Russian Aviation, Space Flight, and Visual Culture

Among the many successes of the Soviet Union were inaugural space flight—ahead of the United States—and many other triumphs related to aviation. Aviators and cosmonauts enjoyed heroic status in the Soviet Union, and provided supports of the Soviet project with iconic figures which could be used to bolster the regime's visions, self-confidence, and the image of itself as forward looking and futuristic. This book explores how the themes of aviation and space flight have been depicted in film, animation, art, architecture, and digital media. Incorporating many illustrations, the book covers a wide range of subjects, including the representations of heroes, the construction of myths, and the relationship between visual art forms and Soviet/Russian culture and society.

Vlad Strukov is an Associate Professor in Film and Digital Culture in the School of Languages, Cultures and Societies at the University of Leeds, UK.

Helena Goscilo is Professor of Slavic at the Ohio State University, Columbus, Ohio, US.

Routledge Contemporary Russia and Eastern Europe Series

58 Fashion and the Consumer Revolution in Contemporary Russia
Olga Gurova

59 Religion, Nation and Democracy in the South Caucasus
Edited by Alexander Agadjanian, Ansgar Jödicke and Evert van der Zweerde

60 Eurasian Integration – The View from Within
Edited by Piotr Dutkiewicz and Richard Sakwa

61 Art and Protest in Putin's Russia
Lena Jonson

62 The Challenges for Russia's Politicized Economic System
Edited by Susanne Oxenstierna

63 Boundaries of Utopia - Imagining Communism from Plato to Stalin
Erik van Ree

64 Democracy in Poland
Representation, participation, competition and accountability since 1989
Anna Gwiazda

65 Democracy, Civil Culture and Small Business in Russia's Regions
Social processes in comparative historical perspective
Molly O'Neal

66 National Minorities in Putin's Russia
Federica Prina

67 The Social History of Post-Communist Russia
Edited by Piotr Dutkiewicz, Richard Sakwa and Vladimir Kulikov

68 The Return of the Cold War
Ukraine, The West and Russia
Edited by J. L. Black and Michael Johns

69 Corporate Strategy in Post-Communist Russia
Mikhail Glazunov

70 Russian Aviation, Space Flight and Visual Culture
Edited by Vlad Strukov and Helena Goscilo

71 EU-Russia Relations, 1999-2015
From courtship to confrontation
Anna-Sophie Maass

Russian Aviation, Space Flight, and Visual Culture

Edited by Vlad Strukov and Helena Goscilo

Routledge
Taylor & Francis Group
LONDON AND NEW YORK

First published 2017 by Routledge

2 Park Square, Milton Park, Abingdon, Oxfordshire OX14 4RN
711 Third Avenue, New York, NY 10017

Routledge is an imprint of the Taylor & Francis Group, an informa business

First issued in paperback 2018

British Library Cataloguing in Publication Data
A catalogue record for this book is available from the British Library

Library of Congress Cataloging in Publication Data
Names: Strukov, Vlad, 1973- editor. | Goscilo, Helena, 1945- editor.
Title: Russian aviation, space flight and visual culture / edited by Vlad Strukov and Helena Goscilo.
Description: Milton Park, Abingdon, Oxon ; New York : Routledge, 2016. | Series: Routledge contemporary Russia and Eastern Eeurope series ; 70 | Includes bibliographical references and index.
Identifiers: LCCN 2016003131| ISBN 9781138951983 (hardback) | ISBN 9781315667881 (ebook)
Subjects: LCSH: Astronautics in mass media. | Astronautics in art. | Astronautics—Social aspects—Soviet Union. | Astronautics and civilization—Soviet Union. | Popular culture—Soviet Union—History—20th century.
Classification: LCC P96.A792 R87 2016 | DDC 700.4/56—dc23
LC record available at http://lccn.loc.gov/2016003131

ISBN: 978-1-138-95198-3 (hbk)
ISBN: 978-1-138-58820-2 (pbk)

Typeset in Times New Roman
by Swales & Willis Ltd, Exeter, Devon, UK

To my brother, a real pilot.
-VS

To Dziudzio, who chose water over air.
-HG

Contents

Figures

Contributors

Anindita Banerjee (Associate Professor of Comparative Literature at Cornell University) explores the interfaces between techno-scientific, cultural, and social imaginations across Russia, Eurasia, and the Indian subcontinent. She is interested in science fictional literature and media, which play a crucial role in negotiating trans-local practices and global understandings of modernity. The subject is explored in her monograph, *We Modern People: Science Fiction and the Making of Russian Modernity* (2013). Banerjee's broader interests are in media studies, critical geography, migration, and border studies. Her articles in these areas have appeared in journals including the *PMLA*, *Clio*, *Science Fiction Studies*, and *Comparative American Studies*.

Birgit Beumers (Professor of Film Studies at the University of Aberystwyth, Wales, UK) specializes in Russian culture, especially cinema and theater. Her most recent publications include *A History of Russian Cinema* (2009) and, with Mark Lipovetsky, *Performing Violence* (2009); she has edited *Directory of World Cinema: Russia* (2010; 2014), *The Cinema of Alexander Sokurov* (2011, with N. Condee), *Russia's New Fin de Siècle* (2013), and (with M. Rouland and G. Abikeyeva) *Cinema in Central Asia: Rewriting Cultural Histories* (2013). She is currently writing a book on early Soviet animation. She is editor of *KinoKultura* and *Studies in Russian and Soviet Cinema*.

Maria Engström (Associate Professor of Russian, School of Humanities and Media Studies, Dalarna University, Sweden) focuses on the role of the Orthodox Church in Russian politics, and contemporary Russian utopian imagination. Her most recent publications include '"Orthodoxy or death!" Political Orthodoxy in Russia' (*Religion, Politics and Nation-Building in Post-Communist Countries*, 2015); 'Contemporary Russian Messianism and New Russian Foreign Policy' (*Contemporary Security Policy*, 2014); and 'Forbidden Dandyism: Imperial Aesthetics in Contemporary Russia' (*Nordic Fashion Studies*, 2012). She is a co-editor of 'Digital Orthodoxy: Mediating Post-Secularity in Russia and Ukraine', a special issue of 'Studies in Russian, Eurasian and Central European New Media' (2015).

John Etty (PhD candidate, University of Leeds) is working on a doctoral thesis on *Krokodil* magazine during the Khrushchev Thaw, considering the journal in relation to theories of carnivalesque humor and satire, transmedia and co-creative production techniques, and performativity. His interests also include flight in visual art in the Soviet period and public art and political graphic satire published on social media in twenty-first-century Russia. He was a co-organizer of the 2010 'Russian Aviation and Space: Technology and Cultural Imagination' conference in Leeds.

Helena Goscilo (Professor of Slavic at The Ohio State University) writes primarily on culture, gender, and visual genres in Russia. She has also published on Polish film and opera, as well as Russian folklore. Her book-length publications in the last five years include *Reflections and Refractions: The Mirror in Russian Culture* (*Studies in 20th and 21st Century Literature*, ed., 2011), *Putin as Celebrity and Cultural Icon* (ed., 2012), *Embracing Arms: Cultural Representations of Slavic and Balkan Women in War* (co-ed., 2012), and *Baba Yaga: The Wild Witch of the East in Russian Fairy Tales* (co-ed., 2013). Among her current projects is the monograph *Graphic Ideology: The Soviet Poster from Stalin to Yeltsin.*

Julian Graffy (Professor Emeritus of Russian Literature and Cinema at University College London) has written widely on Russian film and is the author of *Bed and Sofa: The Film Companion* (2001) and *Chapaev: The Film Companion* (2010). His recent research has focused on Soviet cinema of the 1920s and 1930s, through articles on Igor' Savchenko and Boris Barnet; and on contemporary Russian authorial cinema, through studies of Kira Muratova, Aleksandr Sokurov, Andrei Zviagintsev, Vasilii Sigarev, Boris Khlebnikov, Aleksei Popogrebskii, and Aleksei German Jr. He is currently completing a study of the representation of foreigners in Russian and Soviet film.

Aleksandra Idzior (Associate Professor in Art History and Visual Studies at the University of the Fraser Valley, Canada) is interested in socio-political aspects of art production and reception; artists' responses to social and political crisis; war and the politics of memory; the modern city and urban space; and built and imaginary environments. She has published on a number of topics ranging from the rendition of national politics in nineteenth-century history painting in Germany, Poland, and the USA, to the politics and utopian visualization of urban space in Moscow and New York during the 1920s, and on the war, trauma, and memory in Terasa Żarnower's photomontages.

Mike O'Mahony (Reader in History of Art, University of Bristol) researches the visual and material culture of the Soviet Union and post-Soviet Russia with a particular emphasis on representations of sport, physical culture, and leisure. He has published books and articles on representations of fizkultura, the films of Sergei Eisenstein, the Moscow metro, and the work of Aleksandr Deineka. His most recent publications have addressed cultural representation and the Olympic movement and he is presently researching representations of soccer in Soviet art in advance of the staging of the FIFA World Cup in Russia in 2018.

Vlad Strukov (Associate Professor in Film and Digital Culture, University of Leeds) focuses on visual aspects of contemporary culture in a number of theoretical settings including globalization theory, theories of empire and nationhood, and theories of agency and mediation. He has published on film, animation, social media, art, online gaming, transnational television, and celebrity culture, focusing on Russia in the twenty-first century. He is the principal editor of an international journal, 'Studies in Russian, Eurasian and Central European New Media'. In the last two years he has authored *Contemporary Russian Cinema: Symbols of a New Era* (2016) and co-edited 'From Central to Digital: Television in Russia' (2014) and 'New Media in New Europe-Asia' (2014).

Acknowledgements

We thank (1) our contributors for their work, collegial cooperation, and stoic patience; (2) the various entities that granted permission to reproduce the visuals included in the volume; (3) the School of Languages, Cultures and Societies of the University of Leeds for providing financial assistance at the early stages of this project and for providing Vlad Strukov with research leave in 2013–14, enabling him to complete work on this volume; (4) friends and colleagues who generously sent materials pertinent to our project, including Pedro Hernandez for sharing his technical expertise; and (5) the personnel at RoutledgeCurzon for all stages of the publication process.

We extend a special word of appreciation to Scott Palmer, not only for his excellent book on flight, but also for his thoughtful and generous reading of the manuscript, which enabled us to improve it.

Note on the text

Throughout we have followed the Library of Congress (LoC) transliteration system, except in those cases where names have acquired a standard alternative established by Anglophone publications: for instance, Yeltsin, not El'tsin; Dostoevsky, not Dostoevskii. The scholarly apparatus, however, adheres faithfully to the LoC system.

Introduction

The aerial ways of aspiration and inspiration, or the Russian chronot(r)ope of transcendence

Vlad Strukov and Helena Goscilo

"It's a bird; it's a plane; no, it's Superman!"[1]

"If God had intended man to fly, He'd have given him wings."
George W. Melville[2]

"Space travel nowadays was an escape from the problems of Earth."
Stanislaw Lem, *Solaris*[3]

Celestial space: the heavens in myth

The sky has never been just the sky. It has also—always already—been the heavens, according to the vertical axis extending into the firmament that universally structures metaphysical thought.[4] What for centuries symbolized the identification of the sky with divinity was the transcultural Cosmic Tree of Life (*axis mundi*), which mapped the spatial relationship between the underworld, Earth, and Heaven (Figs. I.1 and I.2). With its roots underground, its branches penetrated into the vault of the sky—the dwelling of the Olympians and their counterparts in other cultures, who governed the fates of the mortals below and communicated their implacable dictates via signs and messengers. Language itself, in tandem with iconography, confirms this paradigmatic conceptual triad: humankind, located in the paradox-riddled middle stratum (of perishable body encasing eternal soul), 'looks **up**' to the **supernal** realm of the divine and **down** upon **sub**terranean Hades/Tarturus/Hell. This symbolic geography, naturalized through the tree as an organic entity, has presided over modes of thought and cultural production for centuries, alternating with the ladder (*scala naturae*) as a means of vertical disposition and negotiation, verified in the Bible, depictions of Jacob's ladder, the philosophy of Neo-Platonists, and Nikolai Gogol's idiosyncratic sense of the world (Fig. I.3). The hierarchical iconography and discourse of ascent/descent persists to this day. Mikhail Bakhtin, citing the metaphysical hierarchy of Dante's *Divine Comedy*, conceived of this vertical temporality as "deep time," wrenched free from linear temporal progression.[5] Fyodor Dostoevsky, in his portrayal of the anti-Christ Nikolai Stavrogin in the novel *Demons* [Besy], similarly suspends time in the chapters titled "Night" [Noch'] to convey the primacy of spiritual hierarchy over narrative continuity.

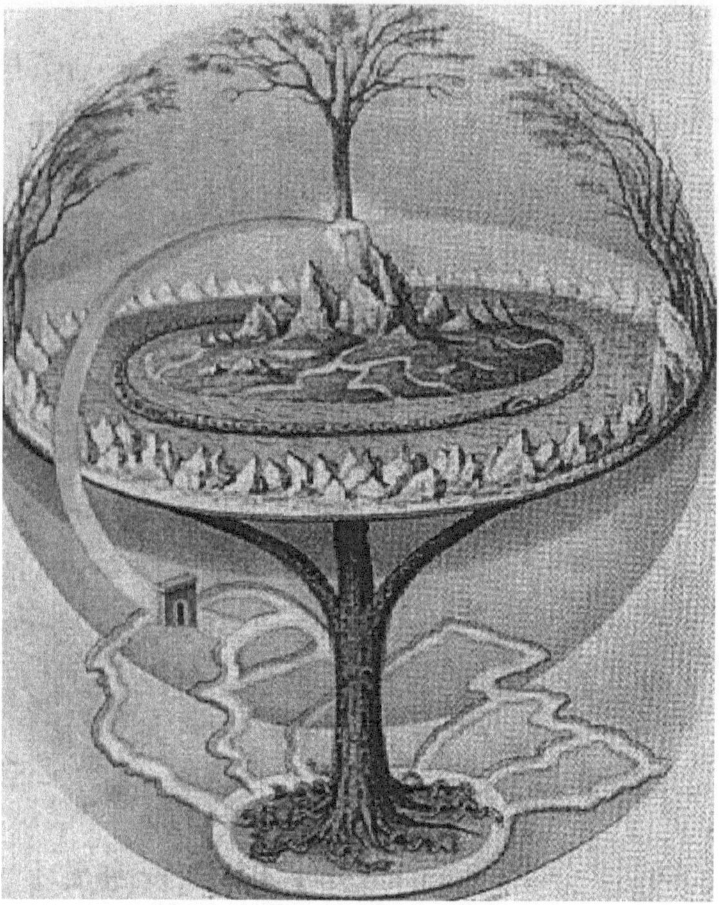

Figure I.1 Yggdrasil, the tree of life central to Norse cosmology, in the translation of Oluf Bagge's *Prose Edda* (1847). Courtesy of Wikimedia Commons

Since ancient times the attempt to reach the firmament's symbolic center or source (i.e., God/sun) has constituted an irresistible challenge to hubristic humanity, which acknowledges, yet aspires to surmount, the established hierarchy or divine order—in the modern world, primarily through untrammeled scientific exploration, for laboratory inventions and discoveries are the modern, secular equivalent of miracles. Throughout centuries efforts to arrogate divine privilege comprise the imaginative scenarios of such works as Goethe's drama *Faust* (1808), Mary Shelley's landmark novel *Frankenstein* (1818), and a host of endlessly proliferating sci-fi and horror films, in which flight figures prominently.[6] These efforts have also resulted in the actual advent of cloning, cryonics, and flight and colonization of space since the late nineteenth century, though famously, as early as during

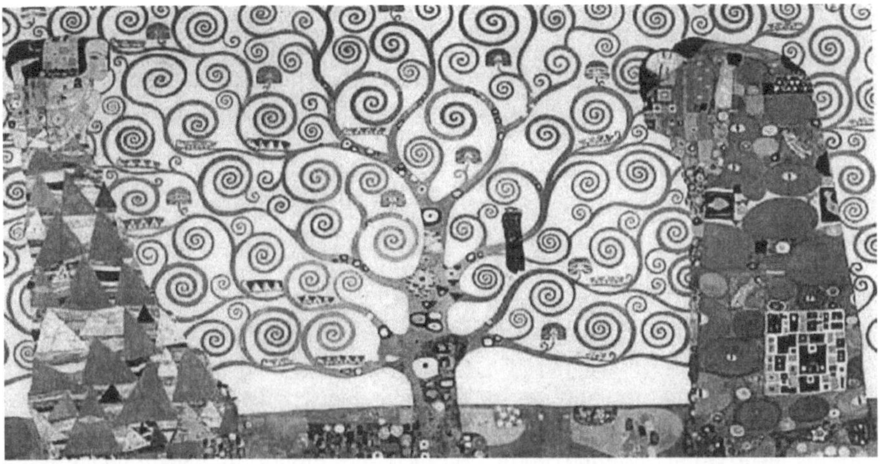

Figure I.2 Gustav Klimt, *Tree of Life* (1909). Courtesy of Wikimedia Commons

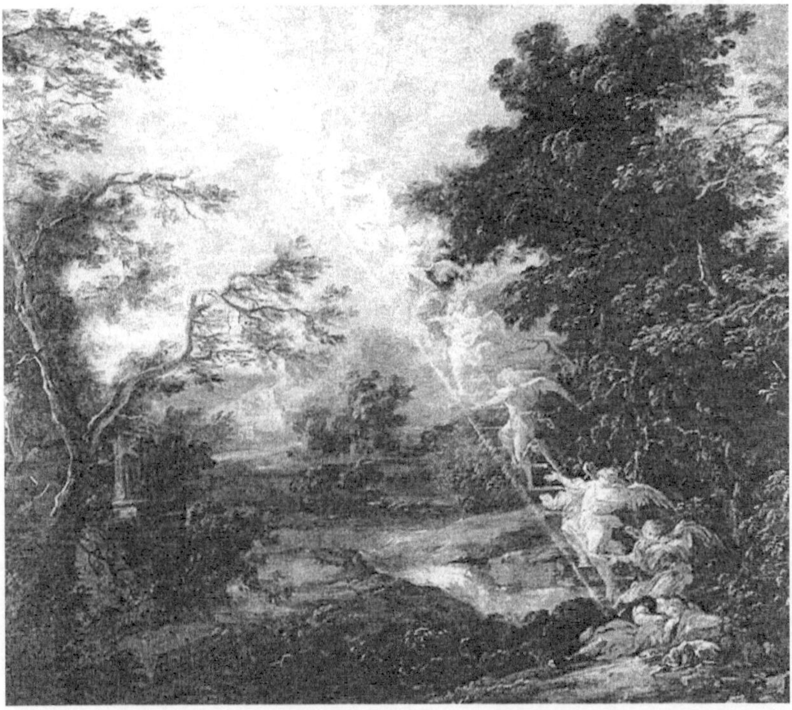

Figure I.3 Michael Lukas Leopold Willmann, *Landscape with the Dream of Jacob* (1691), in which Jacob's ladder reaches up to the heavens, depicting humans' immemorial aspirations. Courtesy of Wikimedia Commons

the Renaissance, Leonardo da Vinci (between 1485 and 1515) studied the flight of birds and created designs for several contraptions intended to enable human flight.[7]

To appreciate the element of metaphysical revolt implied by the birth of aeronautics, we need only recall the famous declaration in 1900 (years after gliders [1890s] and more than a century after the air balloon [1780s] had become reality): "If God had intended man to fly, He'd have given him wings"—a sentiment attributed, not to representatives of any religious institution, but to George Melville, chief engineer of the US Navy. That viewpoint seemed validated by the case of Otto Lielenthal, the German who had pioneered the glider and perished in 1896 following an Icarus-like fall while gliding. It is no coincidence that Mary Shelley subtitled her literary treatment of the Industrial Revolution's experimental excesses *The Modern Prometheus*, after the Titan who stole sacred fire from Zeus for humanity, only to pay for his audacity with his liver, which Zeus' eagle savaged daily. Shelley's cautionary narrative set the stage for earthbound mortals' usurpation of divine prerogative as part of modernity's vaulting ambition to exploit scientifically based modes of creation and exploration.

Icarus as rebellious son or revolutionary hero

Among numerous ancient Greek myths about the human urge to surpass or rank with the gods[8] is the renowned fate, narrated in Ovid's *Metamorphoses*, of the Athenian 'artificer' Daedalus and his son, Icarus, the latter punished when, in defiance of his father's injunction, he overreached by flying too close to the sun. Its heat melted the wax affixing the wings to the youth's body, causing his literal and metaphorical punitive fall[9] (Figs. I.4 and I.5). The medieval world analogized that fall with the Biblical lapsarian myth, according to which Icarus was condignly brought down for his disobedience and *superbia* (vanity): like Adam and Eve, he contravened paternal prohibition—in this case, the warning to avoid the extremes of height and depth (sun and sea). Retribution for flouting authoritative strictures, in other words, both literally and figuratively, adhered to the vertical axis (the 'big F' and the 'little f' fall). Once the Renaissance ushered in the cult of individualism, however, painterly and literary works transvalued both Icarus and Daedalus into symbols for art and freedom (Fig. I.6). After all, Daedalus, the architect of the Minotaur's maze, had devised the wings of birds' feathers as a means of escape from imprisonment in Crete. Hence James Joyce's choice of Stephen Daedalus as the name for the protagonist of *Portrait of the Artist as a Young Man* (1915/16), who, like his Greek forebear, yearns for liberation—from the stifling fetters of Catholic Ireland.[10]

Two features of the Greek myth stand out in the extensive Icarus iconography throughout the ages. First, the patrilinear bond, an official mode of which Scott Palmer deems uniquely characteristic of Russian and Soviet aviation, distinguishing it from the Western tradition, which follows Renaissance values by troping "the occupation of the air" as "individual freedom and personal liberation" (Palmer 2005: 19). Russia's version of the Icarus myth, popularized in the twentieth

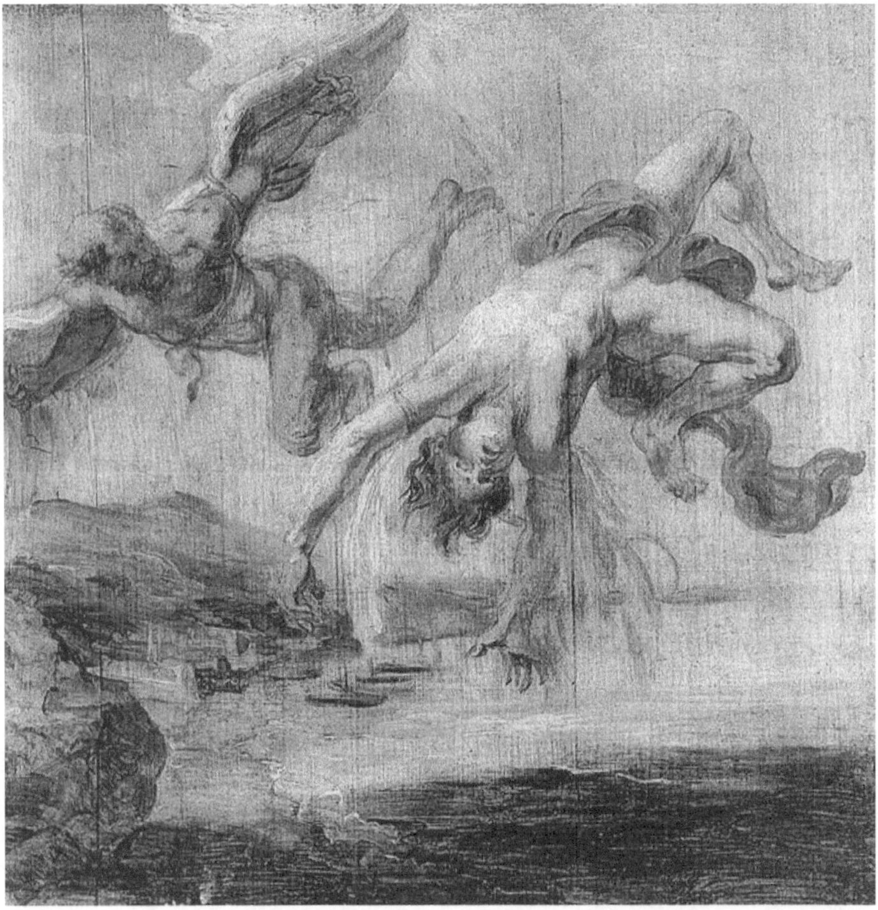

Figure I.4 Peter Paul Rubens, *The Fall of Icarus* (1636), one of numerous visuals capturing the incautious son of a wiser father plummeting to his death. Courtesy of Wikimedia Commons

century—captured on screen in Iurii Tarich's one-hour film, *Wings of a Serf* [Kryl'ia kholopa 1926], and on canvas by Aleksandr Deineka (1940)—abjures individualism for official paternalism, reminiscent of the medieval reading of Icarus. In the sixteenth century, the serf-inventor Nikitka reportedly succeeded in "flying" a short distance (see Chapter 2), but incurred state ire: Ivan the Terrible issued a disquieting edict, which anticipated the American navy engineer's sentiment four hundred years later: "A man is not a bird. He does not have wings. Those who attach wooden wings to themselves do so in opposition to the will of nature. Such is . . . a deed which emanates from unclean powers. For such an association with the forces of darkness, the head of the inventor shall be cut off" (Palmer 2005: 20 citing Borozdin 6). Approximately two centuries later, Peter the

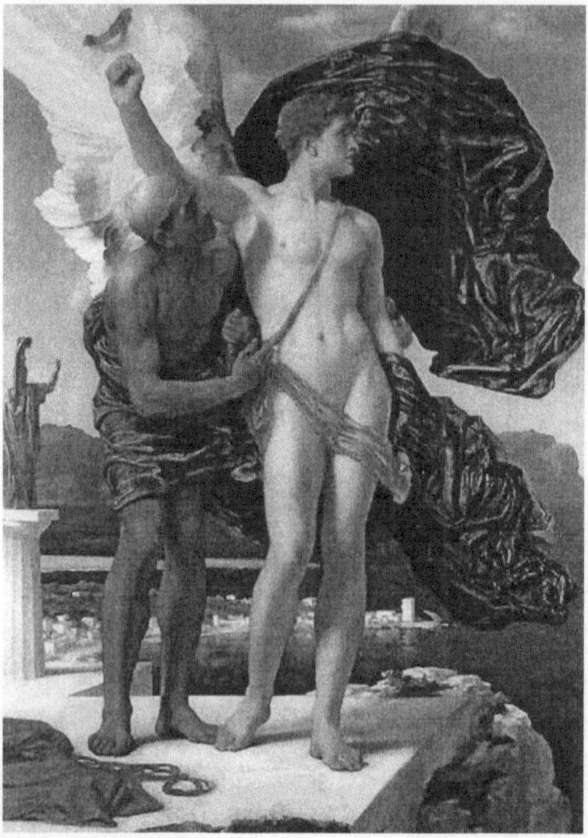

Figure I.5 Frederic Leighton, *Icarus and Daedalus* (c. 1869), showing Icarus in youthful glory before his demise, aided by his concerned parent. Courtesy of Wikimedia Commons

Great underwrote the peasant Emelian Ivanov's construction of a pair of wings with which he intended to fly from the middle of Red Square. After the flight failed, the peasant was beaten and his property sold to repay his debts (Palmer 2005: 20). Apart from spotlighting Ivan the Terrible's medieval superstitiousness versus Peter the Great's pragmatism, both accounts, as Palmer points out, focus on flight "under the watchful eye of state authority and result in the physical punishment of the would-be aeronaut before a collected assembly of his neighbors and countrymen" (Palmer 2005: 20). The Russian Icarus, in other words, "speaks to the virtue of individual obedience and the role of state authority in facilitating transcendence" (Palmer 2005: 21). In short, the state is the paternal eye writ large, as corroborated by Mikhail Kalatozov's cinematic narrative of *Valerii Chkalov*, the pilot who wholeheartedly embraced Stalin's initiatives and off-screen died like Icarus by deviating from a flight plan (1938).

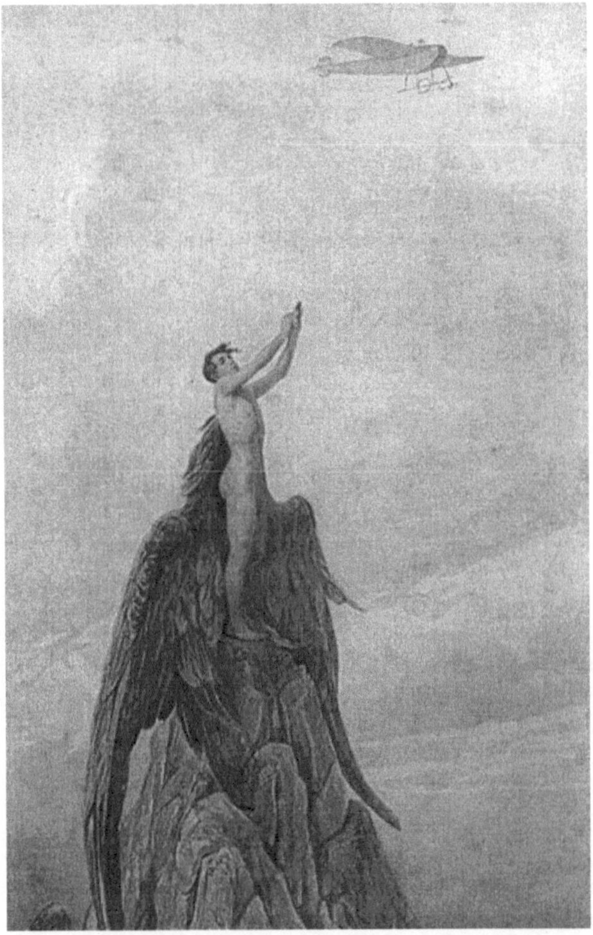

Figure I.6 Sergei Solomko, *Icarus' Dream* (date unknown), which connects human desire to transcend, avian wings, and airplanes in a single, compact image. Courtesy of Wikimedia Commons

Second, Icarus iconography dwells on the final stage of the myth—the poignant drama of the son's demise, echoed in representations of the modern pilot plummeting from the sky, as in Aleksandr Deineka's painting (see Chapter 2). For a variety of reasons, not least the desire to emphasize the wisdom of obedience to state and fatherly precepts, many images on canvas illustrate both paternal(istic) supervision and filial down/fall: Extraordinarily, Sergei Solomko's depiction of Icarus in *Icarus' Dream* (1920s?) links the youth directly with aerial flight. The frequently reproduced image portrays Icarus, poised on a mountain peak and weighed down by huge feather-wings, gazing up at an overhead plane free from all such encumbrances. Solomko makes clear the human desire to equal the airborne machine,

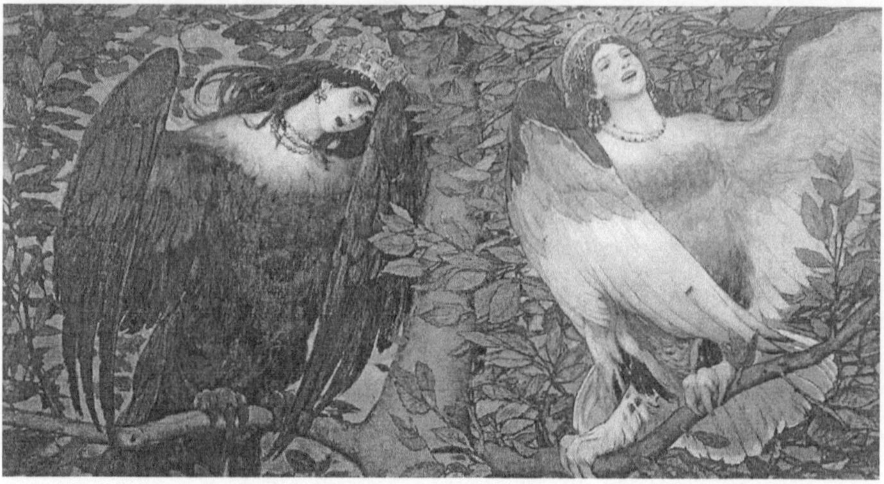

Figure I.7 Viktor Vasnetsov's famous *Sirin and Alkonost* (1896), the two mythical birds whose songs symbolize joy and sorrow. Courtesy of Wikimedia Commons, from A.K. Lazuko's *Viktor Vasnetsov*

moving freely above the earthbound human (Fig. I.6). Solomko's configuration establishes the paradigmatic conceptual triad noted above, with the high mountain evoking remoteness from the earth and its lower depths, so that subterranean, terrestrial, and heavenly map the seemingly deathless metaphysical geography of the Tree of Life.

Wings and avians

Just as the celestial sphere is the divine realm, so flight is synonymous with avians. Practically speaking, Daedalus forged wings from birds' plumage in a spirit of deductive mimesis, as would da Vinci, but Daedalus' stratagem for flight is more important for its metaphorical significance. In a symbolic spatial hierarchy, birds as occupants of the air between humanity and the gods/God are winged intermediaries/mediators—vulnerable creatures on earth yet lofty inhabitants of the skies.[11] Feathered angels of sorts (in Greek, ἄγγελος [*angelos*] = messenger), they represent the passage between the spiritual and the physical worlds, which explains why traditionally they have symbolized the soul liberating itself from the body and flying heavenward.[12] Plato in *Phaedrus* declares: "The natural function of the wing is to soar upwards and carry that which is heavy up to the place where dwells the race of gods. More than any other thing that pertains to the body it [the wing] partakes of the nature of the divine" (Wohl 1994: np). In Russian folklore, the birds of paradise—Finist and Alkonost—perch atop the Tree of Life and through their song also function as divine messengers to those below—variously depicted by Ivan Bilibin, Viktor Vasnetsov, and Tat'iana Tolstaia in the story

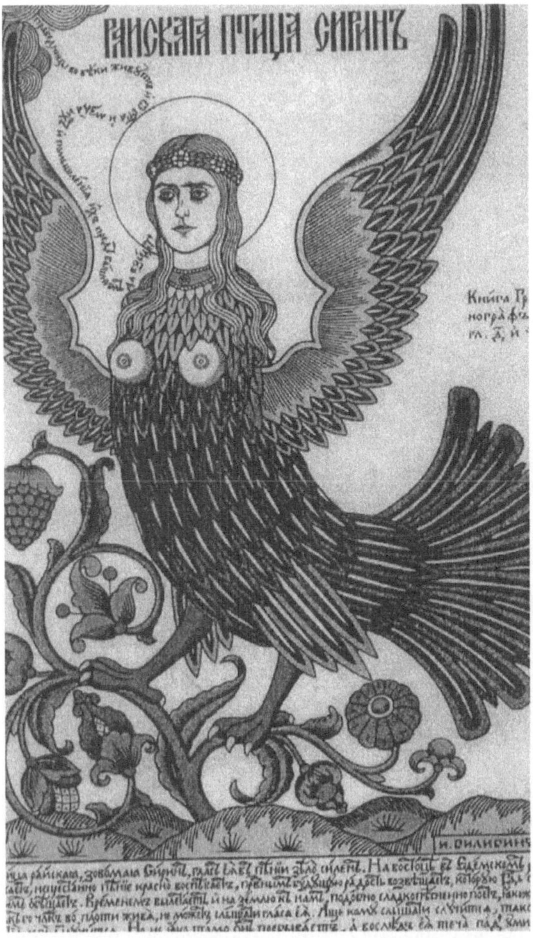

Figure I.8 Ivan Bilibin, *Sirin* (1905), frequently reproduced graphic image of the mythical bird of joy by the premier illustrator of fairy tales in Russia. Courtesy of Wikimedia Commons and Alex Bakharev

"Rendezvous with a Bird" [Svidanie s ptitsei 1980s] (Figs. I.7 and I.8). Saturated in animism, the medieval *Tale of Igor's Campaign* [Slovo o polku igoreve] compares the vatic Boyan to a nightingale and teems with birds who similarly serve as mystic omens. Likewise, in Shakespeare's plays, the owl that appears at noon in *Julius Caesar*, the hoarse raven in *Macbeth*, and so forth all signal disruption in the Great Chain of Being—the *scala naturae* mentioned earlier as a rough equivalent of the Tree of Life, with heaven at its apex.

Birds' traditional association with the transcendent has a rich pedigree: they figure as divine emissaries—Zeus' eagle for the ancient Greeks (Fig. I.9), Jupiter's for the Romans, Perun's for the pagan Slavs, the sun god Tonatiuh's for the

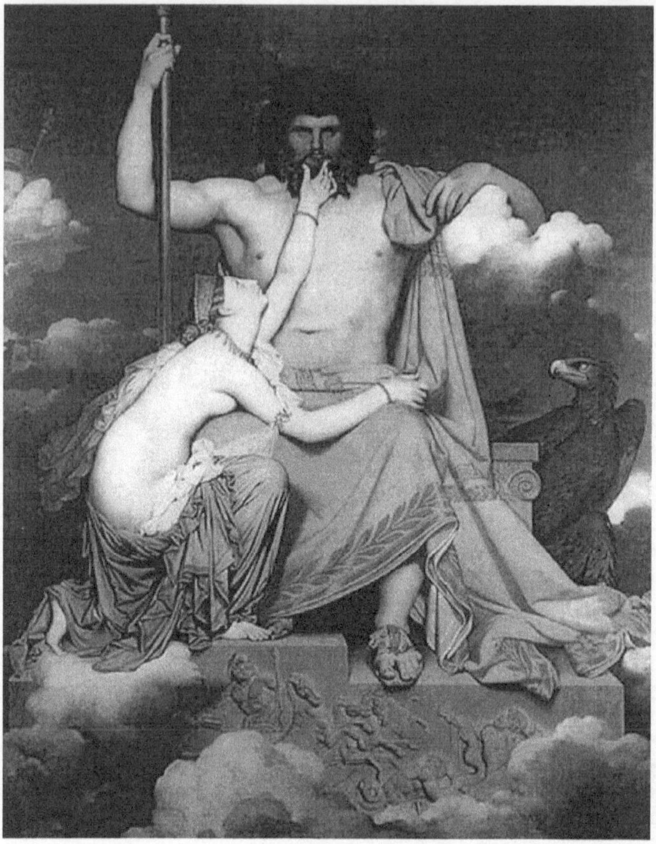

Figure I.9 Jean Auguste Dominique Ingres' *Zeus and Thetis* (1811) visualizes an episode from Homer's *Iliad*, where the nymph Thetis supplicates Zeus to intervene on behalf of her son Achilles. The eagle at the god's left is Zeus' symbol, linking the bird to royal divinity/divine royalty. Courtesy of Wikimedia Commons

Aztecs—while the dove is universally recognized as the Christian symbol of the Holy Spirit[13] (Figs. I.10 and I.11). In ancient Syria the eagle was associated with the sun god and represented the victory of light over darkness. Medieval European bestiaries followed classical traditions in naming the eagle the king of birds, for he was deemed the sole avian endowed with a vision enabling him to look directly at the sun, while in Christian thought the eagle's ability to fly extraordinarily high explains his symbolization of Christ's ascension—hence the bird's frequent depiction on baptismal fonts. This rich heritage of ascendancy accounts for Napoleon's widespread image as an eagle, especially in Romantic verse. The double-headed eagle, as an emblem of Byzantine, Roman, and Russian empires, signaled the ruler's command over both state and church—rendering him just one rank lower than the divine (Fig. I.12). And in Russian literature, kindred symbolism imbues dozens of swans in poetry, Anton Chekhov's seagull, Maksim Gor'kii's stormy petrel, Vasilii Aksenov's steel bird as Stalin (*Steel Bird* [Stal'naia ptitsa 1979]), and a plethora of others.[14] Notable tropological avians onscreen encompass the cranes in Mikhail Kalatozov's *The Cranes Are Flying* [Letiat zhuravli 1957], the various birds in Andrei Tarkovskii's *Mirror* [Zerkalo 1975], and more recently and most pertinently here, the albatross, kites-surrogates, and the seagull in Boris Khlebnikov and Aleksei Popogrebskii's *Koktebel* [Koktebel' 2003], which, not coincidentally, features a former aeronautics engineer and his flight-obsessed son (see Chapter 7 for an astute analysis of the film and its interpolation of birds). The film may be read productively as a contemporary recasting of the Icarus myth, with the father devoid of the son's idealistic notions of flight.

The inherited Greek and Roman double-eagle as a symbol of both regal and divine power—that is, imperial as empyrean—rationalizes the historicized national emblems of numerous countries, including Russia, Poland, and the United States (see Chapter 5 on the recourse to this imperial imagery by the artist Aleksei Beliaev-Gintovt, a proponent of Eurasianism) (Figs. I.13 and I.14). Under

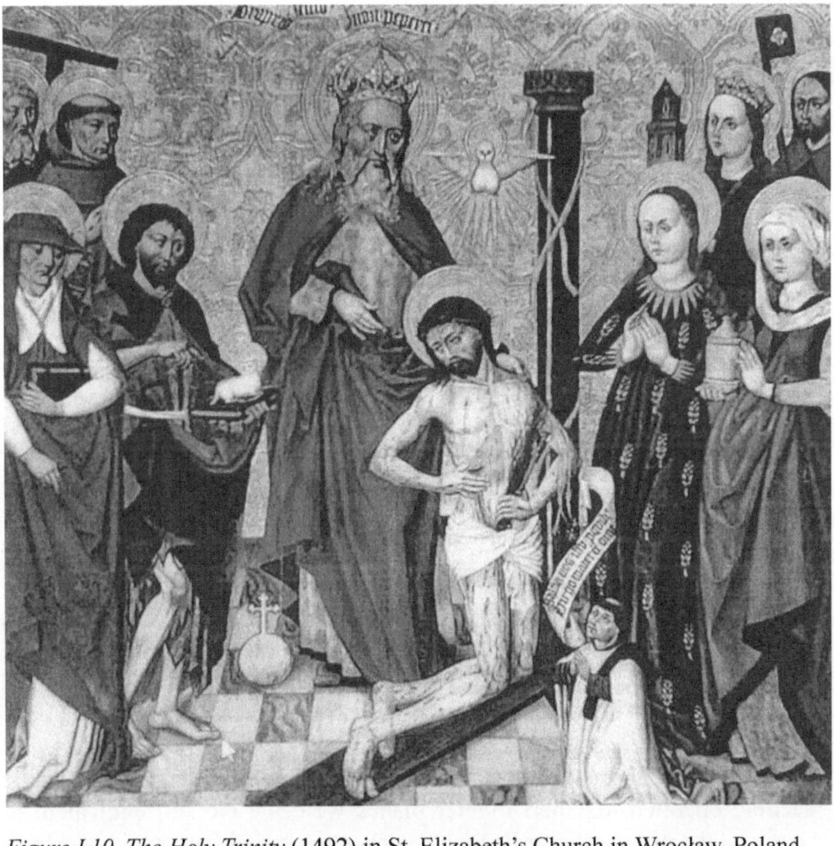

Figure I.10 *The Holy Trinity* (1492) in St. Elizabeth's Church in Wrocław, Poland, represented by the dove. Courtesy of Wikimedia Commons

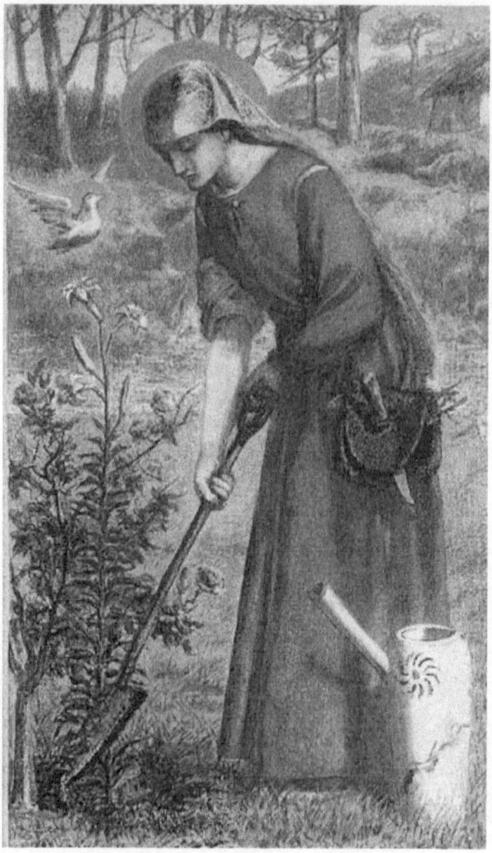

Figure I.11 Dante Dante Gabriel Rossetti's *Mary Nazarene* (1857) symbolizes her spirituality, love, and peace by the dove. Courtesy of Wikimedia Commons

Stalin, iconography linked this symbolism to aviators, in a rhetoric that glorified pilots as Stalin's metaphorical offspring—*sokoly* or *orliata* [falcons or eaglets]—possibly inspired not only by Western classical but also by other mythologies. For instance, Horus, the ancient Egyptian god of the sky and the symbol of kingship, was frequently portrayed as a great falcon with outstretched wings, the sun as his right eye and the moon as his left (again, vision, as in the case of the eagle). His ancient status and exceptional attributes led Egypt Air airline to adopt Horus as its logo.

Stalinist avian nomenclature deified and naturalized technology, while evoking the fantastic transformations in an era that 'made fairy tales come true'; magic and machine operated together. Fighter planes were the modern equivalent of flying carpets, the pilots soaring like the eagles who carried fairy-tale Ivans back to the Russian kingdom. In a historical iteration, the stylized song of today's retro group Liube, "Eaglets Are Learning to Fly" [Orliata uchatsia letat' 1990s],

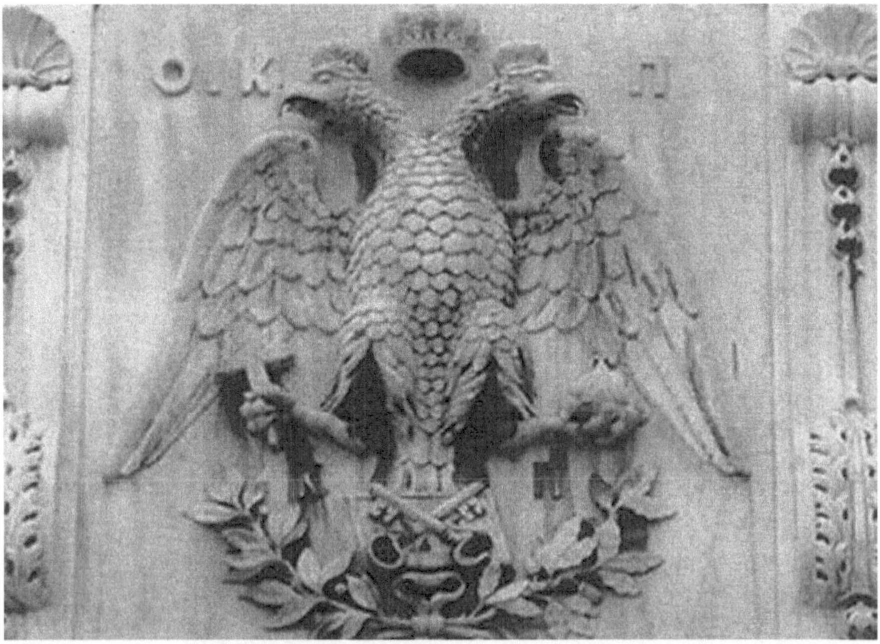

Figure I.12 The Byzantine eagle as a symbol of double-headed power. Courtesy of Wikimedia Commons

celebrates Stalinist and contemporary Russian aeronautic achievements—linking victory over nature with triumph over enemies as a national heritage, summed up in the lines: "They've risen right to the heavens/Just like their forebears near Berlin" [Priamo v podnebes'e vozneslis'/Kak ikh dedy pod Berlinom].[15]

Artistic flights

Long before James Joyce, Charles Baudelaire resorted to both Icarian and avian rhetoric in poeticizing the writer/artist's dilemma: his "Icarus Laments" [Les Plaintes d'un Icare 1862], with the youth plunging seaward as a symbol of attainment fallen short of aspirations, condenses the Faustian drama of constant striving and incommensurability. Five years earlier, his poem "L'Albatross" in the collection *Flowers of Evil* [Les Fleurs du mal 1857] had contrasted the bird's clumsiness on land with its graceful speed in flight—that is, poet as mere man in everyday life versus poet as divinely inspired and semi-divine artist, familiar from Aleksandr Pushkin's poem "Poet" (1827): "Until the poet is called upon . . . " [Poka ne trebuiut poeta . . .]. Both poet and pilot in their different yet related ways aspire to the realm of the divine.

In the late 1890s, flight, no longer confined to trope, became a physical reality. Three years into the twentieth century the Wright brothers made their first patent application for a powered, heavier-than-air flying machine, which they proceeded

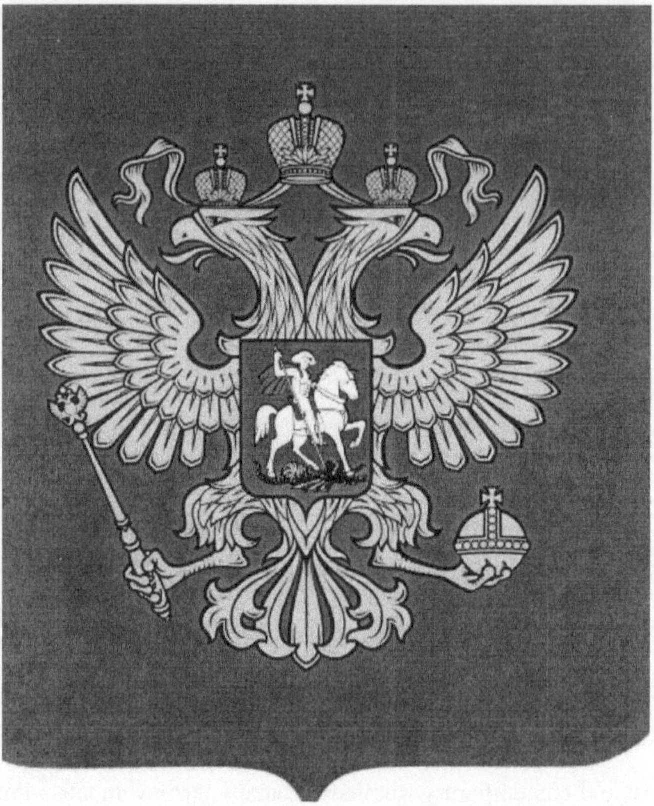

Figure I.13 The coat of arms of the Russian Federation, which combines the two-headed eagle as a symbol of ancient power with the image of St. George, the saint who successfully battled for Russian Orthodoxy, embedded in the eagle. The scepter, orb, and crown inscribe royalty and reflect the excess typical of Russia today. Courtesy of Wikimedia Commons

to operate; in 1906 the Brazilian Alberto Santos-Dumant made the first airplane flight in Europe; in 1908 Wilbur Wright completed 129 flights in France, establishing nine world records for height, distance, and duration; and in 1909 Louis Blériot crossed the English Channel by plane (Paschal *et al.* 2003; Wohl 1994). More importantly for our purposes here, flight also became an obsession and matrix trope for painters and filmmakers during the era of modernity, when technology revolutionized concepts of time and space via accelerated movement through previously unexplored domains. Records from that era indicate that the rapid transition from bicycle to car to airplane created a heady sense of endless possibilities in a world formerly overseen by a circumscribed chronotope that assigned such novelties to the realm of fantasy—or miracle. In short, technologically fueled *bogotvorchestvo* [God-creation] was on its way.

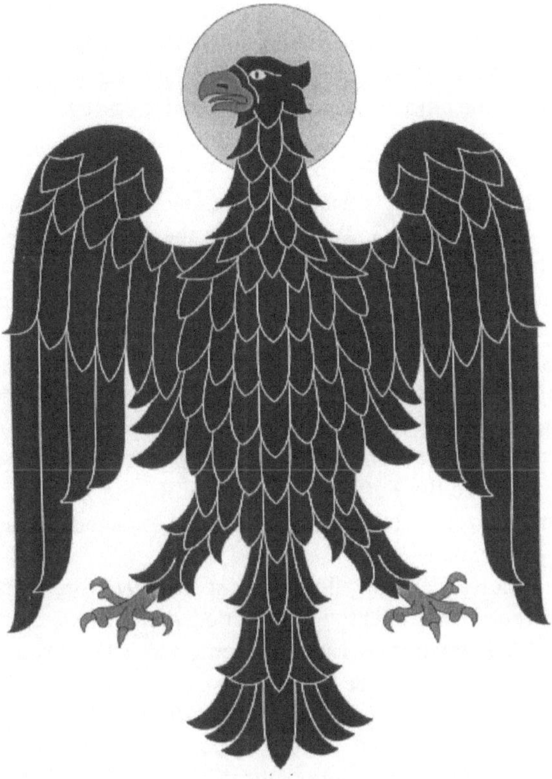

Figure I.14 Heraldic eagle of St. John the Baptist, a symbol of religious power. Courtesy of Wikimedia Commons and Erlenmeyer

The Russian avant-garde responded to aviation with boundless appropriatory enthusiasm, viewing aerial flight as the world of the future, with the artist as aviator/avatar. Tim Harte's stimulating study, *Fast Forward* (2009), accurately characterizes the Russian avant-garde culture of the 1910s and 1920s as inseparable from the aesthetics and ideology of speed—in art, film, and literature. The Italian Futurist Filippo Marinetti, enthralled by the airplane's potential to liberate humanity from time and space, declared in his manifesto of 1909 that the only artists capable of creating a truly Futurist culture would abandon books for technology. The Futurist writer Vasilii Kamenskii concurred, maintaining,

> The airplane—that is the truest achievement of our time. The aviator is the man of worthy heights. Futurists . . . if we are people of the motorized present, poets of universal dynamism, newcomers and messengers of the future, masters of action and activity, enthusiastic builders of new forms of life—then we . . . have no choice but to be fliers.
>
> (Wohl 1994: 145)

Suiting action to word, Kamenskii traveled to France, England, and Poland to learn aviation and managed to become a pilot of sorts, prompting his roommate, the writer Arkadii Averchenko, to inscribe one of his volumes to "celestial Vasilii" from "earthbound Arkadii" (Wohl 1994: 148). That freighted dualism, which coded the discrepancy between the bold, adventurous new and the allegedly retrograde, was inseparable from literati's discourse on aviation—a phenomenon that fascinated intellectuals such as Anatole France, Pierre Loti, Edmond Rostand, Emile Verhaeren, Gerhardt Hauptmann, Gabriele D'Annunzio, Robert Delaunay, and Henri Bergson. Those who, like H.G. Wells, shared the qualms voiced by John Galsworthy about "the prostitution of the conquest of the air to the ends of warfare" composed a minority (Wohl 1994: 153). In terms recalling ancient myths of hubristic aspirations, an outstanding contemporary historian of aviation in culture justly speculates whether Marinetti and others' "desire to conquer the air carried as its unconscious corollary a Luciferian ambition to escape from the last of human limitations . . . and to attain to something approaching godhood through fusion with machines" (Wohl 1994: 154). A partial answer to that question emerged in Russia, where the Futurists Kamenskii, David Burliuk, and Vladimir Maiakovskii wholeheartedly embraced aviation and engaged in various provocative performances in its cause.

The early Futurist promotion of flight culminated in the 1913 premiere of the reported *insuccès de scandale*, the opera signally titled *Victory over the Sun* [Pobeda nad solntsem] in St. Petersburg's Luna [Moon!] Park. With a *zaum'* libretto by Aleksei Kruchenykh, a prologue by Velemir Khlebnikov, music by Mikhail Matiushin, and sets by Kazimir Malevich, the opera featured amateur actors, disconnected pronouncements about violent aggression, sportsmen who proclaimed the existence of Futurist lands where paths lead not only horizontally but vertically, and such sounds as the roar of an engine and a propeller (Fig. I.15). After the broken wing of an airplane falls on the stage, a pilot, heard laughing offstage, walks out and announces that he is unhurt—a reference to Kamenskii's flying accident, which in fact had left him hospitalized and convalescing for months, but which the opera transformed into a demonstration of the aviator's invulnerability. At the conclusion, the actors sing "The world will vanish but we shall have no end!" (Wohl 1994: 158–9). Alienating its uncomprehending audience, the staging displayed in radical form the extent to which the metaphor of flight captured the avant-garde's vision of a brave new future stretching into infinity and the daredevil heroes' intent on exploring this space.

Malevich's Cubo-Futurist sets for the opera reflected his comprehensive shift from representational art to an emphasis on form, which at his pre-Suprematist stage included the mechanization of landscapes and human figures that subsequently would become purely geometric shapes.[16] His fusion of man and machine reified on canvas his fascination with ascent, of escaping the earth. For Malevich flight troped a transfiguration in consciousness, liberation from 3D everyday life, and a redefinition of time and space. In his *Tertium Organum*, he announced: "Art in its highest manifestations is a path to cosmic consciousness" (Wohl 1994: 159, 161), which elevated perception to a higher level of reality

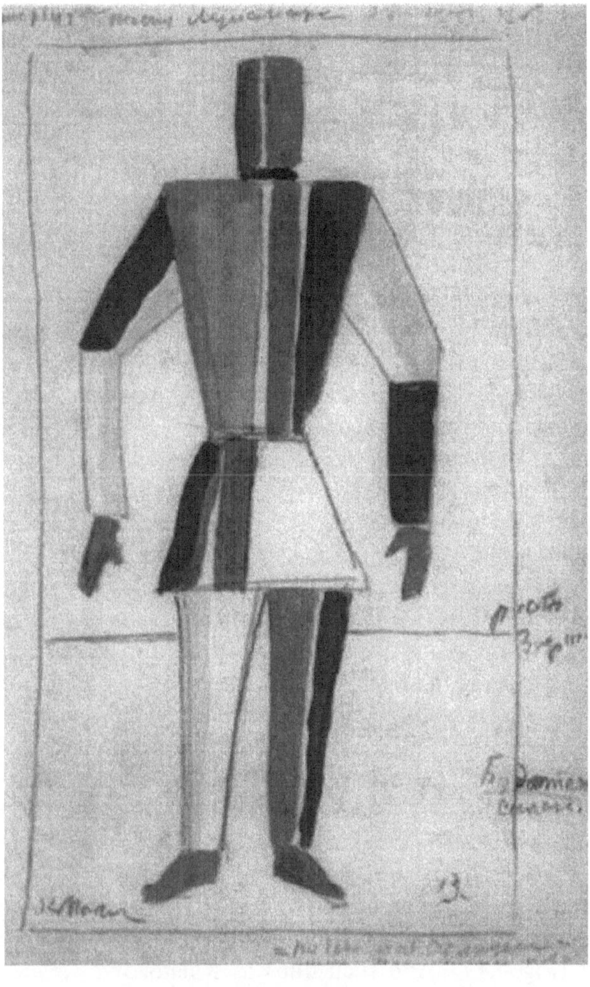

Figure I.15 M. Matiushin's sketch for a costume of the future athlete in the opera *Victory over the Sun* (1913). Courtesy of Wikimedia Commons

(Wohl 1994: 164). Artists now would not **represent**, but would **reveal** that which was accessible only to them (Symbolism + technology). Malevich's *Aviator* of 1915 epitomizes these ideas in their early form, before he embraced Suprematism's complete abstraction. Accordingly, the flat planes of *Aviator* depict a one-eyed man of metal cylinders suspended in Cubist space, his reversed feet alerting the viewer to a hitherto unknown visual world. The head of a transcendently white saw-like sturgeon and an arrow across a whisk both point upward, the latter to the zero emanating yellow light on the man's top hat, "while a green plane, like a flying carpet, floats behind the man" (Wohl 1994: 169). Linked to the fish, the saw dissects the upper register of the painting in half, suggesting the destruction of the past, while the fish, a traditional Christian symbol of resurrection, conveys the new artists' liberation from that past. According to Malevich, artists formerly trapped "in the nets of the horizon, like fish" now could escape by taking off to a new dimension (a reprisal of Daedalus' scenario). Similarly, the fragmentation of "*apteka*" [pharmacy] intimates the dissolution of the petty conventions and commerce of *byt—aptekar',* the avant-garde's pejorative synonym for professionals other than artists or intellectuals (see Maiakovskii's *Chelovek*), while the white A on the saw above the aviator signals his transcendence, likewise symbolized by the zero, which for Malevich meant infinity. In a letter to Matiushin, he explained that he wished to title a new journal *Zero* because "we intend to reduce everything to zero in it. . . . We ourselves will then transcend zero" (Wohl 1994: 171).

One of many works by Malevich hailing the advent of the flying machine, *Aviator* evidences his passionate conviction that the new technology and above all the plane had the capacity to raise humankind above its mundane, terrestrial habits and concerns, fundamentally transfiguring its perceptions with the ultimate goal of achieving a kind of cosmic metaphysical unity. That extraordinary faith in the spiritually liberating power of aeronautics was supplanted during the Stalin era—when the vertical axis reigned absolute—by a determination to impose the Soviet imprimatur on both nature and nations, partly through the colonization of space. **Suprematism** yielded to the telos of political **supremacy**; though still invoking the rhetoric of transcendence, militarization replaced the metaphysics of flight. Deineka occupies an ambiguous place in this shift, for his glorification of the human body's power verges on the metaphysical, though his perception of the pilot's solitude in space intimates not only awe, but also angst (see Chapter 2).

The song "The Aviator's March" [Marsh aviatora] or "Ever Higher" [Vse vyshe], composed in 1920 by Pavel German and Iulii Khait but especially popular in the 1930s, condenses the verbal and visual rhetoric not only of aviation during the Soviet era (van Geldern 1995: 257–8), but also of flight-aspirations throughout centuries. Russian, Soviet, and post-Soviet culture celebrated and continues to celebrate its pilots and its air forces in postcards (Fig. I.16), stamps, and other genres. In fact, some areas in Russia adopted both eagle-like wings and airplanes for their coats of arms (Fig. I.17). And such films as Kalatozov's *Cranes are Flying* and *Koktebel'* by Khlebnikov and Popogrebskii incorporate avian symbolism into the philosophical structure of their narratives, as did Chekhov before them in *The Seagull* [Chaika 1895], and also fairy tales, with casts that included the otherworldly firebird, magical swans, and Baba Iaga's eagles, who transported protagonists across seemingly untransversable boundaries. Though not obviously related to the Roman eagles who functioned as symbolic emissaries of the pagan gods, these supernatural avians nonetheless performed an analogous role as servants of a higher, arbitrary power.

That power is precisely what aviation sought through technology as, in an increasingly politicized and militarized international rivalry, pioneers of aeronautical design struggled to perfect machines that already had revolutionized concepts of both time and space. Politics and the eternal compulsion to discover and colonize the unknown led to the exploration of space, which transformed such science-fiction scenarios as Georges Méliès's historic *Trip to the Moon* [Le

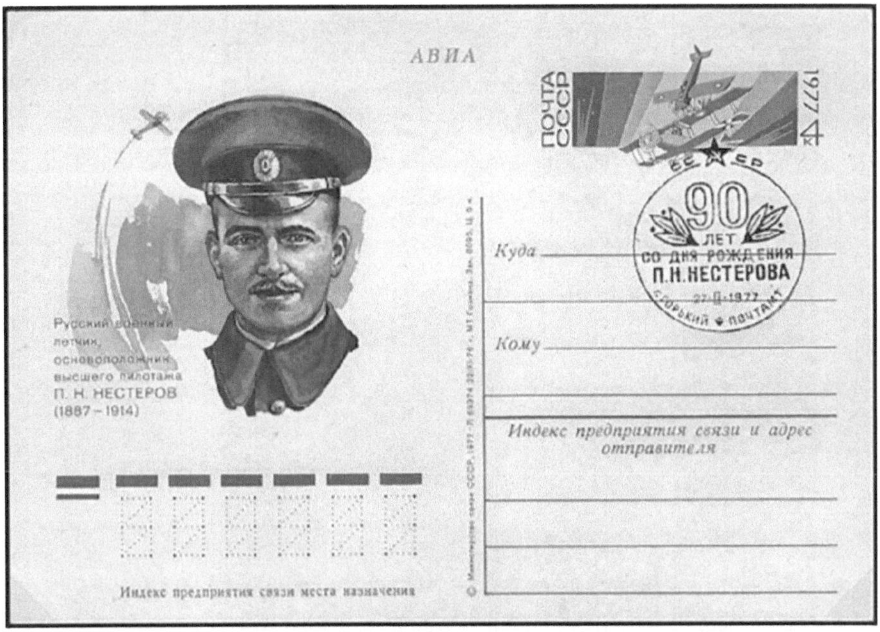

Figure I.16 Postcard commemorating the 90th anniversary of the birth of Petr Nesterov (1887–1914), aircraft designer, pilot, and pioneer of aerobatics

Voyage dans la lune 1902], inspired by Jules Verne's *From the Earth to the Moon* [De la terre à la lune 1865], into lived reality. After WWII and the horrors of Hiroshima and Nagasaki, which demonstrated the catastrophic capacity of aviation for wholesale destruction at a distance that could cushion pilots' ethical concerns about wiping out entire communities, nations' focus transferred to the purportedly peaceful investigation of other planets. At the same time, reprising H.G. Wells's healthy skepticism about vaunted claims of remarkable human progress within the cosmos, fiction by Stanisław Lem, Philip Dick, and Robert Heinlein, and such films as the two versions of *Solaris* (by Andrei Tarkovskii 1972 and Steven Soderburgh 2002), the *Alien* franchise (1979–),[17] and James Cameron's *Avatar* (2009) and its spinoffs, to name but a few, undercut official claims by dramatizing humans' apparently eternal imperialistic drive and their self-absorbed indifference to other civilizations and modes of existence. The hubris of 'conquering' space is eloquently conveyed in not only the name of NASA's Apollo 11, but also its insignia (Fig. I.18). Visual genres especially have tracked technological 'advances' and novelties in the twentieth and twenty-first centuries, disseminating images that prompt both disquiet and delight in viewers faced with the ambiguously seductive spectacle of seemingly endless possibilities unleashed by the historic phenomenon of aviation, which more than a century ago offered a wholly new perspective on earth and on life as it had been lived until then.

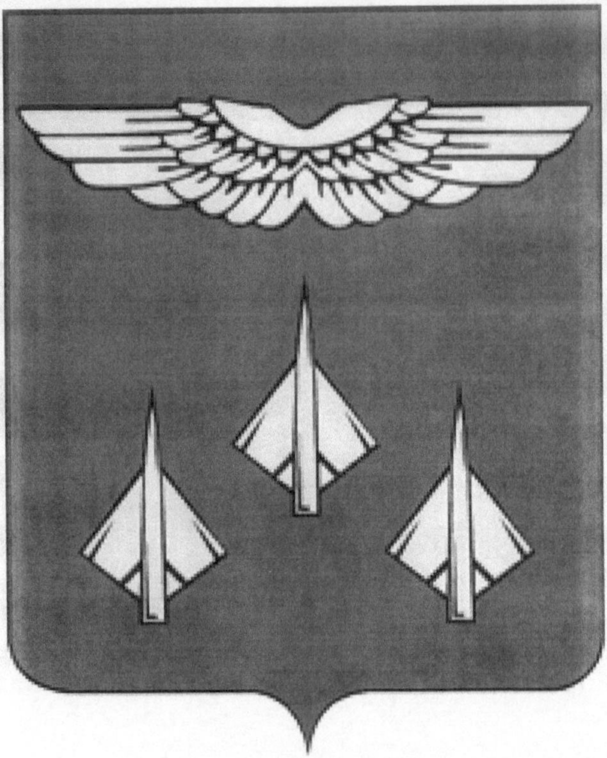

Figure I.17 The coat of arms of Zhukovskii, a town on the Moscow River named after Nikolai Zhukovskii, the founding father of modern aero- and hydrodynamics. The image clearly yokes avian wings to technologically aided flight. Courtesy of Wikimedia Commons

The pragmatics of flying, or the parameters of the Russo-Soviet techno-modernity

The idea of flight has been used in Russian culture not only to define a particular locale of human aspiration—the celestial space—and a condition of the (flying) subject (what we have termed the chronotope of transcendence), but also in a more pragmatic way as a means to expand ideology and represent geopolitical aspirations, especially during the Soviet era. Although aviation took off in Russia well before the establishment of the USSR, the airplane, and later the spacecraft, became the symbol of Soviet modernity, replacing the train as a symbol of late-imperial modernity.[18] Vladimir Tatlin's sculptures and Ilya Kabakov's total installation nicely demonstrate the parameters and direction of Soviet modernity, which evolved from the outward-looking, expansive trajectory of the early Soviet period into the inward-looking, claustrophobic atmosphere of the late Soviet Union.

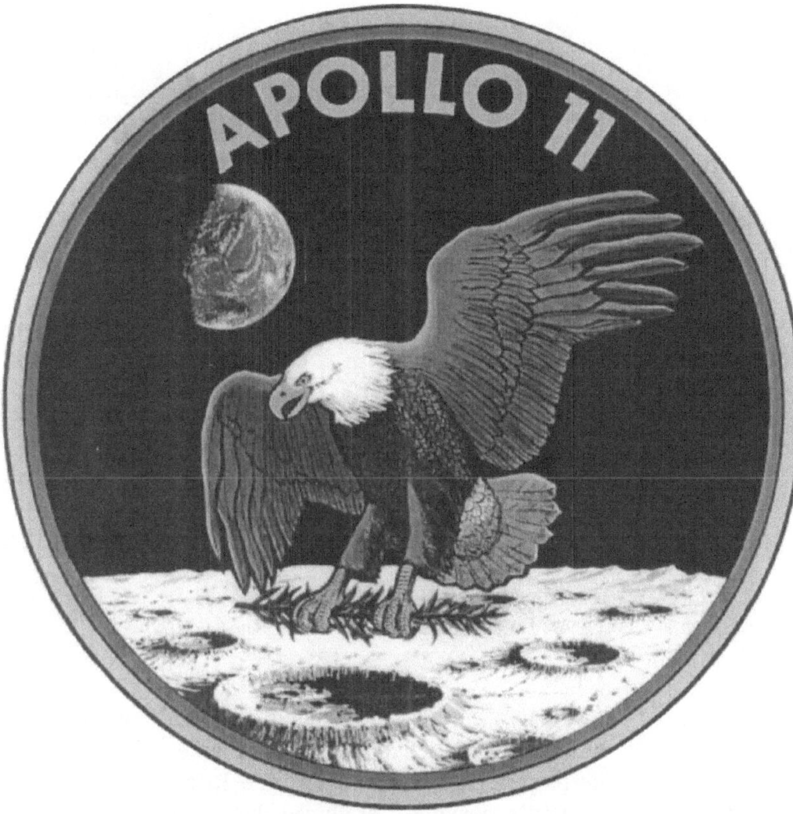

Figure I.18 The NASA logo (1969) identified with its third spaceflight program, Apollo, which successfully landed men on the moon. The image revealingly combines the American eagle with imperialist 'triumph' over other planets. Courtesy of Wikimedia Commons

In 1919–20 the Russian avant-garde artist Tatlin (1885–1993) conceived the *Monument to the Third International* (commonly known as "Tatlin's Tower").[19] It was meant to be a helical tapering metal construction, with an axis running parallel to the earth's axis, with the tower reaching a height of 400 meters. Higher than the Eiffel Tower, the Monument to the Third International was supposed to propel the proletariat above the ground. Inside the tower there were to be three or four rooms, walled in glass and shaped as cubicles, pyramids, cylinders, or hemispheres, which would rotate at different speeds.[20] The rotations of these elements were to represent revolution, both literally and metaphorically. Workers would hold their meetings inside the tower, and from the tower, beams of light would be projected onto the sky while a radio station would continuously broadcast in every known language. Tatlin intended his Tower to represent the Soviet proletariat's aspirations—to broadcast the ideas of the Bolshevik revolution to the workers of the world. It exploited elevation

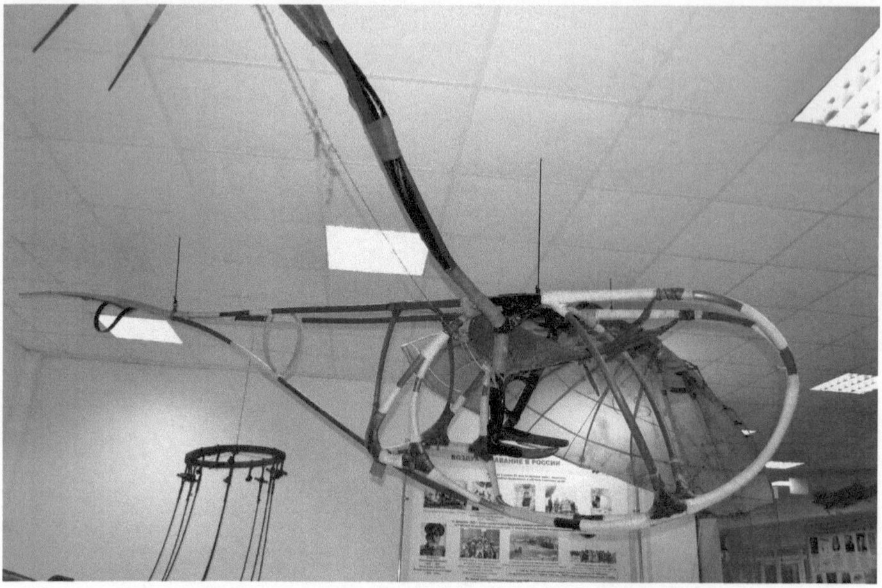

Figure I.19 Letatlin at Central Air Force Museum in Monino, Russia. Courtesy of Wikimedia Commons

and projection not only as structural nodes and elements of design but also as an epistemological undertaking in that it alluded to the Tower of Babel.

Tatlin's next large-scale project, 'Letatlin' (joining the Russian verb 'letat',' meaning 'to fly,' with the artist's surname), produced during his Constructivist period in 1929–32, disentangles the human from the constraints of life on the ground (Fig. I.19). Letatlin is an apparatus with bird's wings that would enable an individual to ascend into the sky on an aerial bicycle, relying solely on the use of muscular power (a type of ornithopter; in Russian, 'makholet') (see Chapter 10 for a critique of these technologies in post-Soviet culture). Letatlin was exhibited in the Italian hall of the State Museum of Fine Arts in Moscow (nowadays the Pushkin Museum) alongside sculptures by Verrocchio and Donatello and next to a copy of Michelangelo's *David*. Such a position for the display of Tatlin's work pointed to the Renaissance idea of a human being as a perfect universal machine and to the newly asserted history of Soviet modernity. Tatlin's work celebrated individual endeavor in the period of Stalin's forced collectivization, thus linking Letatlin to the Russian avant-garde rather than to Soviet official art. At the same time it embraced the aspirations of the new regime in its attempt to record, structure, and catalogue the Soviet realm with the help of aviation. In its organic form and use of natural materials—leather, wood, and baleen—Letatlin eschewed the monochrome, steel, and concrete aesthetic of Soviet art of the period. It conveyed the ancient myth of Icarus (Greece) and Elijah (Russia)—beings rewarded with the gift of flight and occupancy of the sky, albeit temporarily. In its subjugation of Soviet values, Letatlin evoked the rebellious, dissenting nature of flight in European culture.

The nonconformist aspect of flight and aviation as an individual endeavor is particularly evident in the work of late-Soviet dissident artists. In 1984 Kabakov (b. 1933) created an installation about a fictitious character, a lonely dreamer living in earnest solemnity. In his room in a Soviet communal apartment, which Kabakov reconstructs with extraordinary zeal as a total installation, this imaginary man builds a makeshift slingshot that he uses to catapult himself into open space, leaving behind only a hole in the ceiling (Fig. I.20).[21] Flight provides the man with a means of escape from his dilapidated room, the drudgery of everyday life, and the Soviet project altogether. The primitive flight apparatus reveals the paradox of Soviet modernity—its cosmic vision yet shabby materialization. The escape from the communal apartment also symbolizes the paradoxes of Soviet life, philosophy, and practice: in spite of all its limitations the USSR provides the hero with a means of escape, without which he never would have been able to reach the sky. The installation appropriates official discourses around aviation: while an individual endeavor, flight relies on a collective effort; that is, the pilot uses a machine that had been designed and built by many; similarly, Kabakov creates artwork in the name of other, fictitious artists,

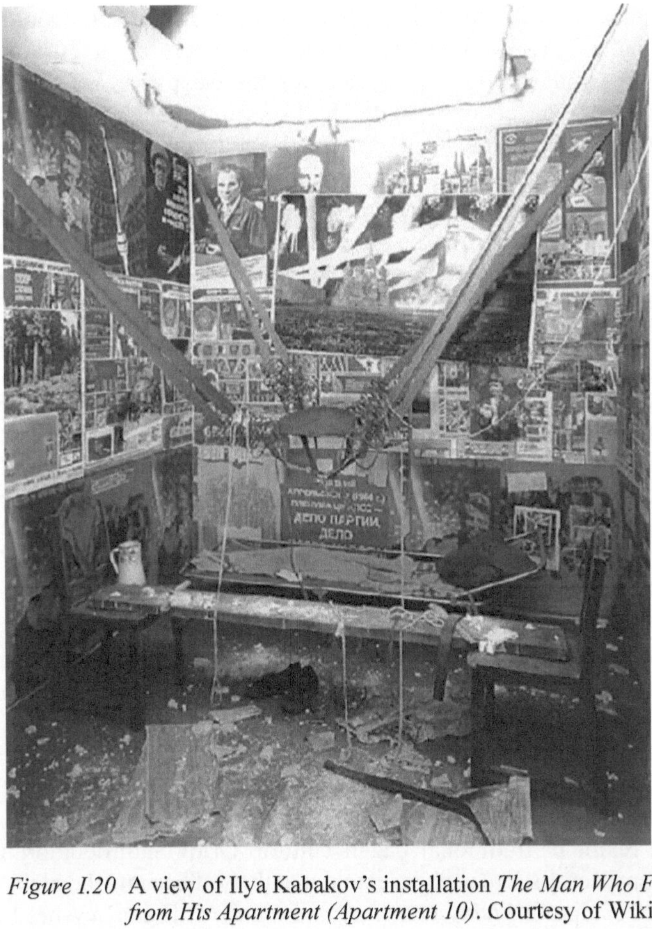

Figure I.20 A view of Ilya Kabakov's installation *The Man Who Flew into Space from His Apartment (Apartment 10)*. Courtesy of Wikimedia Commons

thus suggesting that he is a bearer of an artistic tradition rather than a Romantic type of artist/genius. In other words, Kabakov offers a re-configuration of the chronotope of transcendence. While raising questions of authorship in modernity, Kabakov's installation and especially the technical parameters of the slingshot make a reference to Letatlin, the universal machine of the Russian avant-garde. Thus, the narrative of Soviet modernity and its relationship to flight makes a full circle: it returns to its origins to inspect the genesis of the Soviet project. Like Letatlin, Kabakov's installation conceals the subject of flight, and with it, Soviet subjectivity, transferring the discussion into the realm of mystery. Who is the secretive dreamer? Why did he abandon his apartment? Where did he go?

The postmodern zigzag: flight and the retrospective gaze in contemporary Russian culture

Both Tatlin's and Kabakov's projects in the conspicuous absence of the agent of flight—the pilot—stage the ultimate expression of modernist transcendence insofar as modernity in Tatlin's case and postmodernity in Kabakov's produce the new subjectivity as idea, which is there to be fetishized in the ideological discourse and popular culture (see Chapters 1 and 2 on Soviet art and Chapters 6 and 8 on Soviet film). Indeed, flight and the aviator/space explorer developed into the Soviet grand narrative, according to which the sacred should arise in history at a moment of crisis and destiny. This explains the cult of the aviator and space explorer epitomized, for example, in the figure of Yuri Gagarin, who was the first to journey into outer space when his Vostok spacecraft completed an orbit of the planet on 12 April 1961. In vogue at the time, his name was co-opted for the most cherished projects of Soviet modernity, including the titles of urban areas (e.g. the Gagarinskii district of Moscow), factories, civic centers, cinemas, leisure parks, and so forth, all across the Soviet Union, as if he could provide an authoritative cosmic perspective on late socialism and its struggles.[22] Through the figure of Gagarin, the USSR attempted to transcend the boundaries of human enterprise, offering a high level of generality and an upward trajectory toward the abstract realm of communism. In the late-Soviet period these aspirations, and with them the whole Soviet grand narrative, would become an object of satirical mockery, as in the work of one of the pioneers of Sots-art, Boris Orlov (b. 1941), whose very surname is avian: 'Orlov' means 'of eagles.'

In the late 1980s Orlov produced a series of sculptures and installations that included an ironic treatment of the Soviet preoccupation with aviation and flight. He made drawings and 3D models of planes that would combine elements of different styles and discourses with a parodic transcending effect. For example, his planes featured landing wheels in the shape of chicken legs, breasts, and testicles—a clear reference to the infamous Baba Iaga from Russian folklore, with her mortar and pestle supplying both male and female metaphors. References to Baba Iaga enabled the artist to transcend modernist discourse and take the discussion into the realm of traditional pagan culture. Orlov anthropomorphized planes, and later spacecrafts,[23] to such an extent that the self became the plane. For example, his 1990 sculpture *Flying Onwards and Upwards* [Letat' vyshe! Letat'

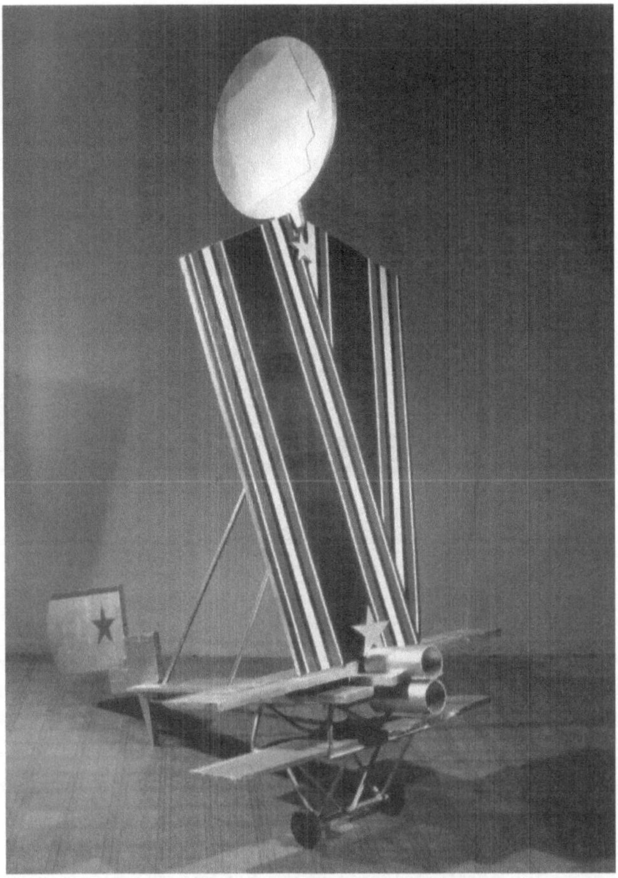

Figure I.21 Boris Orlov's 1990 sculpture *Flying Onwards and Upwards*. Courtesy of the artist

ravnee!] consists of a wooden base shaped like an early plane similar to the WWII IL, painted silver, with an item in the shape of a Soviet ribbon of orders attached to it (Fig. I.21). The ribbon is positioned upside down so that the medal is erected on top of the structure. This attachment is twice the size of the base, and the whole construction resembles a human being, a pilot. The disproportionate size of the pilot part of the sculpture suggests a critique of the cult of aviator and space explorer in Soviet culture: the cult is so grand, so overpowering, that it prevents the plane from flying. Moreover, the structure reveals a significant lack—that of a section or a mechanism enabling a production or capture of the aerial view: instead, it provides a fossilized version of the internal view.

In contemporary Russia the notion of flight returns in a zigzagging, convoluted form as a means of reconciling the contradictions of the new capitalist society. Although the space program is center-staged in Aleksei German Jr.'s *Paper Soldier* [Bumazhnyi soldat 2008], as Julian Graffy argues in Chapter 7 of the volume,

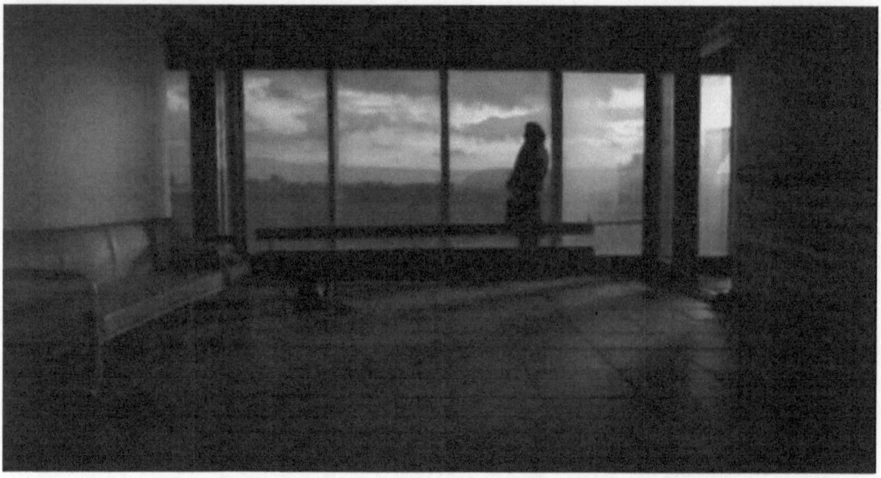

Figure I.22 Lara, a young female flight attendant (Renata Litvinova), waiting for her lover in the terminal building. Vera Storozheva's 2002 *The Sky, The Plane, and the Girl*

the narrative of the film serves to examine Thaw sensibilities. Though the film focuses on the period leading to the first manned space flight, *Paper Soldier* is not concerned with the spectacle of flight. Instead, it focuses on the main characters' personal dilemmas, thus turning the chief Soviet technological achievement into a pretext for reconnecting contemporary subjectivity to its historical predecessors. Similarly, Vera Storozheva's *The Sky, The Plane, and the Girl* [Nebo, samolet, devushka 2002] presents a melodramatic analysis of the love life of a young female flight attendant (Renata Litvinova). Featuring long takes of dark cloudy skies, meant to emphasize the emotional state of the heroine, the film depicts the agents of flight in desolate and claustrophobic spaces, from where, like the character of Kabakov's total installation, they have only one form of escape—death (Fig. I.22). *The Sky, The Plane, and the Girl* makes sense only in relation to the original film on which it is based—Georgii Natanson's *Once More About Love* [Eshche raz pro liubov' 1968]. Here the notion of flight functions as a means of establishing an aesthetic and emotional link between contemporary Russia and the late USSR: by choosing Mikhail Efremov to play the pilot in *The Sky, The Plane, and the Girl*—his father, Oleg Efremov, played the pilot in *One More Time About Love*—Storozheva ascribes a new meaning to flight: the return to and transcendence of history, enabling a retroactive and restorative gaze. The subject of the early Putin era here is on a journey into the space of its own dreams, recollections, and phobias, whereby the cinematic remake functions as a metaphor of the zigzagging path of post-Soviet modernity/Russian postmodernity.[24] (See Chapter 5 on the re-appropriation of the ideology of flight in contemporary art, and also Chapter 9 on the uses of flight to mediate Soviet memory and trauma.)

Especially after the annexation of Crimea and the Malaysia jet crash in the conflict zone of Eastern Ukraine in 2014, which provided Russia with global visibility, there has been an attempt at the governmental level—through the Ministry of Culture—to celebrate and serialize the achievements of the Soviet space program and, as critics would say, to mask the failures of the current program. For example, the government has invested in the production of a film to commemorate Aleksei Leonov's first spacewalk, released in 2016, *First Time* [Vremia pervykh, dir. Iurii Bykov].[25] It will be produced by Timur Bekmambetov, a film director from Kazakhstan with a successful career in both Russia and the US, and, as with Bekmambetov's previous films released in Russia, will feature imagery derived from recent Hollywood blockbusters such as Alfonso Cuarón's 2013 *Gravity*, thus creating another zigzag in the history of flight technology and visual culture whereby a Soviet story will be told using Hollywood techniques for a global audience.

Crowdsourcing in space, or sculpting the future of the Russian space program

In 2009 the Energiia Space Corporation announced an open competition for the best design of work and living environments of spaceships (see Chapter 4 on early Russian designs for living in space). Three prizes were announced at 30,000 ($909), 15,000 ($454), and 10,000 ($303) rubles, respectively. In addition, participants had the prospect of getting contracts with Energiia to work on specific projects. Their portfolios had to include the design concept, a written statement, visualizations, and 3D models. The first prize went to Promgrafika design studio from Nizhnii Novgorod, which developed a redesign of the control panels of the Soiuz-TMA spaceship. Their proposal modernized the appearance of the control panels as well as the design of the company logo, 'The Team of Astronauts,' which would be used on the spaceship and search-and-rescue vehicles on the ground.

The competition was an opportunity for the public to become involved in the Russian space program and a chance for Energiia to acquire new ideas. In this regard, Energiia followed in the footsteps of NASA, which is a pioneer in the field of crowdsourcing, frequently using open innovation to help its teams solve specific challenges, such as its 2013 longeron challenge.[26] The Energiia challenge relinquished the Soviet attitude to space exploration, the technical aspects of which had always been shrouded in secrecy. By inviting the public to respond to real challenges, Energiia attempted to present itself not as an impenetrable bastion of technology, but as an open center of innovation and collaboration (see Chapter 10 on the uses of collaborative play and technologies of simulation in contemporary culture). However, unlike NASA, in the spirit of early Russian space visions, Energiia welcomed holistic elucidations; in other words, the challenge conveners solicited general concepts rather than specific mathematical or engineering solutions. As much is reflected in the decision of the judges to award the first prize to the team that offered not only practical designs for the control panel, but also an Energiia corporate design for all its services and personnel. The Promgrafika project appears to be a concerted attempt to rebrand Energiia and its flagship apparatus. In this regard, the Energiia challenge is characteristic of Putin's corporate capitalism, which, unlike Medvedev's calls for modernization, aims to rebrand and repackage existing industries and products, especially its oil and gas giants, for easier consumption/market flotation both in the country and abroad. This reveals the current government's concerns with re-branding Soviet and Russian technologies to meet the requirements of the new economy of symbols whereby technologies of flight provide Putin's regime with new visual representations of its power.

The Russian government also has burning practical concerns to address. Russia's use of the Baikonur Cosmodrome in Kazakhstan is uncertain because of political instability and financial disputes. As a result, Russia started building a new cosmodrome, Vostochnyi, in the Eastern part of the country in 2010. The construction is on the site of the former strategic missile base. This development is an illustration of the collapse of the Soviet empire and the post-Cold War conversion and re-purposing of military bases and services on a grand scale. Due to be completed in 2018, with the first unmanned rocket launch planned for 2016, Vostochnyi is a massive investment for a country struggling to deliver on its existing commitments. While Russia realigns its technological forces on the ground, the outer space becomes increasingly crowded and competitive, leaving Russia with no option but to optimize its search for talent and innovation. The state no longer has the monopoly on producing the symbolic capital associated with aviation and space exploration or, generally, with the technology of flight. However, Russia remains one of the major global players in terms of producing satellite imagery and other forms of satellite-enabled forms of communication that are paramount in the twenty-first century. For example, GLONASS, or Global Navigation Satellite System,[27] is a space-based navigation system operated by the Russian Aerospace Defense Forces. It provides an alternative to Global Positioning System (GPS) and is the second alternative navigational system in operation with global coverage and of comparable precision.[28] By 2010, GLONASS had achieved 100% coverage of the territory of the Russian Federation, and in October 2011, the full orbital constellation of twenty-four satellites was in position, enabling full global coverage—or a global panorama (see Chapter 10). In addition to supplying imagery from space, GLONASS has a symbolic function of visualizing Russian aspirations on the global stage, which became more articulated with the Sochi Olympics and the political crisis in Ukraine (Fig. I.23) (see Chapter 6 on technological developments, the Cold War, and special effects, and Chapter 3 on political satire and cartoons).

We find a critique of these aspirations in German Jr.'s film *Under Electric Clouds* [Pod elektricheskimi oblakami 2015]. Set in 2017, the centenary of the Bolshevik revolution, the film consists of seven chapters, each telling a story about a particular individual. These individuals are immigrants, engineers, philosophers, cultural entrepreneurs, and so on, each finding him or herself on the margins of contemporary society and all united by their angst about the possible start of a new world war. The film combines two gazes: that of an individual subject who, to borrow from nineteenth-century literature, is a superfluous man, and the other of de-personified, globalized, and omnipresent power. The latter is presented as a series of projections onto electric clouds above the city where the characters live. Streaming advertisements and possibly government messages, the projections are visualizations of a totalizing transcendent gaze from above. This

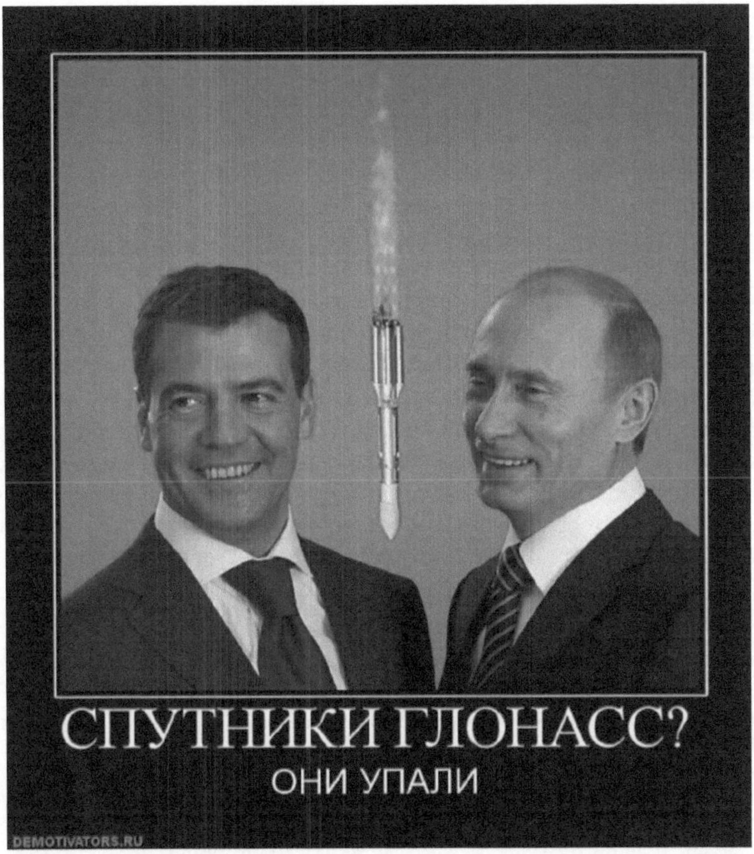

Figure I.23 Anonymous digital poster criticizing GLONASS system by making an allusion to Russian fallen rockets. The phallic symbolism of the falling rocket satirizes the political union of Putin and Medvedev. Courtesy of demativatory.ru

gaze charts the new terrain of human life, where the connection to the sky is not just Romantic, transcendental, but also economic, capitalist. With the sky remaining the only space on the planet not demarcated by the imperatives of capitalism, German Jr.'s film makes a claim about the future of Russian modernity whereby the gaze from above signifies the social and cultural problems of the new century.

***

Unlike previous studies of technologies of flight and Russian/Soviet culture (e.g. Andrews and Siddiqi 2011; Palmer 2006; Siddiqi 2010; Widdis 2003), this volume is not concerned with the history of Russian and Soviet aviation and space exploration. Instead we center on the intersection of the technologies of flight and visual culture in both Soviet and post-Soviet contexts. Equally, unlike the authors of a volume titled *Seeing from Above: The Aerial View in Visual Culture*, edited by Mark Dorrain and Frederic Pousin (I.B. Tauris, 2013), we do not supply a survey of all possible media and forms of representation. For example, we deliberately excluded such important areas as maps, poster art, and photography. This is because our purpose is not to provide a taxonomy of aerial modes of representation but to trace and analyze the impact of the technologies of flight on visual modes of representation in a particular cultural space—Russia, which, as we have demonstrated in our discussion above, on the one hand subscribes to Western European canons, and on the other, charges its energy from its own sources, thus enriching universal visual modes. Our case studies are aimed at theorizing visual culture in the era of transition from analog to digital technologies and also at conceptualizing contemporary Russia in its relations to its own geopolitical predecessors and other countries and their visual traditions. To achieve these objectives, we have organized the volume both chronologically and thematically. Contributions in Part I examine the impact of the technologies of flight in the mediums of painting, drawing (caricature), sculpture, and architecture. Contributions in Part II focus on the moving image by examining representations of technologies of flight in film, animation, and computer gaming (simulation). Chapters 1 to 4 and 6 look at the Soviet period, while Chapters 5 and 7 to 10 reflect on the post-Soviet, contemporary period.

Notes

1 The exclamation invariably greeting the sight of Superman flashing through the sky, in comics, films, etc.
2 Widely attributed to George W. Melville, chief engineer of the U.S. Navy, 1900.
3 Stanislaw Lem, *Solaris.* San Diego, New York, London: Harcourt Brace & Co., 1974. 1.
4 See Mircia Eliade.
5 See Mikhail Bakhtin on Dante's construction of vertical time ("Forms of Time and Chronotope in the Novel") *The Dialogic Imagination*, 1981/1990, 156–8.
6 Scenarios of flight and its attendant concepts of transcendence, 'heightened' imagination, and escape appear not only in screen versions of *Faust*, but also in (1) numerous vampire films, where flying bats symbolize the nocturnal (repressed) urge to defy the cosmic hierarchy (multiple versions of *Dracula* and its derivations), and (2) the philosophical branch of sci-fi visual and verbal texts that equate space exploration with a hubristic self-deification (above all, Stanislaw Lem's classic *Solaris* 1961; Andrei Tarkovskii's film version 1972).
7 See the Syrian rhetorician and satirist Lucian, 120–80 BC—his narrative of travel to outer space, i.e., voyage to the moon as *A True History*.
8 Agamemnon and Apollo, Medusa and Pallas Athena.
9 Icarus' fate shares aspects of the myth of Phaethon, the son of the sun god Helios and the sea nymph Clymene, who likewise plummeted from the sky into the waters below. To prove to his skeptical friends that the sun, Helios, was his father, Phaethon journeyed to the palace of Helios, who confirmed his paternity and promised to grant his son a wish. When Phaethon asked to drive his father's chariot across the sky, Helios tried to discourage him, knowing that no one except the sun could control the horses that pulled the chariot. Like Icarus, Phaethon ignored the paternal warning and took up the reins. Shortly after the chariot ascended skyward, Phaethon lost control of the horses. To stop the fiery destruction resulting from the sun's excessive proximity to the earth, Zeus used a thunderbolt to killed Phaethon, who plunged into the Eridanus River.

10 Stephen was the first Christian martyr, and Joyce's nomenclature slots into the Romantic convention of depicting artists as kenotic geniuses whose gifts are a divine but mixed blessing.

11 More mundanely, terrestrial pigeons actually relayed messages taped to their legs across vast distances, especially during times of travail/war. Hence their specific chronotope, inasmuch as their aerial speed enabled them to traverse humanly non-traversable space along a direct route ('as the crow flies')—avian velocity.

12 It is no accident that in Roger Vadim's comic-based cult film *Barbarella* (1968), action on another planet involves the angel Pygar, who flies through space as a copiously winged creature and mode of transport for the human Barbarella to the city of Sogo.

13 According to Matthew 3:16 and Luke 3:22, the Holy Spirit as dove appears at the baptism of Jesus.

14 See Margaret Ziolkowski's *Literary Exorcisms of Stalinism: Russian Writers and the Soviet Past* (1998).

15 Ptenchiki okrain gorodskikh/Povzrosleli, poleteli./Ptenchiki okrain gorodskikh,/Chto im grozy, chto meteli?// Vy ne sudite strogo ikh,/Vy ne sudite strogo,-/Ptentsov okrain gorodskikh,/Puskai letiat vysoko!// Pripev: Orliata uchatsia letat'!/Orliata uchatsia letat'! Prosti, strana, zabavy ikh,/Prosti, narod rosseiskii,/Ptentsov okrain gorodskikh/Zhdet dal'nii put' armeiskii. Pripev.// Ptenchiki okrain podnialis',/Poleteli stroinym klinom./Priamo v podnebes'e vozneslis',/Kak ikh dedy pod Berlinom!// Khrani, sud'ba, veselykh ikh,/Chtob vse domoi vernulis' -/Ptentsy okrain gorodskikh,/Ptentsy vechernikh ulits . . .

16 For an analysis of the Suprematist response to aviation, including Malevich's, see the excellent article by Christine Lodder, "Transfiguring Reality: Suprematism and the Aerial View," in *Seeing from Above: The Aerial View in Visual Culture*, eds. Mark Dorrian and Frédéric Pousin, London and New York: I.B. Tauris, 2013, 95–117.

17 The original (dir. Ridley Scott) prompted additional, chronologically conscious sequels (all inferior) up to *Alien 5*. The films were directed by James Cameron, David Fincher, Jean-Pierre Jeunet, and Neill Blomkamp (the last yet to be produced).

18 See, for example, the use of trains in Leo Tolstoy's *Anna Karenina*.

19 The imagery of flight and flight technologies was present in all major forms of art during the Soviet period, including canvas and graphic art. Our purpose here is not to provide a survey of these representations, but to theorize flight as a grand narrative of Russo-Soviet modernity.

20 On the ideological uses of Tatlin's sculptures and technologies of flight, see Smith 2014; on the relationship between technology and Tatlin's sculptures, see Cridge 2015; and on the mathematical nature of his art, see Gamwell 2015.

21 For an art-historical analysis of this work see Groys 2006.

22 Similarly, space-inspired architecture dots the space of the former USSR. See Frederic Chauban's outstanding gallery of photographs of these monuments in his *Cosmic Communist Constructions Photographed*, published by Taschen in 2011.

23 See, for example, his drawing titled *The Sirens* (1989).

24 Andrei Konchalovskii's *The Postman's White Nights* [Belye nochi pochtal'ona Alekseia Triapitsyna 2014] documents the divide between the realm of high technologies and the primitive mode of everyday life in contemporary Russia. The film focuses on a community of villagers in the remote area of the Arkhangelsk region, where they survive on poaching and handouts from the state, while a few miles away space rockets are launched from the Plesetsk space station.

25 From the perspective of genre, this film belongs to the category of patriotic blockbuster (see Norris 2012).

26 Longerons are strips of metal used to support solar collectors on the International Space Station. The purpose of the challenge was to find an ultimate angle for longerons to generate as much power as possible during the most difficult orbital positions (see, for example, Bratvold 2013).

27 [Global'naia navigatsionnaia sputnikovaia sistema].

28 Its development began in the Soviet Union in 1976, thus transcending the cultural break associated with the dissolution of the USSR in 1991. Under Putin, the restoration of the system became a top government priority, with relevant funding made available.

Bibliography

Andrews, J. T. and A. A. Siddiqi (eds.) (2011) *Into the Cosmos: Space Exploration and Soviet Culture*, Pittsburgh, PA: Pittsburgh UP.

Bratvold, D. (2013) "NASA Launches Open Innovation Challenge to Improve the International Space Station." http://dailycrowdsource.com/crowdsourcing/projects/1302-nasa-launches-open-innovation-challenge-to-improve-the-international-space-station (accessed 25 August 2013).

Cridge, N. (2015) *Drawing the Unbuildable: Seriality and Reproduction in Architecture*, London: Routledge.

Dorrain, M. and F. Pousin (eds.) (2013) *Seeing from Above: The Aerial View in Visual Culture*, London: I.B. Tauris

Gamwell, L. (2015) *Mathematics and Art: A Cultural History*, Princeton, NJ: Princeton University Press.

Groys, B. (2006) *Ilya Kabakov: The Man Who Flew into Space from his Apartment*, London: Afterall Books.

Harte, T. (2009) *Fast Forward: The Aesthetics and Ideology of Speed in Russian Avant-Garde Culture, 1910–1930*. Madison, WI: U of Wisconsin P.

Higham, R., J. T. Greenwood and V. Hardesty (1998) *Russian Aviation and Air Power in the Twentieth Century*, London and Portland, OR: Frank Cass.

Miller, F. J. (1990) *Folklore for Stalin: Russian Folklore and Pseudofolklore of the Stalin Era*, Armonk, NY: M.E. Sharpe.

Norris, S. M. (2002) *Blockbuster History in the New Russia: Movies, Memory, and Patriotism*, Bloomington, IN: Indiana UP.

Palmer, S. W. (2005) "Icarus, East: The Symbolic Contexts of Russian Flight." *The Slavic and East European Journal*, Vol. 49, No. 1 (Spring): 19–47.

—. (2006/2009) *Dictatorship of the Air: Aviation Culture and the Fate of Modern Russia*, Cambridge: Cambridge UP.

Paschal, H., L. Johnson Dougherty, *et al.* (curators/eds.) (2003) *Defying Gravity: Contemporary Art and Flight*, Raleigh, NC: North Carolina Museum of Art.

Ryan, K. (2002) "Aksenov's ptichii iazyk: Nonsense Reconsidered." *The Slavic and East European Journal*, Vol. 46, No. 1 (Spring): 29–46.

Siddiqi, A. A. (2010/2014) *The Red Rockets' Glare: Spaceflight and the Russian Imagination, 1857–1957*, Cambridge: Cambridge UP.

Smith, M. G. (2014) Rockets. *Revolution: A Cultural History of Early Spaceflight*, Lincoln, NE: Nebraska UP.

von Geldern, J. and R. Stites (1995) *Mass Culture in Soviet Russia; Tales, Poems, Songs, Movies, Plays, and Folklore 1917–1953*, Bloomington, IN: Indiana UP.

Widdis, E. (2003) *Visions of a New Land: Soviet Film from the Revolution to the Second World War*, Yale: Yale UP.

Wohl, R. (1994) *A Passion for Wings: Aviation and the Western Imagination 1908–1918*, New Haven, CT and London: Yale UP.

Ziolkowski, M. (1988) "The Reversal of Stalinist Literary Motifs: The Image of the Wounded Bird in Recent Russian Literature." *The Modern Language Review*, Vol. 83, No. 1 (Jan.): 106–120.

Part I

Art and architecture

1 Ever onwards, ever upwards?

Representing the aviation hero in Soviet art

Mike O'Mahony

"[T]he invention of the airplane was at first perceived by many as an aesthetic event with far-reaching implications for the new century's artistic and moral sensibility."

(Wohl 1994: 1)

Introduction

The emergence and development of aviation in the years between 1903 and 1927 undoubtedly marked one of the major transformations of the early twentieth century. In less than a generation, powered flight rapidly advanced from the first tentative experiments of the Wright brothers at Kitty Hawk, North Carolina, to Charles Lindbergh's epoch-shifting crossing of the Atlantic. Yet early aviation achievements did more than transform the technological landscape of the modern world, heralding a new age of movement and speed. They also facilitated a significant transformation within the broader field of cultural production. Thus while technicians and designers were striving to develop the capabilities of the earliest airplanes, writers and artists were similarly striving to find new ways to articulate, in both word and image, the shifts in consciousness brought about by flight.

Painters were among the earliest cultural practitioners to explore and embrace the implications of aviation. In Paris, for example, then the epicenter of artistic experimentation, aviation's capacity to alter the way in which both time and space conventionally had been conceived had a direct impact upon the most innovative artists of the era. Among these, both Pablo Picasso (1881–1973) and Robert Delaunay (1885–1941) consciously exploited the implications of powered flight in their search for new visual vocabularies to express the experience of modern life.[1] In Italy, too, the heady, intoxicating speed, energy, and sheer danger of early aviation similarly inspired Futurists such as Carlo Carrà (1881–1966), whose belligerent, nationalist, and militaristic *Patriotic Festival* of 1914 drew explicitly upon the vortex-like, spinning form of an airplane propeller.

Given the Russian Empire's industrial backwardness in the early years of the twentieth century, it is perhaps surprising that this rich dialogue between technological advances in powered flight and avant-garde cultural developments was established from the earliest days of aviation history. For example, one of the nation's earliest aviators, Vasilii Vasil'evich Kamenskii (1884–1961), was also a major figure in both the verbal and the plastic arts. Working alongside such influential pre-revolutionary members of the Russian Futurist circle as David Davidovich Burliuk (1882–1967), Velimir (Viktor Vladimirovich) Khlebnikov (1885–1922), Aleksei Eliseevich Kruchenykh (1886–1968), and Vladimir Vladimirovich Maiakovskii (1893–1930), Kamenskii explicitly drew upon his personal experiences as a pilot in much of the work produced during this period. Defining himself as a "poet-aviator," Kamenskii reinforced the strong bond between the technological and aesthetic dimensions of flight in his declamatory poems and public performances, seeking to evoke the sensations of flight in both verbal and visual form.[2] Other pre-revolutionary artists similarly attempted to re-conceptualize the physical world in the wake of the aviation breakthrough, as evidenced in Kazimir Severinovich Malevich's (1879–1935) *Simultaneous Death of a Man in an Airplane and at the Railway* (1913) and *The Aviator* (1914), both of which can be included among the artist's early attempts to reconfigure the representation of time and space in art (Wohl 1994: 166–71). Further, as the vital role that aviation would come to play in the military sphere became more evident, Natal'ia Sergeevna Goncharova (1881–1962) also invoked aviation—for example, in *Angels and Airplanes*, one of fourteen lithographs collectively published under the title *Mystical Images of War* in 1914. Here, Goncharova intriguingly embraced the apocalyptic potential of flight as she sought to articulate visually a link between the traditional folk culture of ancient Russia and the modern world of aviation. It is thus clear that, from the outset, aviation provided much more than merely a new subject matter for artists to exploit. Its potential to invoke such metaphorical concepts as ascension and descent, or fall, and conflate these with a modern reconceptualization of time and space ultimately signaled that aviation could indeed provide a rich seam for artists to mine.

Imaging aviation after October

Despite the endeavors of the pre-revolutionary avant-garde, early Bolshevik artists initially proved slow to respond to the challenges posed by aviation. As Scott Palmer has demonstrated, the new regime certainly sought to exploit the symbolic significance of aviation to promote Soviet technological advances (Palmer 2006: 85). However, artistic attention at this time was focused predominantly upon the production of official posters celebrating what Palmer has referred to as "air-mindedness" (Palmer 2006: 2). In this context it is perhaps surprising that throughout the 1920s and early 1930s, relatively few paintings or sculptures on the aviation theme found their way into major exhibitions and the growing public collections of official Soviet art. The emergence of Socialist Realism, however, would begin to alter this pattern, not least as the first tentative steps in this new, if ill-defined, artistic method coincided with what was to become the golden age of Soviet aviation. Accordingly, in this chapter I want to examine representations of aviation produced during the mid- to late 1930s, specifically under the aegis of Socialist Realism. My purview includes both painting and sculpture, and highlights works that were proposed and rejected as well as those that were successfully executed and achieved popularity. As key products of Socialist Realism, representations of aviation are frequently dismissed as little more than officially endorsed paeans to state policies. What I aim to demonstrate, however, is that these works engaged in complex and diverse ways with the ever-changing significance of both aviation and artistic developments. As such, they can be regarded as fascinating and intriguing documents providing visual evidence of the fluctuating, rather than stable, political and cultural environment of the time. In particular, I focus on three forms of artistic response to the aviation theme: portraits of the famous pilot, Valerii Chkalov; monuments to the aviation heroes of the Cheliuskin mission; and representations of other major successes and failures in Soviet aviation.

Imaging the heroic pilot: portraits of Chkalov

The early days of Soviet aviation can be characterized by the fetishization of the machine. Here, the airplane, rather than its operator, was prioritized as a dominant symbol of Soviet technological and industrial progress. While the pilot was recognized as a valuable cog in the greater socio-political machine, his or her role, like that of the worker, remained essentially subordinate to the collective will of the masses. By the mid-1930s, however, Socialist Realism's promotion of the positive hero was beginning to leave an indelible mark on Stalinist culture. Celebrating the exemplary achievements of individuals, the most notable example being the famous shock-worker Aleksei Grigor'evich Stakhanov (1906–77), was now given precedence, deflecting attention away from the technological to the human. Aviation similarly witnessed a shift of emphasis toward the celebration of the heroic pilot. Record-breaking flights were planned and executed, as much, it seems, for their publicity as for their scientific value, and the pilot rapidly came to symbolize the archetypal Soviet New Person [novyi chelovek]. Despite the fact that the record-breaking flights of the late 1930s were all achieved through the efforts of a team of engineers, aircraft designers, back-up and auxiliary staff, and flying crews of at least three, most of the media adulation was reserved for the individual. This provided both an opportunity and a challenge for artists to develop a new mode of portraiture focused on a new kind of hero. Here, the artistic representations of one pilot in particular, Valerii Pavlovich Chkalov (1904–38), offers an interesting insight into contemporary attitudes toward the aviator-hero and the artistic conventions adopted to represent this new character.

Chkalov was the best known, and most frequently represented, of all Soviet pilots. Born to a humble peasant family in the town of Vasil'evo, he rose to become one of the Soviet Union's most famous citizens. During his lifetime Chkalov received two Orders of Lenin and the Order of the Red Banner, and became an elected Deputy of the Supreme Soviet of the USSR. He even had two regions renamed in his honor.[3] Chkalov's aviation career began while he was still in his teens, working as an airplane assembler for the Red Army during the Civil War. He soon graduated to flying school, joining the Red Banner Fighter Squadron in 1924. By the end of the decade he had become a fully fledged test pilot for the Soviet Air Force. Chkalov first achieved widespread public recognition in 1936, when, together with co-pilot Georgii Filippovich Baidukov (1907–94) and navigator Aleksandr Vasil'evich Beliakov (1897–1982), he completed a record-breaking long-distance flight that captured the headlines in the Soviet press. Taking off from Moscow in an ANT-25 aircraft strategically named *Stalin's Route* [Stalinskii marshrut], the crew flew along the edge of the Arctic Circle and down over the Kamchatka peninsula before landing on the island of Udd. Airborne for a total of 56 hours and 20 minutes, Chkalov's airplane covered a total distance of 9,374 km (5,825 miles). Yet it was not simply the achievement that catapulted Chkalov into the public eye. On his return to Moscow, the successful pilot was greeted by none other than Stalin. The intimate embrace of the Soviet leader and hero-pilot, captured by a press photographer and subsequently published on the front page of *Pravda,* rapidly became an iconic image symbolizing the close bond between Stalin and Chkalov as the most favored of all Soviet aviators (McCannon 1998: 107). The following year, Chkalov and his team undertook an even more daring mission, flying directly over the North Pole to the United States. Although the distance covered was less than the flight of the previous year, the nature of this achievement, traversing the most inhospitable terrain on the globe, confirmed for Chkalov a celebrity status subsequently matched in the Soviet Union only by the achievements of cosmonauts.

By now Chkalov's media popularity was fully established. He was fêted in the Soviet press, and his life story, largely conforming to the standard precepts of an emerging Socialist Realist biographical literature, made frequent appearances in articles and books. In 1941, Lenfilm studio released a dramatic movie recounting his exploits.[4] While such literary and cinematic representations of Chkalov have attracted scholarly attention, portraits of the pilot by contemporary artists have largely been overlooked. Yet these compelling images, in their specific attempt to find a visual vocabulary to convey Chkalov's public reputation and to epitomize his capacity to overcome all obstacles in his unswerving dedication to the state, can reveal a great deal about the Chkalov myth and contemporary attitudes to aviation. They also offer fascinating insights into the artistic representation of Soviet heroes in the early period of Socialist Realism.

Among the numerous contemporary painted portraits of Chkalov, the most famous is perhaps that produced by Mikhail Shteiner in 1937 (see Fig. 1.1). Shteiner notably represented Chkalov as a rock of a man, exuding both physical and mental strength. The monumental proportions and the adoption of an upright posture and stern expression, focused beyond the viewer, certainly conform to conventional Socialist Realist modes of portraiture, invoking heroic purposefulness. Yet there is also something individualistic and slightly unconventional about the representation of this figure. He is depicted leaning slightly backwards, defying military deportment, one hand casually shoved in his pocket, the other holding a pipe. Significantly, Chkalov's reputation to a large extent had been forged on precisely such unconventionalities. As a young pilot he had been grounded for performing daredevil maneuvers without official authority, including, most famously, flying beneath the Trinity Bridge in Leningrad (Baidukov 1991: 54–9).

Interestingly, this very unconventionality was often presented in a positive light, as a sign of the pilot's youthful energy, his indefatigability in constantly testing the limits of his skill and endurance. In this way, as Katerina Clark has pointed out, aviation heroes were typically presented as "sons" to "father" Stalin in the great metaphorical Soviet family (Clark 1981: 124–9). Their youthful exuberance was thus seen not as a sign of challenging discipline and authority, but as a spontaneous expression of impatience, a desire to commence their ritualistic trial of passage to manhood, to fulfill their obligations to the state.

In Shteiner's portrait, Chkalov's costume is similarly unconventional. A uniform collar and tie appear beneath a fur-lined jacket, suggesting the pilot's endurance of the cold. Yet this costume also resembles a more informal track suit, an allusion to Chkalov's physical fitness and his widely publicized participation in sport and physical culture. As aviation historian Robert Wohl has pointed out,

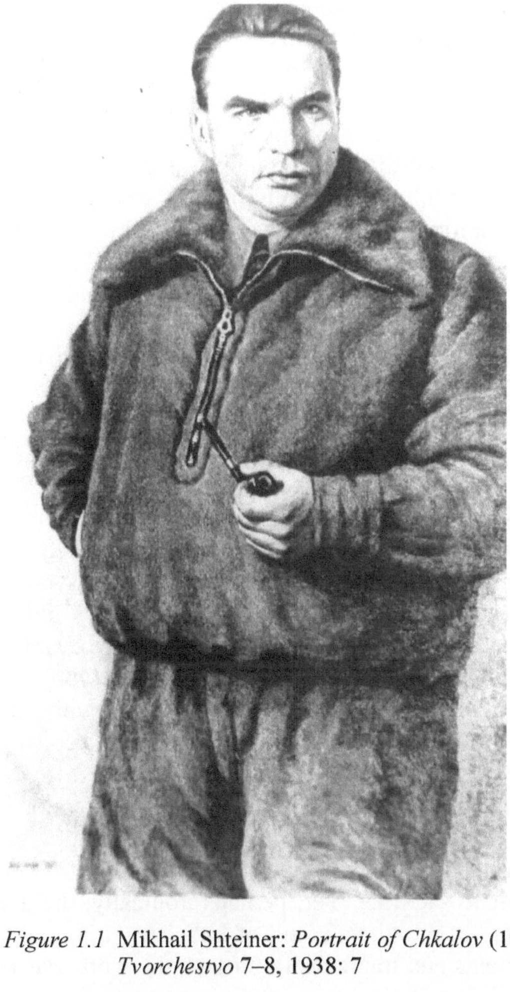

Figure 1.1 Mikhail Shteiner: *Portrait of Chkalov* (1937), oil on canvas. From *Tvorchestvo* 7–8, 1938: 7

the links between aviation and sport had been well established throughout Europe since the invention of the aeroplane, perhaps epitomized in the figure of French sportsman and WWI ace Roland Garros (1888–1918) (Wohl 1994: 203–5). In both Tsarist Russia and the Soviet Union, this link was similarly embraced, with aviation achievements frequently appearing on the pages of the specialist sports press. Chkalov, notably, was adored by the sports media. In his youth he had been an avid participant in soccer, tennis, swimming, and gymnastics; later on he developed a love for motorcycling. He used his fame to establish motorcycle competitions and encourage record-breaking achievements, published articles on his involvement as a referee at major sports events, and was renowned for encouraging children to take up sport (Baidukov 1991: 282). Shteiner's presentation of Chkalov as simultaneously a heroic individual of great authority and discipline, and a slightly unconventional man of the people, thus drew together both labor and leisure, directing both toward serving the state. In this context, his portrait of Chkalov can be seen as exemplary in establishing a conventional Socialist Realist mode for the representation of the new Soviet citizen.

In December 1938, while on a routine test flight, Chkalov encountered disaster. After experiencing engine difficulties, he refused to abandon his plane and instead attempted a crash landing. He was killed instantly. Tributes to Chkalov soon appeared in all the major journals and the aviation hero was given a prestigious burial spot in the walls of the Kremlin, with Stalin himself acting as principal pall-bearer. As part of the memorialization process, the art journal *Iskusstvo* reproduced Shteiner's portrait alongside the great pilot's obituary.[5] Within a year, several more portraits of Chkalov, including Shteiner's, were put on display at the 1939 exhibition *Stalin and the Peoples of the Soviet Union* at the Tretyakov Gallery in Moscow. This exhibition also introduced the public to a work that subsequently became the most famous representation of the pilot; a bronze bust of Chkalov by Sarra Dmitr'evna Lebedeva (1892–1967) (see Fig. 1.2).

Awarded a state prize by the Committee for Art Affairs in 1940, the bust captures the very essence of physical endurance that typified Chkalov's flights (Ternovets 1940: 39). Focusing solely on the head, with no costume details to suggest the pilot's profession, Lebedeva produced a loosely finished, rough-surfaced bronze cast, still emphasizing the clay from which it was originally molded. The facial features are ill-defined, marked out with scratches and indentations reminiscent of the natural erosion of the landscape, and the eyes remain unformed. Here, Lebedeva represented a pilot whose hardened and worn features compellingly reflect the very harshness of the Arctic landscape with which he was associated in his lifetime. Yet the very passivity of his expression also suggests a calm, almost natural, acceptance of his fate. Chkalov's capacity for extreme physical endurance is thus marked both externally and internally. As a monument to the dead hero, Lebedeva's work aptly captures the essence of Chkalov's existence, his willing engagement with the extremes demanded of a record-breaking Soviet test pilot and his ultimate sacrifice. Yet, perhaps ironically, the appearance of the finished work was not the direct result of Lebedeva's interpretation of the pilot's heroic achievements and tragic death. In fact, the work was begun

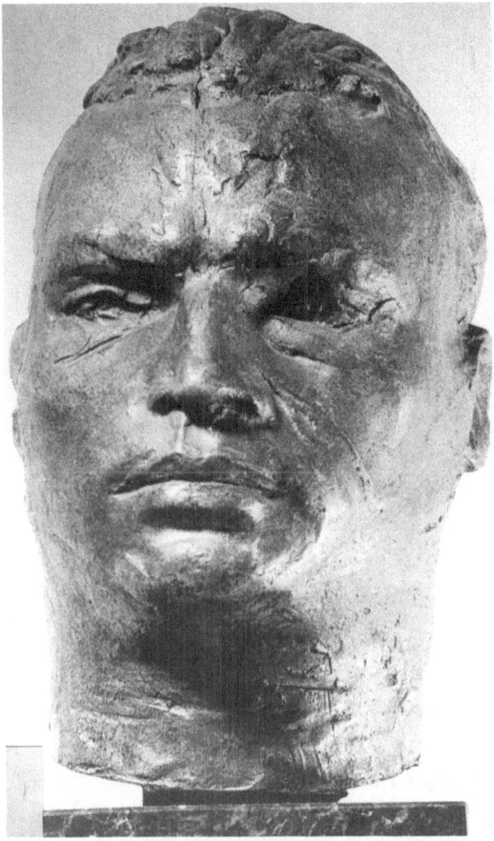

Figure 1.2 Sarra Lebedeva, *Chkalov* (1936), bronze. State Russian Museum, St. Petersburg

in 1936, following Chkalov's first long-distance Arctic flight. According to Boris Ternovets, Chkalov posed for Lebedeva on two separate occasions (Ternovets 1940: 16). The huge demands made upon his time, however, made it difficult to arrange further poses and Lebedeva left the clay model unfinished in her studio. Over the next two years the clay dried out, shrinking a little in the process, but not cracking. When the news of Chkalov's death broke, Lebedeva was asked to complete her sculptural portrait as a monument to the aviation hero. She refused, seemingly unable to work on it in the light of such a great loss to the nation. It would be another year before Lebedeva was able to return to the work. Rather than complete it, however, she now had it cast in bronze in this unfinished condition, no doubt recognizing how the rough surface and details of the original clay model now contributed to the symbolic rendering of Chkalov's life.

Shteiner's and Lebedeva's representations of Chkalov, both originally produced within the pilot's lifetime, notably received official approval as posthumous monuments to the aviation hero. Each represents Chkalov, metaphorically, as an implacable rock capable of withstanding any onslaught. Both foreground physical and mental endurance as prime qualities for the Soviet pilot—the very qualities that the state and military institutions hoped would be induced in the broader population. As key representations of Soviet aviation, they celebrate the achievements of the nation, though notably by reference to the physical body of the individual, which emphasizes Chkalov's masculinity.

Vera Mukhina and the monument to the Cheliuskin heroes

Other artists, however, adopted a different language to express the nation's progress in aviation. Here, interestingly, the success of Shteiner and Lebedeva can be contrasted with the official rejection of a project executed by the Soviet Union's best-known monumental sculptor, Vera Ignat'evna Mukhina (1889–1953).

In 1937, on a visit to an airplane factory outside Moscow, Mukhina was invited to be a passenger on a training flight. While in the air the anonymous pilot performed a 'death-loop,' taking the plane through 360 degrees in the air, and it was only on landing that Mukhina discovered that the pilot for her very first flight was none other than Chkalov (Voronov 1989: 192). Saddened by his subsequent death, Mukhina proposed that a monument be erected both to Chkalov and to aviation heroes in general. The monument she proposed was to be a domed pantheon containing urns with the ashes of past aviation heroes. At the center of the pantheon, Mukhina planned to install a huge sculpture representing the mythological image of *Icarus* [Ikar] in the final instance before his crash into the ocean (Zamkov 1989: 72–73) (Fig. 1.3). This monument, symbolically associating Chkalov with Icarus, was always unlikely to gain state approval. Problematically, Icarus' demise was as a consequence of flying too close to the sun, against the advice of his father, Daedalus. While this could have been read as an extension of Chkalov's unconventionality, the consequences hardly subscribed to the heroic image of the fearless but wise pilot. Moreover, Mukhina's proposal focused exclusively on the tragedy of Chkalov, his final fatal moment, rather than on his heroic achievements, thus prioritizing the failures rather than the successes of the Soviet aviation program. And finally, Mukhina's adoption of a mythological subject matter specifically rejected the image of the modern pilot. Mukhina's pantheon would never be built.[6]

Mukhina's choice of a mythological subject to express flight is perhaps surprising, not least because earlier that very year she had encountered official resistance to such a form of representation. In the summer of 1938, the Moscow City Soviet launched a competition to produce a monument to the heroes of the Cheliuskin expedition, the first event in Soviet aviation to attract worldwide publicity (Abolina 1954: 50). In February 1934, the entire nation was gripped by news of a major disaster. The scientific exploration ship, the *Cheliuskin,* while on a mission to find a safe passage along the northern coast of Russia, became trapped in the icy waters of the Arctic Ocean, was crushed, and sank to the bottom of the sea. The ship's crew of one hundred and four, including a six-month-old baby, managed to escape the vessel but were now stranded on a moving ice-floe.[7] With the spring thaw approaching, and food supplies limited, the fate of the crew

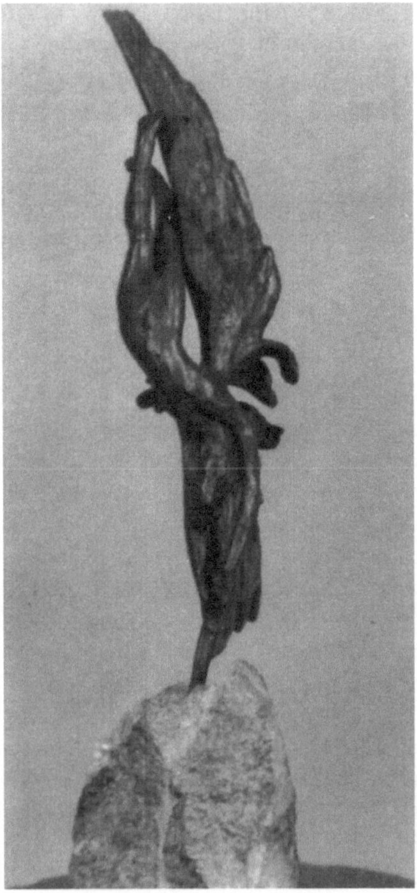

Figure 1.3 Vera Mukhina, *Icarus* (1938), bronze. State Tretyakov Gallery, Moscow

looked bleak. The Soviet authorities, desperate to save both the crew members and an international reputation, launched a daring rescue mission. Several aircraft flew to the polar region, pinpointed the crew, landed on the melting ice, and successfully evacuated them. The Cheliuskin rescue was celebrated both nationally and internationally. It even resulted in the establishment of a new honorific title, Hero of the Soviet Union, bestowed upon seven of the pilots.

The subsequent proposal to erect a monument to the Cheliuskin heroes was launched as part of the Reconstruction of Moscow project. The recent completion of the Moscow-Volga Canal had connected the capital to the principal water route to the sea, thus necessitating the rebuilding of several bridges across the Moskva River to allow for the passage of larger vessels. To symbolize both the naval context of the Cheliuskin mission and the air link between the capital and the far north that had enabled the rescue, the Moscow City Soviet proposed that the monument be built as part of a bridge (Voronov 1989: 178–9). Although linking so many disparate factors proved to be a difficult task, Mukhina willingly took up the challenge. Doubtless she was hoping to reinforce her reputation as the pre-eminent sculptor of the age following her recent success in Paris with the monumental *Worker and Collective Farm Woman* [Rabochii i kolkhoznitsa] crowning the Soviet Pavilion at the 1937 *Exposition Internationale.*

Initially, Mukhina produced two plaster models, no longer extant, based upon a simple obelisk form and replicating the verticality of the Palace of the Soviets planned to be built nearby (Voronov 1989: 179). These, however, were soon abandoned. Teaming up with her brother-in-law, the architect Sergei Andreevich Zamkov, she now proposed a third project comprising three separate sculptural groups spread horizontally across the span of the bridge. The central supporting pylons were each to act as pedestals for two of these sculptural groups; on the left, the isolated crew of the Cheliuskin, as if still stranded on the ice; on the right, an airplane and its crew, the rescuers, having landed. In these two groups Mukhina stuck rigidly to a literal narrative representation, expressing the two key elements within the Cheliuskin episode. However, these groups were essentially to form a backdrop to the principal feature of the bridge design. From these pylons, two concrete piers extended out into the river to form a triangle, at the apex of which, rising as if from the river itself, Mukhina incorporated a colossal figurative sculpture of *Boreas* [Borei], the mythological personification of the north wind (see Fig. 1.4). Represented as a scowling, bearded old man, Boreas soars through the air with arms outstretched. His long hair and beard are hurled around in a maelstrom while his bearskin cloak acquires a life of its own in the gale-force winds. Despite his aged appearance and rugged features, Boreas' naked, lithe body conveys the very power and strength of raw nature, reinforcing a sense of the physical endurance required of both crew and rescuers in the Cheliuskin mission. To overcome the problem of representing movement and dynamism, Mukhina replaced the traditional stone pedestal with one made of colored and faceted glass.[8] Resembling an iceberg rising directly from the river, and thus signifying the Arctic conditions of the rescue mission, this design would certainly have acquired an extra effectiveness in Moscow's frozen winters. Above this structure, as if carved from the frozen landscape itself, the figure of Boreas rises into the sky.

Mukhina's adoption of a figure from classical mythology notably contrasted with her recent emphasis on modern Soviet citizens, the urban worker, and 'kolkhoznitsa' deployed in Paris. Perhaps unsurprisingly, the Moscow City Soviet commissioners rejected the Boreas subject matter.[9] As Mukhina later complained:

> People were not aware of the mythological tales from antiquity. Boreas, or Zephyr, or any of the personifications of the winds and natural forces, as understood in the past, were not close to the masses; for a modern monument it was essential to use real forms taken from real life.[10]

Still determined to win the commission, Mukhina produced yet another version of the monument, this time working with the architect Aleksandr Vasil'evich Vlasov (1900–62). In this fourth version Mukhina incorporated a pilot standing four-square upon a stone pedestal and gazing purposefully forward (see Fig. 1.5).

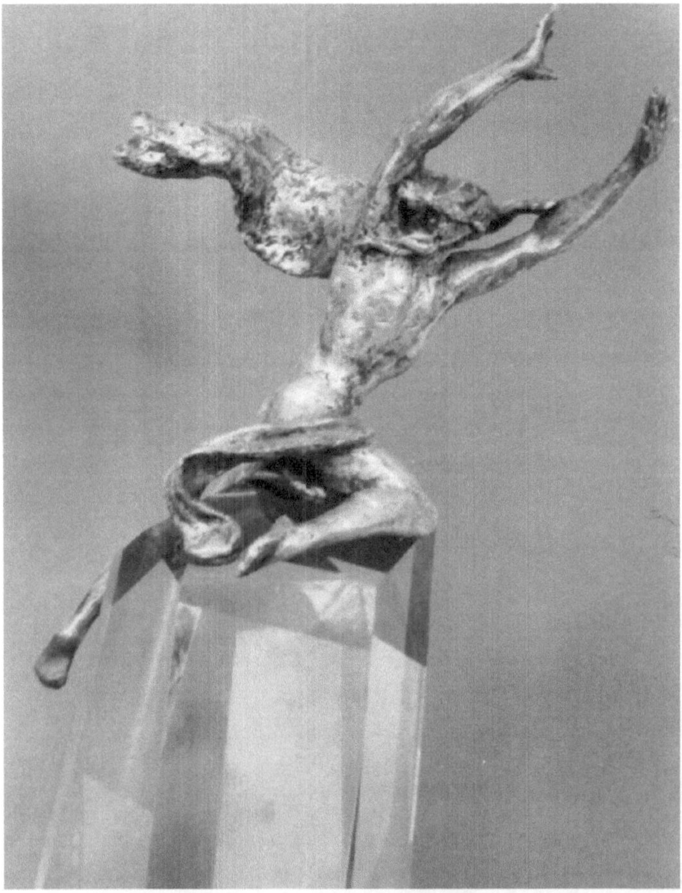

Figure 1.4 Vera Mukhina, *Boreas* (1938), bronze and glass. State Tretyakov Gallery, Moscow

The figure is dressed in a modern air force uniform, with an enormous propeller poised on one of its three points in his outstretched arms. Despite its huge weight, the propeller is held lightly in the pilot's arms, which extend like the wings of an aircraft. This gentle grip invokes the skill and control required of pilots in flight, the propeller effectively standing in for the joystick of a metaphorical aircraft. Here, man and machine are conflated into one monumental representation. An existing drawing by Vlasov further reveals that the pilot was originally designed to stand on the prow of a ship, cast from concrete and rising from a staircase leading down into the Moskva River (Abolina 1960: 179) (see Fig. 1.6). In this position he resembles a ship's figurehead, thus referencing the naval context of the Cheliuskin. The Moscow City Soviet commissioners found this version of the design for the Cheliuskin monument much more acceptable, but delays and the commencement of war prevented the project from being fulfilled (Abolina 1954: 50).

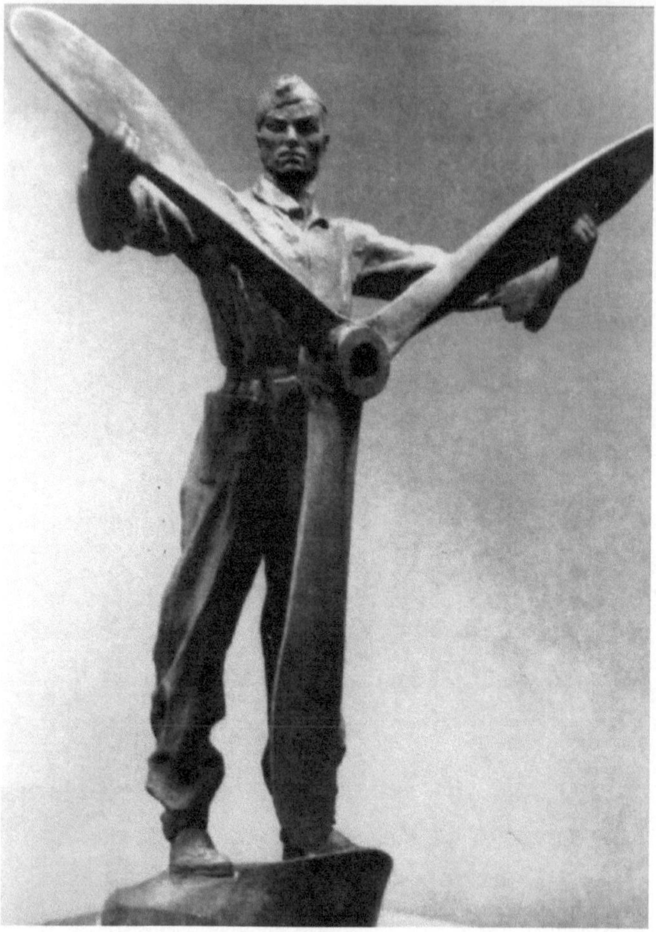

Figure 1.5 Vera Mukhina, *Pilot* (1938), bronze. State Tretyakov Gallery, Moscow

The official rejection of the mythological subject matter in both the *Icarus* and *Boreas* monuments is indicative of the overriding importance attached, at this time, to the representation of contemporary figures. Thus, in her final version Mukhina rejected the allegorical personification of the north wind, reverting to the more heroic figure-type typical of Socialist Realism. Yet, in contrast to the previously mentioned portraits of Chkalov, Mukhina presented an anonymous pilot, a generalized everyman symbolizing the potential for great deeds, not only of individuals, but also the population at large.

Visualizing the successes and failures of Soviet aviation

While Soviet aviation experienced many major successes during the 1930s, these were inevitably accompanied by a series of tragic accidents. Perhaps the most

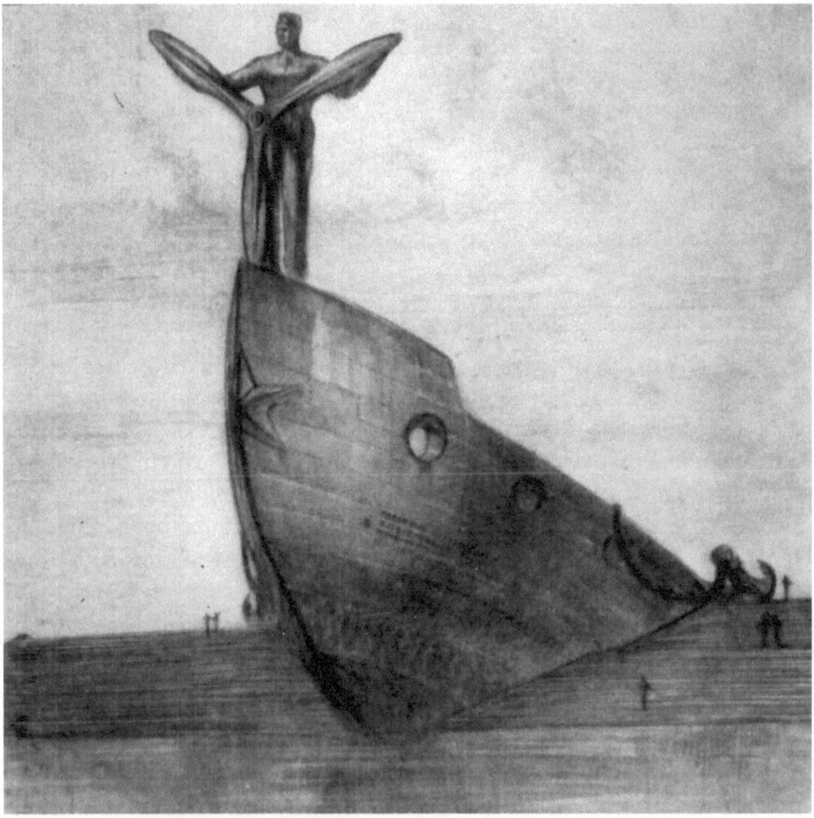

Figure 1.6 Aleksandr Vlasov: *Pilot* (1938), drawing for Cheliuskin monument. R.A. Abolina, *Mukhina*, Moskva: Iskusstvo, 1960. Pl. 193

famous concerned the rise and fall, quite literally, of the famous air colossus, the ANT-20, named the *Maksim Gor'kii.* Claimed to be the largest aircraft in the world, the *Maksim Gor'kii* was designed and constructed in just eighteen months and made its first public appearance in June 1934, when it was flown directly over Red Square in Moscow as part of the celebration for the return of the *Cheliuskin* crew (Palmer 2006: 208). Despite its record-breaking size, the *Maksim Gor'kii* was little more than a propaganda weapon designed to publicize and promote Soviet aviation achievements. As such, it was widely recorded in photographs and on film. It also attracted the attention of contemporary artists. Shortly after its maiden flight, Vasilii Vasil'evich Kuptsov (1899–1935), a student of the out-of-favor painter Pavel Nikolaevich Filonov (1883–1941), produced a large-scale work representing the aircraft flying high above the city of Leningrad on a festival day (see Fig. 1.7). Kuptsov's meticulously detailed and coloristic handling lends a fairy-tale quality to the painting. This aspect is further reinforced by the adoption of a viewpoint suggesting an omniscient spectator looking down from the heavens upon this magical marvel of Soviet aviation. Below, minuscule figures can be seen marching across Vasil'evskii Island toward Palace Square, where the Winter Palace and gardens are decorated in a style reminiscent of the early days of the Bolshevik State. The festive mood is further enhanced by colorful parachutes dropping, confetti-like, what appear to be propaganda messages, doubtless extolling the virtues of Soviet achievements. The presence of a flotilla of sailboats on the Neva simultaneously reinforces the notion of both leisure and transportation. Here, however, the inclusion of a battleship and the accompanying smaller aircraft reminds the viewer of the militaristic objectives of the ANT-20, designed, if never deployed, as a long-range, heavy-duty bomber (Palmer 2006: 206). Produced in the same year as the declaration that Socialist Realism would now be the only accepted method for artists, Kuptsov's overt celebration of the *Maksim Gor'kii* provides visual evidence that while the focus on officially sanctioned subject matter was integral to the emergence and consolidation of a Socialist Realist aesthetic, stylistic variation remained a significant component within many artistic practices in the immediate wake of the Soviet Writers' Congress.

In May 1935, less than a year after Kuptsov completed his canvas, the significance of the *Maksim Gor'kii* for Soviet aviation history was altered forever. A mid-air collision with one of the smaller airplanes that habitually accompanied

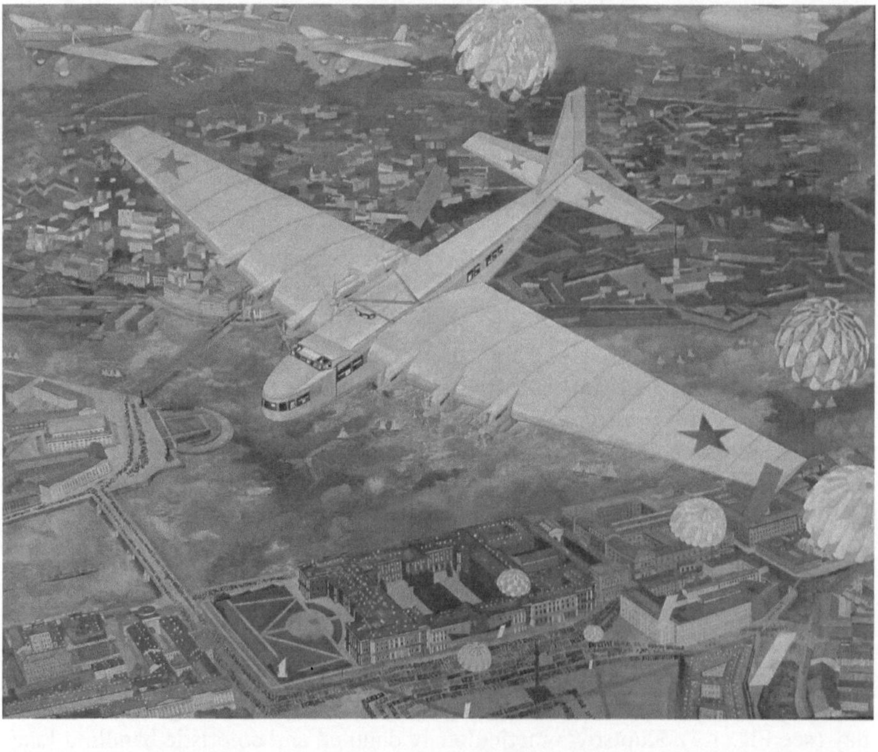

Figure 1.7 Vasilii Kuptsov, *The Maksim Gor'kii* (1934), oil on canvas. Russian Museum, St. Petersburg

its flight caused the ANT-20 to crash, killing its crew of eleven. To make matters worse, a further thirty-six passengers, shock-workers invited to join the special flight, were also killed. Overnight, Kuptsov's work was effectively transformed from a celebratory paean to Soviet aviation successes into a visual requiem for the lost crew and airplane. As a consequence of this tragedy, Kuptsov's work doubtless acquired an additional sense of nostalgic pathos for some viewers. For others, perhaps, its naïve optimism may also have implied a sense of hubris.

While subsequent and unforeseen events contributed to an alteration in the potential meaning and reception of Kuptsov's image, for other painters such incidents provided a starting point. In January 1934, a team of stratospheric balloonists headed by Pavel Fedorovich Fedoseenko (1898–1934) undertook their maiden flight in a specially constructed craft named the *Osoaviakhim-1.* The hydrogen-filled balloon had been designed to reach an altitude of approximately 20,500 m (67,250 ft), thus establishing a new world record. However, having achieved this height, Fedoseenko then implemented the high-risk strategy of ascending still further, climbing to 22,000 m (72,000 ft). Tragically, during the descent, Fedoseenko lost control of the balloon, which subsequently crashed to earth, taking the lives of all three crew members. Although the wisdom of Fedoseenko's decision to overstep the design limits of the *Osoaviakhim-1* was certainly questioned, this did not prevent him and his crew from being posthumously awarded the Order of Lenin.

Shortly after the accident, the painter Georgii Nikolaevich Bibikov (1903–76) began work on a large-scale painting immortalizing the event (see Fig. 1.8). Unlike Kuptsov, Bibikov produced his image with the full benefit of hindsight, thus allowing the artist to incorporate subtle clues to the tragedy about to unfold. Bibikov's work is clearly derived from contemporary photographs that show the crew of the *Osoaviakhim-1* proudly standing on the spherical gondola in which they would make the ascent. Similarly, the compositional focus on the Red Army soldiers surrounding the craft and holding the ropes prior to the take-off is also a direct quotation from these documentary sources. Other details, however, add a notable sense of unease to the scene. The mood of the image, for example, despite representing the dramatic moment of take-off, is remarkably subdued. A sense of reflective silence, rather than fanfare, pervades the scene, and the majority of facial expressions are represented as more pensive than animated. One figure alone, at the apex of the group of pilots, is shown smiling, and this is the instrument operator and physicist Il'ia Usyskin, the youngest of the group by at least a decade. Here, Usyskin's youthful optimism serves to reinforce the conventional sense of a positive hero, though that attitude starkly contrasts with that of the other figures present. In the lower-right foreground, for example, a figure of authority exudes uncertainty as he raises his right hand in a gesture that is part salute, part wave of farewell. This hesitancy is further reflected in the questioning glance of one of the foreground soldiers who appears to be awaiting the delayed command to release the ropes. Even the muted tones of Bibikov's image, though accurately reflecting the winter conditions of a January ascent, seem simultaneously to stand metaphorically for the thin atmosphere into which the pilots will climb. The tall, vertical format of the work, echoed in Fedoseenko's raised right arm, emphasizes the imminent upward thrust of the balloon, while the composition itself is dominated by a vast, empty, gray sky. Yet perhaps the most telling inclusion of all is that of the figure at the very center of the composition. Rather than look toward his colleagues or the ground crew, flight engineer Andrei Vasenko directs his gaze toward the suspended weights designed to control the buoyancy of the craft. He, alone, in a moment of seeming premonition, concerns himself with the stability devices that would eventually contribute toward the flight's failure and the demise of the crew.

Bibikov's *Osoaviakhim-1* might best be read as an ambivalent response to the fatal flight. On the one hand, the work certainly celebrates the deeds of these aviation heroes, whose compositional centrality and elevation ascribes to them the status of extraordinary individuals. Yet at the same time, Bibikov does not shy away from visual allusions to the potential flaws, both human and mechanical, that ultimately led to the mission's failure.

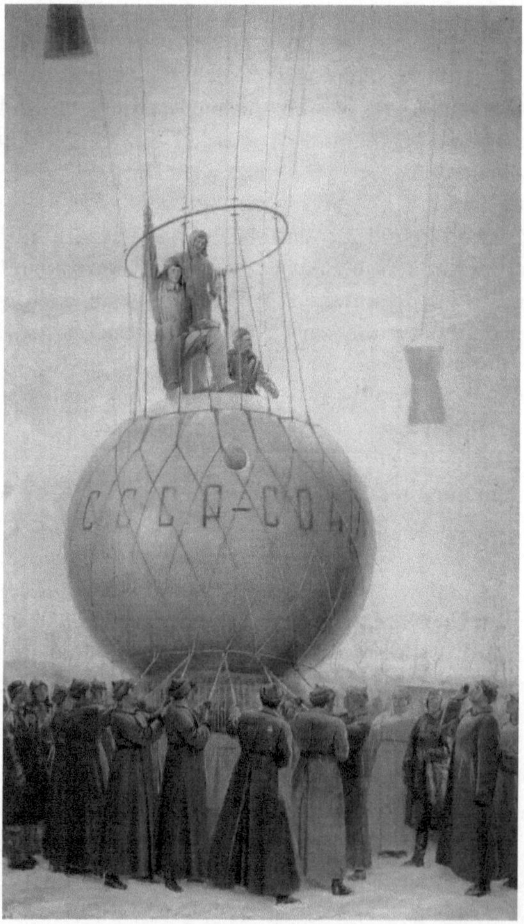

Figure 1.8 Georgii Bibikov, *The Osoaviakhim-1* (1935), oil on canvas. Russian Museum, St. Petersburg

The end of an era

By the late 1930s, the golden age of Soviet aviation record-breaking was coming to an end as the imminence of war loomed large. In 1939, however, the authorities would launch one last international propaganda mission promoting Soviet flight. On 28 April, Vladimir Konstantinovich Kokkinaki (1904–85), with co-pilot Mikhail Gordienko, took off from Moscow on a long-distance flight bound for the United States. Here their landing was scheduled to coincide with the opening of the Soviet Pavilion at the New York World's Fair. All was going well until fog over the north Atlantic caused the pilots to lose their bearings, forcing them to come down safely, if somewhat ignominiously, in a quiet stretch of rural Canada. From here, they were transported to New York, where, on entering the Soviet Pavilion, they were greeted by the monumental 10 x 16.5 m panel representing *Famous Peoples of the Soviet Union* [Znatnye liudi strany sovetov]. Produced by a team of artists headed by the painter Vasilii Prokof'evich Efanov (1900–78), this panel represented forty-eight of the Soviet Union's most famous and high-achieving sons and daughters, descending a *trompe l'oeil* staircase directly toward the viewer. Among several aviators included in the panel was none other than Kokkinaki himself.[11] Given the international publicity surrounding this failed flight, Kokkinaki at the very least must have felt conflicted by the contrast between his heroic painted presence in the Soviet Pavilion and his somewhat less elevated personal arrival. Yet, in a broader sense, this literal conflation of the realities of Soviet aviation and its representation in visual culture perhaps encapsulates the significance and interrelatedness of the two practices during this period. After all, pilots and artists were ultimately fulfilling similar roles both as participants in, and simultaneously manufacturers of, official Soviet mythology.

Notes

1 While the extent to which awareness of aviation developments influenced the reconceptualized spaces articulated in early Cubism and Orphism is disputed, there is little doubt that flight fascinated artists such as Picasso and Delaunay. Moreover, explicit references to flight can be identified in such works as Picasso's *The Scallop Shell* [*Notre avenir est dans l'air*], 1912, and Delaunay's *Hommage à Blériot*, 1914.

2 For an extended account of Kamenskii's experiences as a pilot and air-poet, see Markov 1968.

3 In 1937, Chkalov's hometown of Vasilevo was renamed Chkalovsk and the island of Udd, the destination of the pilot's first major long-distance flight, was renamed Chkalov Island.

4 Scripted by Chkalov's co-pilot, Baidukov, *Chkalov* was a major hit, successfully following up on the earlier released aviation classic, *Flyers* [Letchiki 1935]. The film was subsequently released in the United States as *Wings of Victory*.

5 *Iskusstvo* 6, 1938 unpaginated.

6 Mukhina did, however, produce a small version of the *Icarus* statue for an exhibition in Moscow in 1944, which was stolen. Plans to produce a larger version of the work for the American art collector and wife of the United States envoy to Russia, Marjorie Post, sadly never materialized, but a small bronze version was cast in 1953 toward the end of Mukhina's life, and was acquired by the Tretyakov Gallery (Zamkov 1989: 72–73).

7 One of the crew, Quartermaster Mogil'evich, was drowned during the evacuation procedure, but all other crewmembers were ultimately saved. See Otto Schmidt, 'The Chelyuskin Expedition' (Brown 1935: 1–22).

8 In the 1930s Mukhina developed an interest in glass as a medium for both decorative and sculptural works (Popova 2009: 33–37).

9 Mukhina's original belief that an allegorical treatment might be acceptable was probably based upon the success of her earlier allegorical personification of nature in a small bronze entitled *The Wind,* 1926–7.

10 [Narod ne znaet antichnoi mifologii, ni Borei, ni Zefir, ni olitsetvorenie drugikh vetrov i sil prirody,.poniatnye drevnim, ni stanut blizkimi massam; dlia sovremennogo pamiatnika neobkhodimy real'nye obrazy, vziatye iz zhizni.] Voronov 1989: 179.

11 See *New York Times*, 17 May 1939: 20 and *Pravda*, 18 May 1939: 6. A small portrait bust of Kokkinaki was also produced in 1939 by Vera Mukhina.

Bibliography

Abolina, R.I. (1954) *V.I. Mukhina*. Moscow: Iskusstvo.

Abolina, R.I. (1960) *Mukhina*. Moscow: Iskusstvo.

Baidukov, G. (1991) *Russian Lindbergh: The Life of Valery Chkalov*. Trans. Peter Belov. Washington: Smithsonian Institute Press.

Brown, A. (1935) *The Voyage of the Chelyuskin by Members of the Expedition*. New York: Macmillan.

Clark, K. (1981) *The Soviet Novel: History as Ritual*. Chicago, IL: The University of Chicago Press.

Davies, J.E. (1943) *Mission to Moscow*. London: Victor Gollancz Ltd.

Heyman, N.M. (1978) "NEP and the Industrialization to 1928." *Soviet Aviation and Air Power: A Historical Overview*. Ed R. Higham and J.W. Kipp. London: Westview Press.

Markov, V. (1968) *Russian Futurism: A History*. Berkeley and Los Angeles: University of California Press.

McCannon, J. (1998) *Red Arctic: Polar Exploration and the Myth of the North in the Soviet Union 1932–1939*. Oxford: Oxford University Press.

Odom, W.E. (1973) *The Soviet Volunteers: Modernization and Bureaucracy in a Public Mass Organization*. Princeton, NJ: Princeton University Press.

Palmer, S.W. (2006) *Dictatorship of the Air: Aviation Culture and the Fate of Modern Russia*. Cambridge, MA: Cambridge University Press.

Popova, I. (2009) "Proizvedeniia V. I. Mukhinoi v Dekorativno-Prikladnom Iskusstve." *Vera Mukhina, 1889–1953*. Ed. E. Petrova. St. Petersburg: Palace Editions.

Ternovets, B.N. (1940) *Sarra Lebedeva.* Moscow, Leningrad: Gosudarstvennoe Izdatel'stvo "Iskusstvo."

Voronov, N.V. (1989) *Vera Mukhina*. Moskva: Izobrazitel'noe iskusstvo.

Whiting, K.R. (1978) "Soviet Aviation and Air Power under Stalin, 1928–1941." *Soviet Aviation and Air Power: A Historical Overview*. Ed. R. Higham and J.W. Kipp. London: Westview Press.

Wohl, R. (1994) *A Passion for Wings: Aviation and the Western Imagination, 1908–1918.* New Haven, CT: Yale University Press.

—. (2005) *The Spectacle of Flight: Aviation and the Western Imagination, 1920–1950.* New Haven, CT: Yale University Press.

Zamkov, V. (1989) *Vera Ignat'evna Mukhina 1889–1953.* Moscow: Gosudarstvennaia Tret'iakovskaia Galereia.

2 Deineka's heavenly bodies

Space, sports, and the sacred

Helena Goscilo

I love for things to be on a large scale and with a lifelike quality of form, which is why I'm in love with Greek sculpture, where the harmony of life's truth and an organic perception of the beautiful are identical.

A.A. Deineka, "From an Autobiographical Sketch" (1946/1947)[1]

If there is a dream that is capable of showing the *vectorial nature* of the psyche, it is certainly the *dream of flight*.

Gaston Bachelard, *Air and Dreams* (1943)[2]

Introduction: moving bodies

Dubbed the 'Painter of Dancers,' Edgar Degas (1834–1917) reportedly explained his renowned partiality to ballet subjects as merely "a pretext for depicting movement"[3]—a declaration substantiated by his *Young Spartans Exercising* (1860), *The Start of the Hunt* (1863–5), and countless paintings of horse-races and women busy at their toilette.[4] A kindred fascination with movement held the Soviet 'Painter of Athletes,' Aleksandr Deineka (1899–1969), in thrall to motion throughout his life, irrespective of the zigzags in his professional fortunes and the country's political climate.[5] His contemporary, Rafael Kaufman, quickly registered as much: "The leading theme in Deineka is the person in action, in movement" (Kaufman 1936: 88). Deineka's lifelong infatuation with the dynamism of sundry forms started in childhood, when "his predilection for action and his desire to depict motion" materialized in sketches of "marching soldiers, . . . galloping horses, running and jumping dogs or wild animals, birds in flight, and moving carriages" (Sysoyev 1982: 5). During his studies in VkhUTEMAS,[6] where he enrolled in 1921, Deineka focused on the rules advocated by the rigorous graphic artist Vladimir Favorskii (1886–1964) for depicting movement, and in 1923 he produced a series of engravings and etchings addressing principles of dynamic composition: *Acrobats*, *Foxtrot*, *Tennis*, *Football*, and *Boxing* (Sysoyev 1982: 6). Tellingly, the universally admired canvas automatically associated with his name, *The Defense of Petrograd* [Oborona Petrograda 1928], depicts armed

Soviet citizens on the march,[7] while his most characteristic paintings render young Soviets engaged in sundry athletics and sports, from sprinting to skiing and football.[8] Deineka himself acknowledged, "I like seeing people running and jumping. Static situations are by no means the best in a lot of cases. Sometimes the composition of a picture absolutely demands strong dynamism."[9]

Sport, in fact, dominated his prolific artistic output during the 1930s, when he also painted more than a dozen works devoted to airplanes—the early phase of which, according to Scott Palmer, aroused Americans' (and avant-garde Russians') "unbridled enthusiasm" (Palmer 2006: 2). Though Deineka found flight as a mode of transportation enthralling, largely because of its novel, elevated perspective on the world below, which inevitably transforms one's apperception of phenomena and therefore one's relationship to them (Sysoev 1989: 33),[10] his response to it seems to have been more complex. Intriguingly, the terms of his reaction evoke the sensations attributed to encounters with the sublime in Edmund Burke's classic formulation: "The speed makes your heart beat faster, and a lump of *delight* and *horror* gets stuck in your throat. The sun is above you, and above you melts the *endless expanse of space* with a lustre like that of the sea. . . . The *endless expanses* under me change from turquoise to violet in the distance. . . . The world becomes *beautiful* and *infinite*" (emphasis added).[11] Flight was inseparable from Deineka's lifelong concerns, which embraced versatile bodies in motion, perspectival transformations,[12] and an aesthetic associated with both, which he deemed critical for artists. "Our horizons have broadened extraordinarily," he announced. "We've got used to seeing the earth from an airplane. That means we look not only into the distance, but also *up and down*, which, of course, forces an artist to decide issues of space in a new way" (emphasis added).[13] Such a conviction allied Deineka with Futurist Aeropainting [*aeropittura*], which in 1929 Filippio Marinetti, the Italian grand master of manifestos, declared the new phase of Futurist aesthetics, with aerial phenomena its guiding precept.[14] According to Futurist Aeropainting, flight offered new points of view that demanded and inspired novel expressive means and placed the viewer in constant motion, encouraging "a suppression of detail" and promoting "a synthesis and transformation that brings forth universals" (Silk 256). Though the painterly idiom Deineka elaborated had little in common with Futurist praxis, the principles advocated by Marinetti and his acolytes—focus on new (aerial) perspectives, motion, and suppression of detail—coincided with Deineka's general sense of aviation's impact on art and his own praxis.

In analyzing Deineka's artistic exploration of flight, my chapter swims against the current of scholarship on Deineka, which dwells chiefly on his political significance—his connection to the officially promulgated enthusiasm for aviation during the 1930s–40s.[15] My interest lies elsewhere: in his avowed, frequently iterated passion for the "harmony" and "beauty" of mobile human bodies and their continuity with airborne planes, both of which Deineka associated with the numinous. The symbolic significance of flight derives from ancient concepts of the cosmic structure in world mythologies and theologies—most relevantly here, the Tree of Life/Sacred Tree/Tree of Knowledge (the *axis mundi* connecting earth and heaven[16])—inscribed in visual and verbal texts that provide a basis for the modern world's vision of avian flight and aviation as symbolic modes of transcendence. At the heart of this vision resides the verticality of religious thought (instanced in the Neo-Platonist strain of poets such as Shelley and Russian Symbolism's concept of *a realibus ad realiora*), which in a quasi-secularized key fueled the avant-garde's faith in human rebirth through aviation and Soviet (as well as fascist) notions of winged Nietzschean supermen in space.[17] According to Gaston Bachelard, for Nietzsche, as for William Blake (1988: 77), air was a higher, subtler matter—"the very substance" of human freedom and intimately associated with the imagination (Bachelard 1988: 136). Joseph Corn's tellingly titled study *The Winged Gospel* (1983) reports that in the early days Americans treated airplanes as an object of reverence bordering on worship and referred to themselves as "disciples," "apostles," and "prophets," with aviation a "winged gospel" or "holy cause" (Corn 1983: viii). The influence of traditional religions on this discourse is self-evident, for in religious thought salvation of the soul (and, in some strands of belief, the body) entails its ascent to heaven, as illustrated in various texts, paintings, operas, and film. Hailed as a means of technological and cultural salvation, the airplane transferred focus from earth to sky and the latter's longstanding spiritual associations.

Deineka's numerous depictions of the body developed and revised the avant-garde obsession of the 1910s and early 1920s with the modernist streamlined, efficacious physique as a visible marker of immortality.[18] For Deineka, that immortality was inseparable from areté (ἀρετή),[19] the pursuit of perfection, which he found embodied in Greek sculpture. To convey that unity, Deineka displaced his subjects—both organic and mechanical/industrial—from the horizontal to the vertical axis, investing them with spiritual significance. This axial shift, when oriented upward, reflected his conviction that "[f]or an artist our era gives birth to subjects in heaven."[20] In this spatial symbolism of transcendence analogizing human bodies with avian and aviatory flight, their elevation emblematizes a striving for the ultimate/infinite—immemorially synonymous with divinity. Accordingly, athletes and aviators are the supreme legatees of the mythological Icarus, whose hubristic flight to The Source was translated by the utopia-bound Soviet New Man and Woman into meritorious self-perfection in an officially atheistic state that nevertheless appropriated religious symbolism for its often grandiloquent verbal and especially visual rhetoric.[21]

Curiously, not Deineka, but his close OST (Society of Easel Painters [Obshchestvo stankistov])[22] associate and collaborator Iurii Pimenov (1903–77) most eloquently formulated the transcendental potential of sport as a supreme mode of distilled movement that is fundamentally aesthetic: "Contemporary life is full of movement. Sport is the culmination of movement—in it, movement is liberated from the incidental; it becomes refined, pure, and therefore beautiful. The beauty of sport is the beauty of movement advancing toward perfection. That's why I love sport, the high culture of human movement" (*Sport v sovetskom* 1980: page facing plate 20). The concept of "movement advancing toward perfection" articulated by Pimenov informs almost all of Deineka's oeuvre, particularly his renditions of athletes, and his valorization of velocity links him to modernity and, in complex ways, to its modernist artistic representation. Profoundly in tune with his era, Deineka attempted to capture its accelerated rhythm, initially in images of work, executed via avant-garde techniques influenced by graphics (flat surfaces, primary colors, minimalist delineation of figures),[23] before elaborating a modified version of Socialist Realism to represent sports and aviation—two activities that widespread convention during the 1930s collapsed into sameness.

Deineka's dynamic bodies at work and play

> Sport has one exciting feature: it fits into the most diverse frameworks of art. As a subject it's inexhaustible, for it's democratic and popular. . . . Sport accommodates all nuances of feeling. It's lyrical. There's a lot of optimism in it. It's the source of the heroic.
>
> A.A. Deineka[24]

> I love sport. . . . It has always seemed to me that sport ennobles man, as does everything beautiful. I love the will of athletes, which they can govern. I began to love the Greeks even more as I became better acquainted with the world of sport.
>
> A.A. Deineka[25]

Numerous canvases and illustrations executed by Deineka in the 1920s (especially for the mass propaganda magazines *At the Factory Workbench* [U stanka], *Atheist at the Factory Workbench* [Bezbozhnik u stanka], and *Daesh'* [Let's Produce]) spotlight bodies animated by physical labor, moving in harmony with the machines they operate, bent over or extended to reach equipment, but rarely static. Probably the best-known examples in this key, *Building New Factories* [Na stroike novykh tsekhov 1926] and *Textile Workers* [Tekstil'shchitsy 1927], feature lightly garbed, barefoot female workers amid newly constructed technology who in *Building New Factories* are positioned in semi-open space high above ground level. More than thirty years later, Deineka recalled, "Rhythm and a certain ornamentality lie at the basis of the composition of my painting *Textile Workers*; it is the rhythm of the ceaseless circular motion of the looms. I almost mechanically subordinated to this rhythm the weavers with their smooth, melodious movements."[26] At the same time, his earlier professional trip to the Donbass stimulated not only depictions of miners such as *Before Descending into the Mine* [Pered spuskom v shakhtu 1925], where "the entire picture is constructed along the vertical,"[27] but also several paintings during the 1930s that intersect with his signature subject of sports. Though the artistic genre of the nude, seminal in Western Europe for centuries, largely bypassed both Russian and Soviet-exhibited art, scenes of naked swimmers were not uncommon during the Soviet era,[28] and Deineka's *Lunch Hour in the Donbass* [V obedennyi pereryv v Donbasse 1935] belongs to this category (Fig. 2.1).[29]

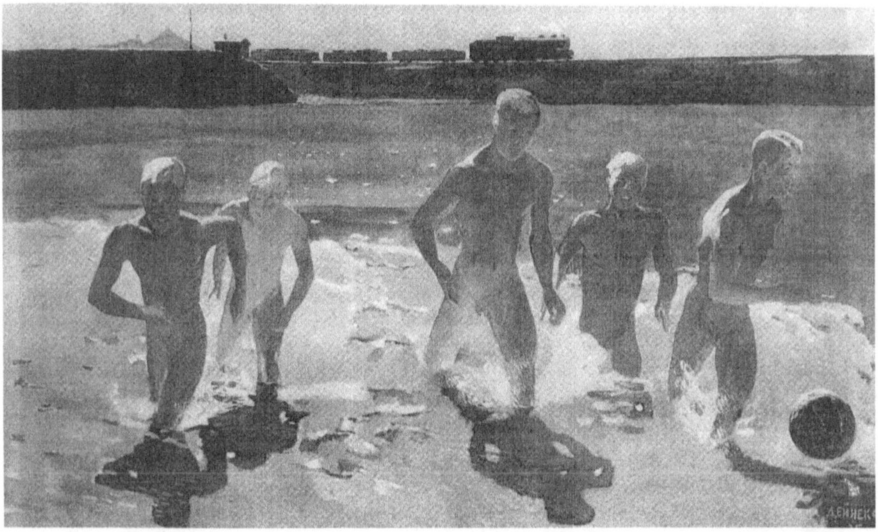

Figure 2.1 *Lunch Hour in the Donbass*, Deineka (1935). Courtesy of Art © Estate of Aleksandr Deineka/RAO, Moscow/VAGA, New York

A gendered companion piece to his *Midday* [Polden' 1932] (Fig. 2.2), which captures five young women emerging from an unidentified body of water against an industrial backdrop, *Lunch Hour* . . . depicts in a similar environment five equally naked young men leaving the shallow water where they presumably have been playing with the ball that now floats in front of the youth to the right. His gaze, focused on something outside the painting's border, implies motion beyond what the canvas actually depicts. Smiling and splashing, the blond quintet personifies spontaneity, youthful energy, and innocence, as if the water has both cleansed and baptized them. Though, unlike many of his contemporaries, Deineka did not hesitate to sketch in male and female genitalia, here and elsewhere he blatantly associates nakedness with an innocence that evokes an Edenic interlude, for his bared subjects completely lack self-consciousness and guile. That innocence likewise obtains in *On the Balcony* (1931), *A Mother* (1932), *Sleeping Boy with Cornflowers* (1932), *Children Resting* (1933), *Young Girls Bathing* (1933), *Gymnastics on the Balcony* (1935), and *Future Pilots* (1937), which all portray 'natural states'—of sleep, sport, and maternity.

Sport, as copious scholarship maintains, is rarely an innocent, autonomous, and self-oblivious activity. Leni Riefenstahl's (1902–2003) choice of the Olympics for her technically acclaimed cinematic tribute to fascism in the 'documentary' *Olympia* (1938) was singularly apposite,[30] for the games in honor of Zeus that originated in Greece in approximately 776 BC from the outset fulfilled a political function—that of forging political alliances and establishing dominance over rival city-states. Moreover, the concept of competition,[31] like the rhetoric of setting records, attaining goals, and defeating the clock, points to a cardinal feature of

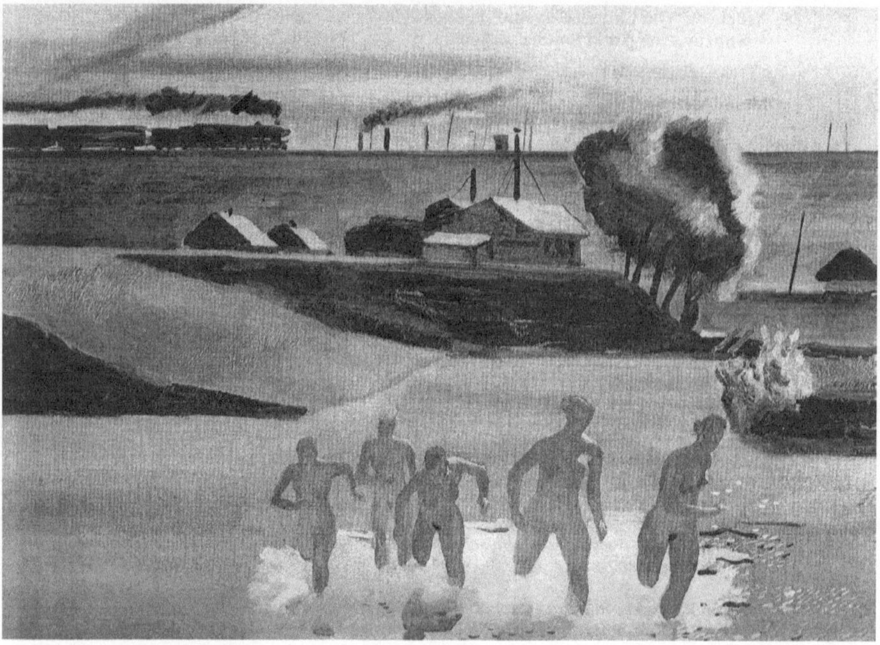

Figure 2.2 *Midday*, Deineka (1932). Courtesy of Art © Estate of Aleksandr Deineka/ RAO, Moscow/VAGA, New York

Soviet political discourse that inhered in its iconography: namely, the perceived equivalence between battle and sport, in which sport is conceived as battle—a posited equivalence that explains why both frequently function as tests of manhood.[32] Though history eliminates all doubt regarding the symbiosis of aviation and warfare, and in the Soviet Union aviation undisputedly straddled sport and military aggression, I contend that Deineka's depiction of both sport and aviation is anomalous. While he was hardly ignorant of or alienated from the official imperatives of his era, for him, as for the more philosophically inclined ancient Greeks, physical fitness and skill denoted mental balance, ethical probity, and above all a potential for *aesthetic transcendence*. Accordingly, his numerous portrayals of runners, cyclists, swimmers, etc. frequently render the human form in action as an extension of the self-sufficient world of nature, such as the famous *Freedom* [Razdol'e 1944] (Fig. 2.3), where trim young female runners in gym outfits seem to flow through an unpopulated verdant landscape of rolling hills, river, trees, haystacks, and largely sunny skies. Even a cursory glance at his *Defense of Sevastopol* [Oborona Sevastopolia 1942] reveals his essentially athletic treatment of war (not warlike treatment of sport[33]): the postures and body contours of Russians defending their nation duplicate those of shot-putters, javelin-throwers, tennis-players, and specifically the American wielding a bat in the painting *Baseball* (1935), which resulted from Deineka's 1935 visit to the United States.[34] That visit also yielded a series, all belonging to that year, represented by *An American Cabaret Show*, *A Negro Show*,

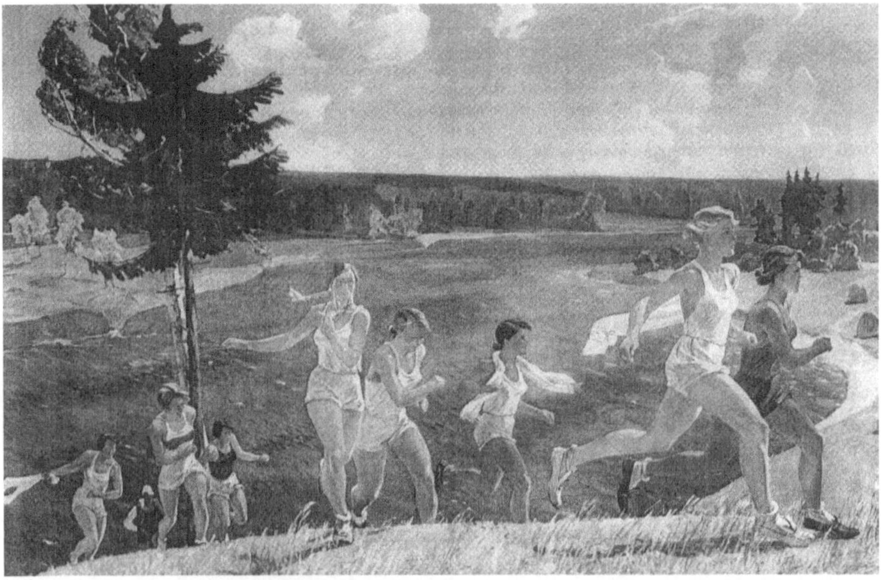

Figure 2.3 *Freedom*, Deineka (1944). Courtesy of Art © Estate of Aleksandr Deineka/RAO, Moscow/VAGA, New York

and *Negroes Dancing*, which explored a mode of movement new in Deineka's repertoire: dance. Unlike Aleksandr Gerasimov, the other artist privileged to travel abroad in the mid-thirties, who had nothing positive to say about the contemporary West (Bown 1993: 129–30), Deineka responded creatively to what he encountered. As always, apart from a few landscapes, he targeted bodies in motion.

As the aspiring Olympians of the new era, Deineka's young athletes defy conventional limitations of time and space through the beauty of their striving for perfection, and he renders their efforts in a style not fully consonant with the dictates of Socialist Realism. *Football Player* [Futbolist 1931/2], for instance, constitutes an extraordinary and original glorification of the sportsman through its dynamism and above all its striking composition. Relying on the bold device of visually literalizing the symbolic, the painting portrays transcendence along the vertical axis, in the process abandoning verisimilitude and accurate proportions (Fig. 2.4). The emphatically horizontal strokes of the paintbrush used to depict the sky create the impression of forces originating beyond the left edge of the painting that propel the young football player forward. Miraculously suspended in the sky, he has kicked the football, which hangs in the air above him, to an altitude comparable to the top of the church tower providing the backdrop for his action. The youth's foregrounded body is dramatically out of proportion in relation to that tower and the churches behind it, signaling his primacy in a humanized empyrean realm. In addition to the buildings, which loom tall, the football player's gaze, left arm, and right leg are all oriented upward, toward the ball, which in shape, texture, and location could double as the planet earth, sun, or moon. Furthermore,

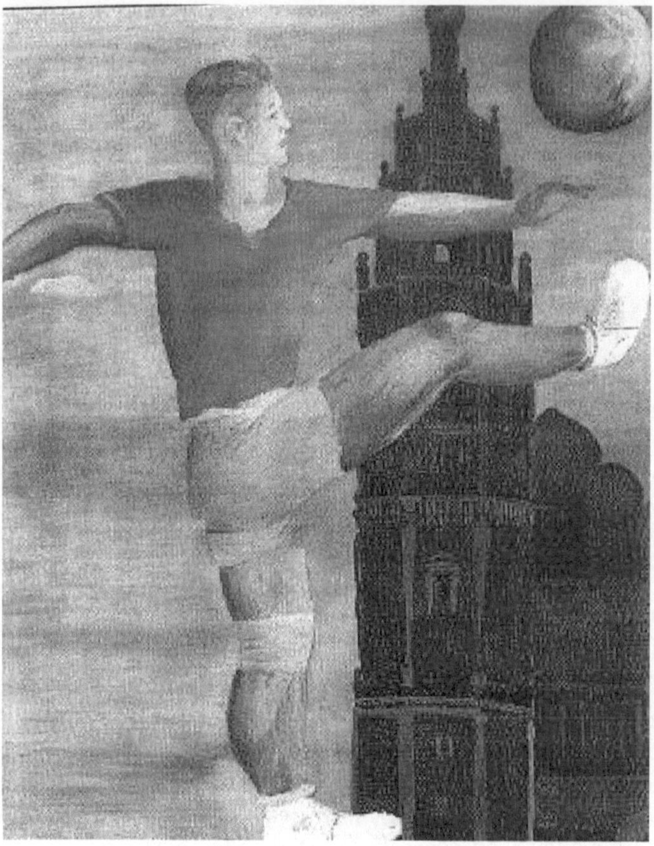

Figure 2.4 *Football Player*, Deineka (1931/1932). Courtesy of Art © Estate of Aleksandr Deineka/RAO, Moscow/VAGA, New York

since his shorts and what on first glance appears to be a bandage on his left knee are actually brushstrokes rendering the sky, he partially merges with it, having attained supranatural heights. All aspects of the painting contribute to the image of an athletic young superman who defies gravity to dwell with the gods. And one has only to compare *Football Player* with Serafima Riangina's *Higher and Higher* [Vse vyshe i vyshe 1934], a transparent hymn to Stalin's Second Five-Year Plan, to gauge the distance between loyal Socialist Realists and Deineka at his original best.

Relying on a somewhat different technique in addition to symbolization, Deineka's *Ball Game* [Igra v miach 1932] similarly elevates three young women to exalted status as they ostensibly play catch or volleyball amid nature (Fig. 2.5). Naked, with strongly built, rosy, sun-kissed bodies, the threesome form an Edenic trinity reaching for the heavens, quite literally in the case of the largest figure, in the foreground, with her back to the viewer. The hand of her raised right arm

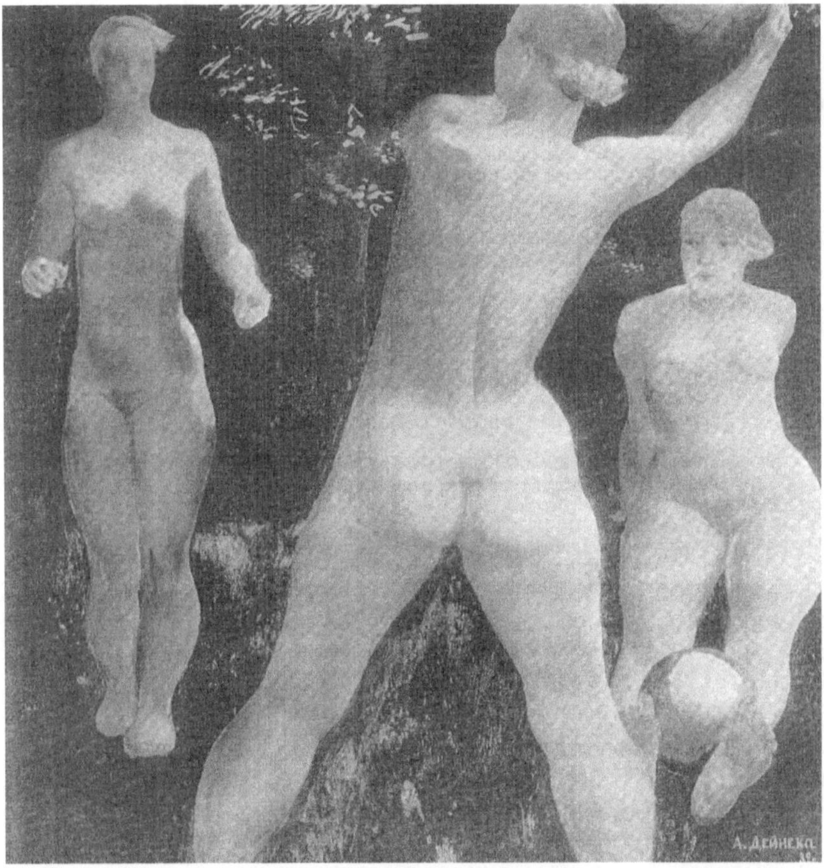

Figure 2.5 *Ball Game*, Deineka (1932). Courtesy of Art © Estate of Aleksandr Deineka/ RAO, Moscow/VAGA, New York

appears to be touching a bronze sphere sooner reminiscent of a setting sun than a ball, especially since the playmate in a reclining position to the right balances what is unmistakably a (superfluous?) ball between her feet. In an unspecified setting, which lacks detail and boasts only roughly delineated grass and a vaguely defined tree in the rear, the three are not firmly planted on the ground, but appear suspended above it, as if dwelling in the air. The imprecise setting and the non-particularized treatment of the figures instance Deineka's attachment to the silhouette, the properties of which he conceived in the following terms:

> [B]y silhouette I don't mean simply black painted on white, but rather, a large formt that absorbs everything that's essential for a close perception of detail.
>
> Successful examples of the unique artistic form are Michelangelo's *Adam* and Falconet's *Peter I.* A silhouette is not only a flat surface, but includes volume and color. . . . A clear arrangement of silhouettes in a painting of mass

> subjects splendidly guarantees visibility from a long distance, and their contouring [outlines traced in black, HG], observed in ancient frescoes, is one of the means of preserving a large form.[35]

As elsewhere, the silhouette in *Ball Game* allows Deineka to circumvent the meticulous detail of naturalistic representation and to preserve the 'large form,' i.e., it enables comprehensive generalization and elevates the female trinity (a modern, semi-secularized Troitsa) to the realm of myth. The painting makes clear why Evgenii Kibrik claimed that Deineka's art was "characterized above all by two main traits—broad generalization [the technique of the silhouette, HG] and dynamism, the intense rhythm and tempos in which the characters of Deineka's paintings live and function" (Kibrik 2012: 403).

Whereas these two works exploit the vertical axis, two sports paintings produced almost three decades apart operate on a paradoxical fusion of axes that overtly rather than implicitly aligns them with Deineka's aviation images. *Goalkeeper* [Vratar' 1934] (Fig. 2.6) typifies Deineka's proclivity to create the pictorial equivalent of cinema's open form, whereby the partial presence of an item leads the gaze beyond the perimeter, as in *Lunch Hour in the Donbass*, examined above. Part of a goalpost close to the right and two small figures (projected against a wood in the distance) balancing it to the left are the only items apart from the focal figure of the goalkeeper, again outsized and in motion. Presented from a lateral-cum-rear angle, he is leaping, arms outstretched, toward the ball that is half-visible at the left border. On the one hand, the man's body is maximally stretched along the horizontal axis; on the other, he is reaching up for the ball and is, yet again, fully suspended in the air. His body mimics an airplane starting a gradual ascent. In like vein, a late painting of a female athlete, *Youth* [Molodost' 1961–2] (Fig. 2.7), catches her as she is about to clear the bar on a high jump. Once more, the female spectator on the left is counterweighted by the forest of firs to the right, with grassy hills and indistinct houses behind the sports stadium in the background. Slender and tanned, the toned body of the young athlete, clad in a red top and shorts, hovers momentarily above the bar, her arms at her sides evoking folded wings. As in the case of the male goalkeeper, however, her body

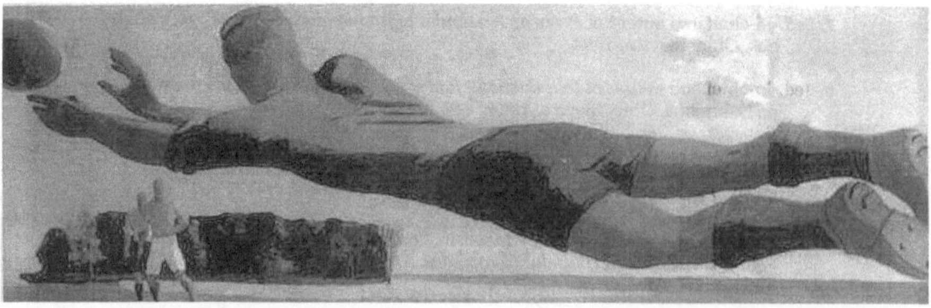

Figure 2.6 *Goalkeeper*, Deineka (1934). Courtesy of Art © Estate of Aleksandr Deineka/RAO, Moscow/VAGA, New York

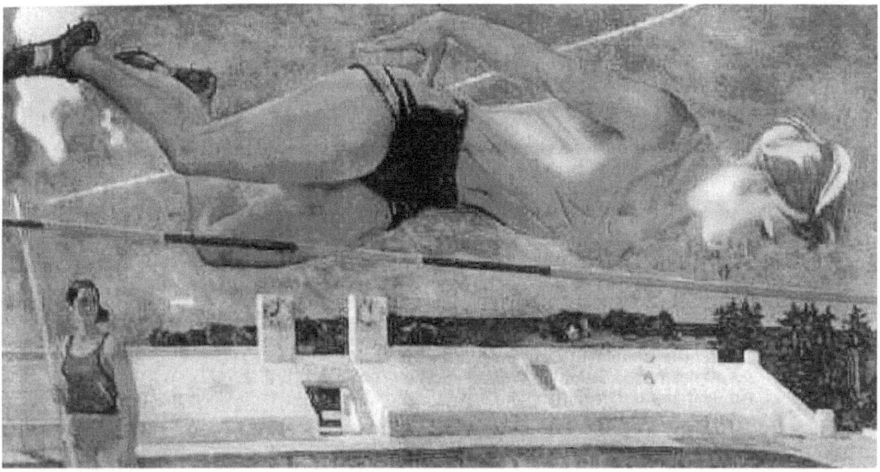

Figure 2.7 *Youth*, Deineka (1961–2). Courtesy of Art © Estate of Aleksandr Deineka/RAO, Moscow/VAGA, New York

is extended horizontally, occupying the center in the upper half of the painting. Since she is portrayed from a low angle, the viewer sees her feet as if kicking at the clouds in the upper-left register. Moreover, an ascending white line across the sky (along the dynamic diagonal) evokes the exhaust track normally left by an airplane, inviting the viewer to draw a parallel between the airborne athlete and the aircraft—both defying the laws of gravity by soaring heavenward. What symbolically allies these athletes with the divine is the combination of vertical and horizontal materialized in the cross representing Christ the man-God and his sacrifice—a configuration that could be symbolically mapped not only onto the structures of cathedrals and churches, but also onto winged aircraft.

Winged woes and wonders

> The love of our time for aviation, for people with designers' talents, with the courage and persistence of a Chkalov, provided artists with the unusual theme of soaring gliders or parachutists falling like fireworks from the sky.
>
> A.A. Deineka, 'Echoes of the Past' (1941)[36]

Inaugurated in the 1930s, Deineka's paintings of aviation could hardly differ more from such bombastic stereotypes of Soviet supremacy in the air as Vasilii Kuptsov's *The Maxim Gorky* [Maksim Gor'kii 1934], where the huge white airplane with red stars on its wings and other, lesser aircraft and parachutes in the distance are observed from overhead ('God's vantage point') and dominate the landscape beneath.[37] By contrast, Deineka's images of aviation belong to two sui generis categories in which remoteness is paramount: the far-off airplane without any human presence and sundry modes of aircraft observed by boys. To the first group belongs *Airborne* [V vozdukhe 1932] (Fig. 2.8), a quintessentially Lermontovian depiction of a lone plane with bright red wings flying high above the clouds against a background of towering mountains, backlit by the sun, that dwarf the aircraft. The painting is remarkable for putting the isolated plane in perspective vis-à-vis the endless billowing clouds (along the horizontal axis) and the massive structure of the crags (along the vertical). What results is an image of existential solitude that evokes Lermontov's scenarios of solitary elevation, exemplified in the desolate protagonist's shuttling between heaven and earth in his narrative poem *Demon* (1841), and anticipates the sense of humans as minuscule specks within unknowable cosmic expanses conveyed in Stanislaw Lem's philosophical novel *Solaris* (1961). And while Deineka's *The Red-Winged Giant* [Krasnokrylyi gigant 1937] (Fig. 2.9) alters the proportions, it retains the identical subject matter: here seen from the rear, the airplane is larger and resembles an albatross with widespread wings soaring aloft amid snowy mountain peaks. The upper two-thirds of the painting consist of the endless vista of the sky, disrupted solely by the aircraft. No sign of civilization punctuates the eerie sense of infinite, indifferent space—what Deineka referred to, with both fascination and trepidation, as "endless expanses."

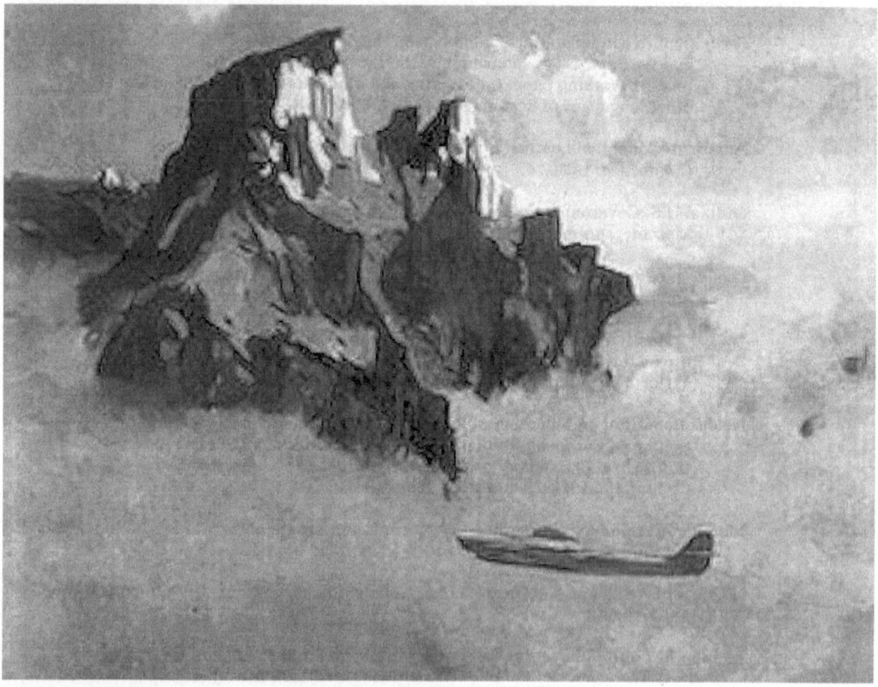

Figure 2.8 Airborne, Deineka (1932). Courtesy of Art © Estate of Aleksandr Deineka/RAO, Moscow/VAGA, New York

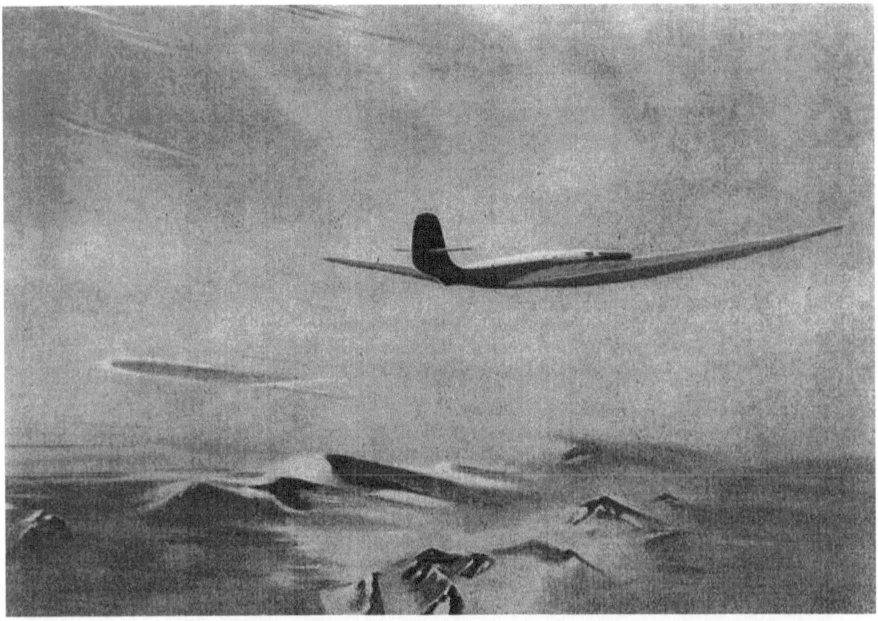

Figure 2.9 *The Red-Winged Giant*, Deineka (1937). Courtesy of Art © Estate of Aleksandr Deineka/RAO, Moscow/VAGA, New York

By contrast, *Hydroplanes* [Gidroplany 1934] depicts from a high angle five hydroplanes of various sizes flying close to one another above some land and buildings on the upper right of the canvas, as well as, on the bottom right, part of a wharf, on which three tiny human figures bespeak the remoteness of earth. Characteristically, these two sections balance each other, and propel the viewer's gaze off the right border of the painting.[38] Peopled aviation receives scant attention in Deineka's oeuvre, however, apart from a 1934 rendition of a parachutist that rehearses the wartime *Downed Ace* [Sbityi as] nine years later. In style and mood *Parachutist over the Sea* [Parashiutist nad morem 1934] (Fig. 2.10) is of a piece with his Lermontovian solo planes, inasmuch as it charts the ascent of a hydroplane from which the parachutist has just leaped, while the man, his body paralleling the dynamic diagonal of light above him, is suspended in the sky, whose boundless space is broken only by part of a mountain. Plummeting earthward headfirst, he has yet to open his parachute, and the disturbing impression is of a forced and fatal descent. The image recalls above all the plethora of visuals exploring the dire fall of Icarus. Compositionally, the parachutist occupies the center of the pictorial space, with the plane above and the rock formation jutting out from the lower-right corner. His solitariness within the "endless expanses" that Deineka remarked on while flying could hardly contrast more emphatically with other painters' celebratory depiction of parachutists, such as Georgii Nisskii

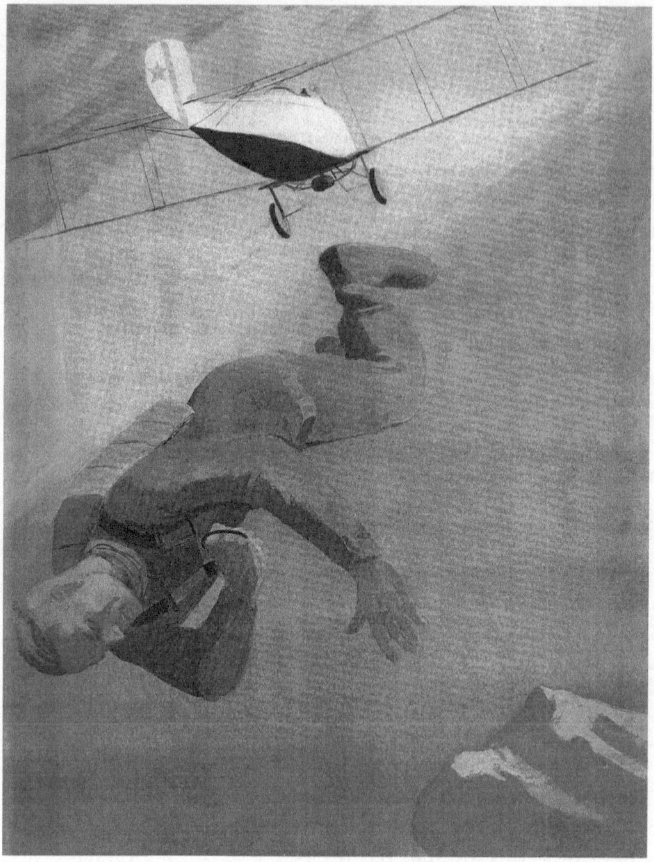

Figure 2.10 *Parachutist over the Sea*, Deineka (1934). Courtesy of Art © Estate of Aleksandr Deineka/RAO, Moscow/VAGA, New York

and Vladimir Odintsov's *Parachuting as a Sport* [Parashiutizm 1938] and Igor' Pekun's *Parachutists* [Parashiutisty 1972]. Conceived in a festive key, both populate a bright blue sky with multi-colored, open parachutes descending like inflated, vivid mushroom caps. Incongruously, one of the female parachutists in the former painting even grips a bouquet of flowers in her hand.

Such sunny mass enjoyment of parachuting makes Deineka's wartime *Downed Ace* (1943) (Fig. 2.11) all the more bleak and jarring. Disaster saturates both the upper and lower register of the painting, with the remains of abandoned, shelled buildings and vehicles merely skeletons of their former selves in the distance, and lopsided, rusty metal bars extruding from the earth in the foreground—a landscape of death and destruction that awaits the body of the titular ace as, parachute and eyes closed, he is projected against the sky during his headlong fall. Deineka provides only a partial glimpse of his plane at the right edge of the painting, the

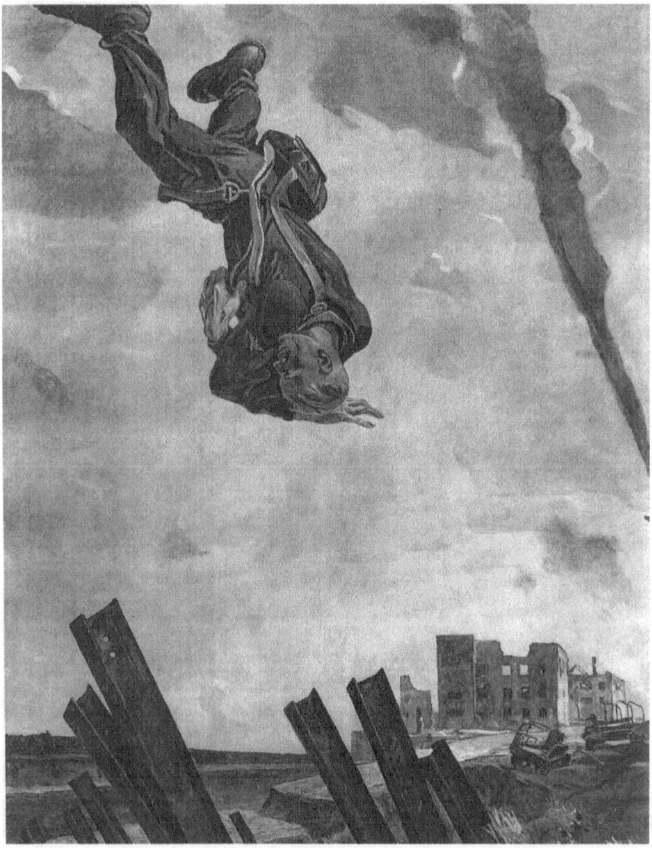

Figure 2.11 *Downed Ace*, Deineka (1943). Courtesy of Art © Estate of Aleksandr Deineka/RAO, Moscow/VAGA, New York

smoke billowing from it paralleling the angle of the pilot's body, both depicted from a low angle that intensifies the aura of calamity. Almost exclusively dull tones—grays and browns—dominate, with the airman's uniform merely a darker shade of the clouds from which he appears to have dropped. The only light emanates from the sky, especially in the remote rear, where flames shoot upward. A German, judging by his uniform, the luckless pilot is portrayed with a pathos all the more unexpected in the context of the devastated Soviet landscape that he, as the arch enemy, presumably helped to bomb and that now will be his grave. It is as if war has realized the ominous potential of aviation and parachuting merely implied by Deineka's earlier *Parachuting over the Sea.* Certainly his subsequent *Parachute Drop on the Dnepr* [Parashiutnyi desant na Dnepre 1944] lacks the disquieting originality of *Downed Ace*, as does *Berlin. The Day the Declaration Was Signed* [Berlin. V den' podpisaniia deklaratsii 1945], in which a triumphant

Russian plane flies over a partly ruined edifice. Both merely mark, in rather conventional terms, historic moments during WWII, and lack the existential dimensions of flight lyrically communicated elsewhere. Deineka's unique angle of vision, like his artistic fecundity, diminished with age, which probably explains why straightforward optimism thrives in his *Moscow. Landscape* [Moskva. Peizazh 1952], where the airplane soaring over the capital is secondary, possibly referencing the war from which, as confirmed by the panorama of the city, it has fully recovered.

Those paintings of the 1930s by Deineka that pair aviation with small boys likewise reveal no existential angst, but sooner incline to inspirational prophecy. The 1934 *The Pioneer* [Pioner] (Fig. 2.12) foregrounds a darkly tanned boy, naked except for skimpy briefs, seated on a low stool and engrossed in watching two planes in the distance. The painting establishes a division between the boy's current world—an unidentified, flat, slate-blue surface, as if perched on air, disrupted only by the vividness of the red-bound book lying at his feet[39]—and the light pink of the brightly illuminated, silhouetted mountains, which harmonize with the white and light gray of the planes. Yet composition and color link the

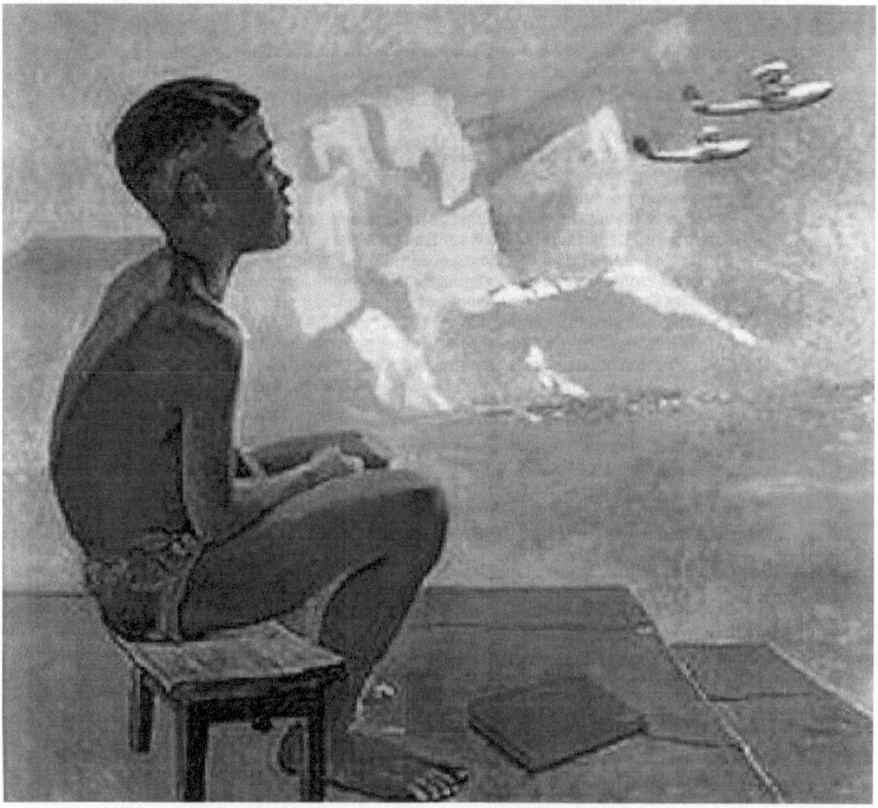

Figure 2.12 *The Pioneer*, Deineka (1934). Courtesy of Art © Estate of Aleksandr Deineka/RAO, Moscow/VAGA, New York

two spheres: the planes, however faraway, fit into the curve of the boy's body and their hues echo those of his shorts. Likewise the sky to the upper left iterates, in a paler tone, the gray-blue of the surface beneath the seated boy. Spatial here signals temporal, for the connection established through colors and angles holds the promise of attraction blossoming into an eventual identification: in time, the Pioneer will become a pilot.

A kindred scenario unfolds in *Future Pilots* [Budushchie letchiki 1938] (Fig. 2.13), which has prompted various political readings (Kiaer 2012, O'Mahony 2006). Given the year, it is possible but unconvincing to interpret the painting as anticipating Germany's aggression three years later (Sevastopol was attacked by Nazi forces in 1941), when the leisure activity of flight would become a civic duty.[40] But in 1938 the threat was overwhelmingly domestic, given Stalin's paranoid purges (1936–9), which peaked that year, when Deineka's first wife, the artist Pavla Freiburg, and the poet Osip Mandelshtam died during their imprisonment

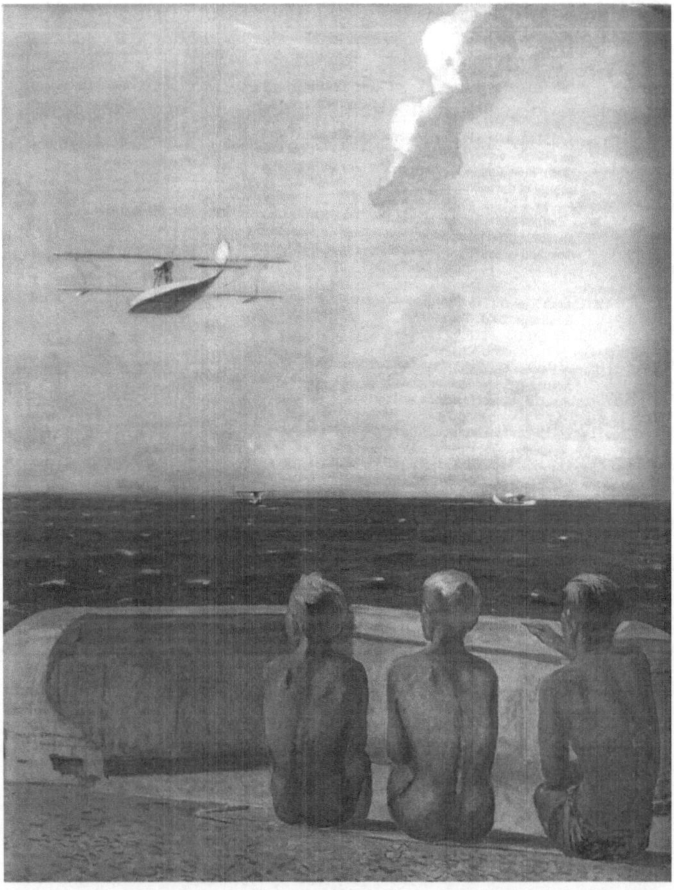

Figure 2.13 *Future Pilots*, Deineka (1937). Courtesy of Art © Estate of Aleksandr Deineka/RAO, Moscow/VAGA, New York

and the graphic artist Gustavs Klucis, a former fellow member of October [Oktiabr'], was summarily executed on Stalin's orders. These events are what doubtless prompted Deineka, who never incorporated Stalin into his works and joined the Party only in 1960, to succumb to embarrassing sycophancy with a stereotypically improbable painting of the purportedly demotic Great Leader, titled *Lenin on an Outing with Children* [Lenin na progulke s det'mi].[41] Certainly, both the title and content of *Future Pilots* court prophecy, and not necessarily in a monitory mode, for the work compactly holds out the exciting promise of a maturation that projects the youthful spectators in the painting, relaxed and devoid of responsibilities, as the next generation of adventurous, skillful fliers—the adult subjects they survey. The painting overlaps in several respects with the earlier *In Sevastopol. The Hydro Station 'Dynamo'* [V Sevastopole. Vodnaia stantsiia 'Dinamo' 1934], where young men and women in swimwear, their backs to the viewer, likewise observe aeronautical activity overhead and in the distance, with no hint of impending war.

Future Pilots belongs to the series of works Deineka devoted to the Crimea, the charms of which he frequently metonymized as young boys relaxing under the warming sun at Sevastopol. As he recalled in later years, "I loved that city from my young days, loved its cheerful inhabitants, its blue sea, verdant parks, [and] climbing, winding streets."[42] The painting's title makes explicit what *The Pioneer* intimates more subtly. On the Crimean coast of the Black Sea three tanned, blond boys of different ages, their backs to the viewer, sit behind a concrete breakwater, gazing out to sea, where three hydroplanes of correspondingly different size are discernible. One seems to have just landed in the water, another hovers close to the sea's surface, probably about to land, and a third is flying in the direction of the horizon. The composition directs the attention of the viewer and that of the seated trio toward the distant horizon—symbolizing aspirations and the future. The three figures are arranged and dressed according to age/experience, with the smallest, on the left, naked (innocent), the middle clad in abbreviated trunks, and the oldest, to the right, sporting regular trunks and presumably explaining something to his two younger companions as he gestures with his right hand at the airborne hydroplane. Carefree and protected by the breakwater, the three are too young to engage in the open aerial activity they witness, which inevitably entails risk, but also tests one's mettle and through the freedom of flight liberates one from earthbound, limited perspectives. 'Future' instates a tacit *Bildungsroman*, affirming that, like the Pioneer, once grown into adulthood, the boys will realize their dreams and take to the skies.

Nikitka, the Russian Icarus or prophetic pioneer

> Today Nikitka [the Russian Icarus] is seen as a symbol of Russian wit, daring, and will.
>
> A.A. Deineka, 'Echoes of the Past' (1941)[43]

Whereas aviation derived from modern technology, miraculous flight has been a diverse and richly elaborated phenomenon in Russian folklore for centuries, ranging from the magic carpet that transports characters through the air to such supernatural personae as winged dragons, speaking eagles that carry humans on their backs, Finist the Bright Falcon, who doubles as a prince, and the most formidable 'aerodynamite' of them all, Baba Iaga, with her trademark mortar and pestle. As Scott Palmer rightly argues, in Russian fairy tales, "Flight divides the world of the divine from the world of man [sic] and, in doing so, underscores the cosmic hierarchy within which the individual must act on a daily basis," which explains why "folkloric flight symbolically represents the hero's movement towards a higher, more enduring consciousness" (Palmer 2005: 25). Analyzing Russia's myths about flight, Palmer adduces the little-known story of Nikitka, a serf during the rule of Ivan the Terrible who "devised a mechanism" intended for aerial transportation. Despite his alleged success (!), the superstitious Ivan decreed his decapitation for having violated nature and allied himself with dark forces: "A man is not a bird. He does not have wings. Those who attach wooden wings to themselves do so in opposition to the will of nature. Such is not a godly deed, but a deed which emanates from unclean powers" (cited in Palmer 2005: 20). However irrational, that condemnation accurately recognized Nikitka's yearning for a superhuman feat that transcended then current earthly knowledge and capabilities.

As Deineka's admiring words cited in the epigraph above confirm, his opinion of Nikitka diverged radically from the punitive tsar's. And in 1940 Nikitka became the subject of one of his most unusual paintings, pointedly titled *Nikitka, the First Russian Flyer* [Nikitka, pervyi russkii letchik] (Fig. 2.14). It comprises three elements: the church tower with prominent bells from which Nikitka may have leaped; the crowds of spectators in the snow-covered square below (divided into two groups, on the left and the right);[44] and Nikitka himself. The vertical axis is emphasized by the tall tower (recalling *Football Player*) that stretches to the top of the painting and beyond, and by the uplifted heads of the spectators, who stare at Nikitka with various expressions—some with astonishment, others with laughter, and yet others with fear. One person, presumably a woman, has covered her face with her hands. The most noteworthy figure, riveting everyone's attention, unquestionably is the visionary Nikitka, distinguished from his audience not only by his action, but also by his appearance: while everyone else is bundled up in winter clothes and headgear, he is bareheaded and wears a light caftan, with the sleeves rolled up, exposing his arms. Gripping the huge, wide-span avian wings attached to his body by a rope, he is not flying up, but descending—a trajectory that leads one to question whether he has arrived from elsewhere instead of having launched himself from the bell-tower.

The "picture," according to Deineka, "was constructed vertically. . . . To depict the highest point in the city at that time I set the bell-tower upon a high river bank."[45] If Deineka adhered wholesale to the myth and set out to portray the success of Nikitka's bold venture, then the painting dramatizes the moment before his safe landing, with the tower merely serving as a visual reminder of verticality. Otherwise, Deineka shows the catastrophic consequences of the enterprising peasant's failure—an unlikely scenario in light of Deineka's admiring words. Which version one favors determines the significance of the flock of birds flying around the

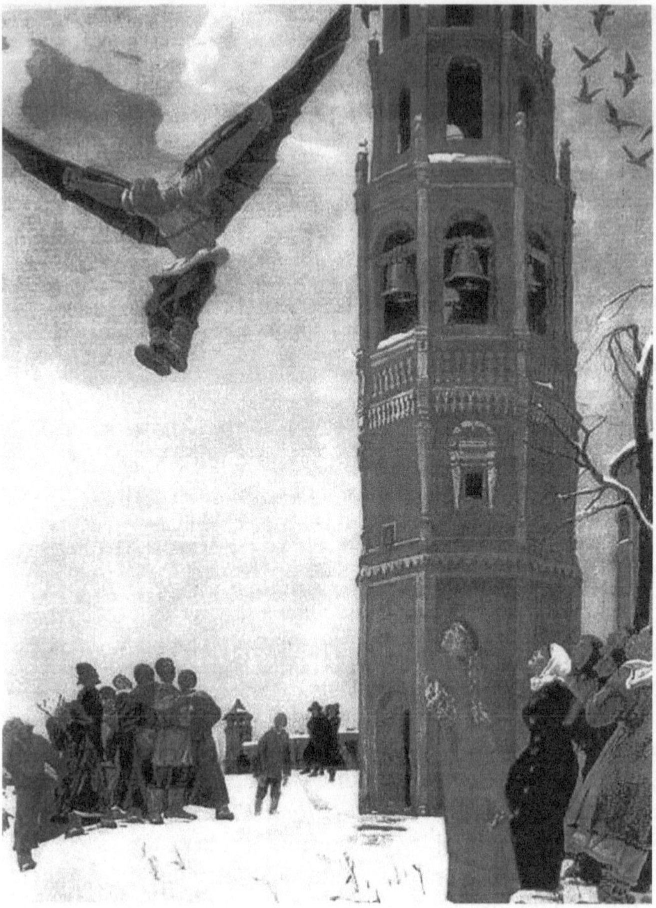

Figure 2.14 *Nikitka, the First Russian Flyer*, Deineka (1940). Courtesy of Art
© Estate of Aleksandr Deineka/RAO, Moscow/VAGA, New York

bell-tower, their wings a miniature version of those 'worn' by Nikitka. As denizens of the sky that has always been their natural habitat, through their fluid locomotion they may underscore by contrast the futility of man's hubristic determination to strive heavenward; or they may simply (and more probably) function as a symbolic analog for the airborne Nikitka. After all, in its earliest stages aeronautics (following da Vinci's sketches of 1505, 'Codex on the Flight of Birds,'[46] almost five centuries before the world's first successful powered aircraft by the Wright brothers) was inspired by the aerodynamic principles behind avian flight. In either case, the painting directly documents what Deineka's depictions of athletes merely intimate—the human drive to 'reach for the sky,' the tropologically loaded title of a British film (1956) devoted to the biography of the WWII flying ace Douglas Bader.

Heady altitude in the subterranean cathedral

> Art is human conscience.[47]
> I consider sincerity to be the very foundation of art.
> This sincerity means, first and foremost, the ability
> to express the ideals of one's own time.[48]
> Aleksandr Deineka

Notwithstanding the peculiar originality of *Nikitka* . . . , no work by Deineka articulates the continuum of sport-airplane-heaven as explicitly and sumptuously as his ceiling designs for the Central Soviet Army Theater in Moscow (1937)[49] and two of the city's early major metro stations: Maiakovskaia (1938) and Novokuznetskaia 1941–3).[50] For centuries the ceilings of cathedrals, churches, and palaces had functioned as conduits to and facsimiles of heaven through the religious figures and scenes adorning them. During the Early Renaissance Andrea Mantegna's designs for the Camera degli Sposi in the Ducal Palace (1472–4) in Mantua, Italy set the tone for illusionistic ceiling frescoes. His deployment of a radical perspective known as *di sotto in sù* ('from below upward'), his placement of figures in convincing settings, and his incorporation of naturalistic details initiated a hardy tradition of illusionistic ceiling painting, with Michelangelo's frescoes at the Sistine Chapel in the Vatican (1508–12) and Antonio Correggio's *Assumption of the Virgin* in the dome of the Parma Cathedral (1530) marking milestones in the efflorescence of that tradition (Hunt 2001–2). The illusion of 3D space on an otherwise 2D or basically flat surface above the viewer directs the raised eyes to a persuasively materialized spiritual world of immaculate forms made accessible through the artist's visionary insight. This complex genre, anchored in *trompe l'oeil*, is what offered an artistic solution for Deineka, a fervent admirer of the Renaissance, when he turned to the assignment of creating appropriate ceiling designs for the huge public underground venues in question.

As documented by the numerous scholars examining the design and decoration of the Moscow metro, this massive-scale undertaking—the brainchild of Stalin and Lev Kaganovich—had a pragmatic, ideological, and psychological agenda. It was intended to provide an efficient system of transportation throughout the capital (the horizontal plane), to market Soviet prestige at home and abroad, and, through artistic means, to create the aura of an illuminated cathedral (the vertical axis).[51] Among manifold problems facing the architects and artists commissioned for the project, those of space and light predominated: how to infuse huge stations (some of them 500 feet long) located approximately 100–200 feet below ground level with sufficient brightness and, moreover, to incorporate decorations inspiring enough to transmit the sense of a sunlit paradise—the radiant future [svetloe budushchee] promised by Soviet ideology.[52] The task required a fundamental inversion, to be executed on an enormous scale and at an accelerated pace. Few collaborators solved the gargantuan problems of such an endeavor as successfully as the team of architect Aleksei Dushkin (1904–77), "star-designer of pre-war Moscow metro stations" (Cooke 1997: 139), and his friend Deineka at what would prove to be the spectacular, prize-winning station of Maiakovskaia.[53]

Dushkin's imaginative design divided the central vestibule into thirty-five evenly spaced bays by columns lined in marble and industrial stainless steel, both known as effective reflectors of light. For the inordinately high ceiling decorations that topped the bays culminating in cupolas—the height of the ceiling was approximately twice that of the New York and Berlin subways (Wolf 1994: 305)—Deineka resorted to dramatic foreshortening and chose the light-reflective material of mosaic.[54] To attract and sustain the upward gaze of passengers he devised a celestial gallery of scenes conceived as glimpses of 'uplifting' facets of contemporary Soviet life spanning twenty-four hours. These elliptical ceiling panels, illuminated by special lights and projected against a background of open sky, exalted sundry 'bodies in motion': the accomplishments of working youth and, more pertinently to my discussion, athletes and aircraft—the heavenly bodies constellated and cast in a mode that rendered them virtually interchangeable dwellers in the firmament (Fig. 2.15).[55] Maiakovskaia station, in short, offered Deineka the appropriate venue for situating his long-held ideals in the empyrean realm—those worthy of "being looked up to" both literally and symbolically in an environment approximating the religious enclaves in which illusionistic ceiling frescoes urging the ocular trajectory of *di sotto in sù* had originated and flourished.[56] Deineka's recollections of the intricacies involved in the artwork at Maiakovskaia address precisely the problems attendant upon the height of the ceiling, which demanded extreme foreshortening and militated against visibility of the images (Sysoev 1989: 2, 52–5)—concerns typical for the genre.

Proponents of an unwaveringly political interpretation of Deineka's works predictably view the preponderance of planes in his design as Deineka's collusion with or obedience to the state's mandate that citizens be reassured of the nation's capacity to defend its borders from the air (see Cooke 1997, Friedman 2000, Groys 2012, O'Mahony 2006). Given the primacy of the vertical axis and the celestial background of the ceiling, as well as the official requirement to select a theme for each station, it is unsurprising that approximately two-thirds of Deineka's thirty-three images are wholly or partially aeronautical,[57] for skies by definition suggest flight, which is why birds also grace two of Deineka's panels (Fig. 2.16). According to the temporal progression captured in the twenty-four hours represented by the visuals, the color of the sky lightens or darkens and projects the brightness of sunshine or is punctuated by clouds. Singly, severally, or in group formation, a wide variety of aircraft (including hydroplanes) and parachutists (all with their parachutes open) appear in one panel after another, some designed to be viewed horizontally, and others vertically.

Less logically dictated but appreciably more interesting and in keeping with Deineka's lifelong attachments and earlier works are the four panels featuring athletes who presumably are performing at the peak of their abilities (*aretê*): the ball players, skier, high jumper, and divers. Reminiscently of *Midday*, three young men in shorts, their arms outstretched, are reaching for a ball located at the center of the image and evoking a fiery, spinning solar sphere in the bright blue vault

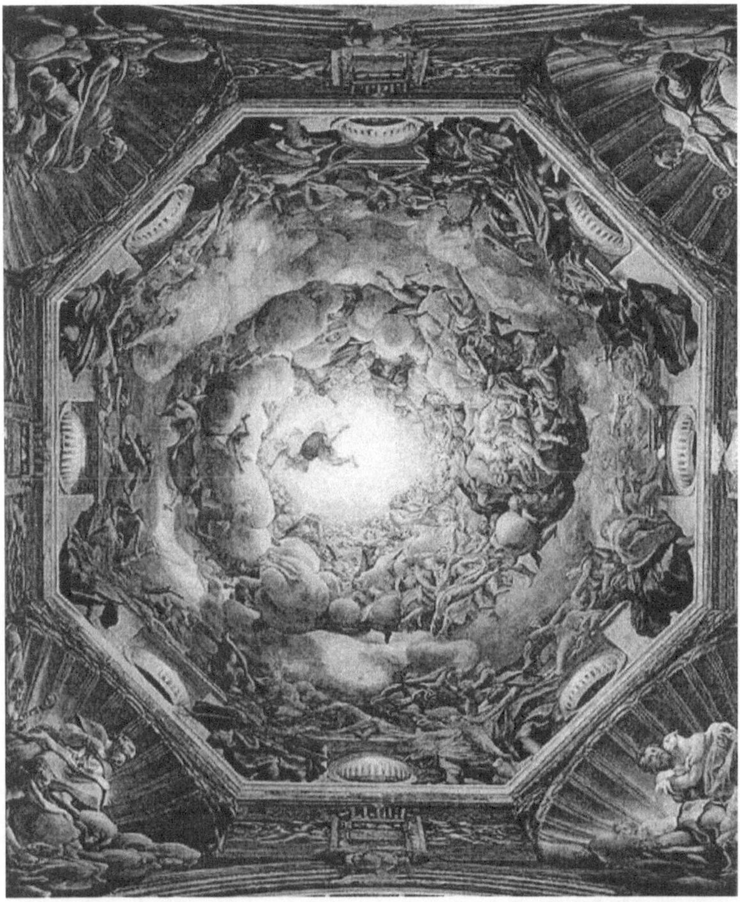

Figure 2.15 Antonio da Correggio, *Assumption of the Virgin* (between 1526 and 1530). Courtesy of Wikimedia Commons

above them (Fig. 2.17). Like countless Deinekian athletes of the 1930s, they appear suspended in the air, floating "weightlessly among the clouds," as Jane Friedman notes (2000: 61), and the instant initial impression is of young gods toning their perfect physiques through play. Elsewhere Deineka's earlier *Goalkeeper* springs to mind in the treatment of the solitary skier, outfitted in red and projected against a pale sky, for his body stretches along the horizontal plane, though the two treetops behind and below him index the enormous, empirically impossible altitude he has attained in his ski jump (Fig. 2.18). By contrast, the high jumper (anticipating the later *Youth*), his body occupying the horizontal plane and one arm pointing upward, has leaped so high that, viewed from the panel's low angle, nothing but the heaven above is visible—presumably, through his athletic prowess he has left all things terrestrial behind (Fig. 2.19). Indeed, all these panels contain no sense of ground

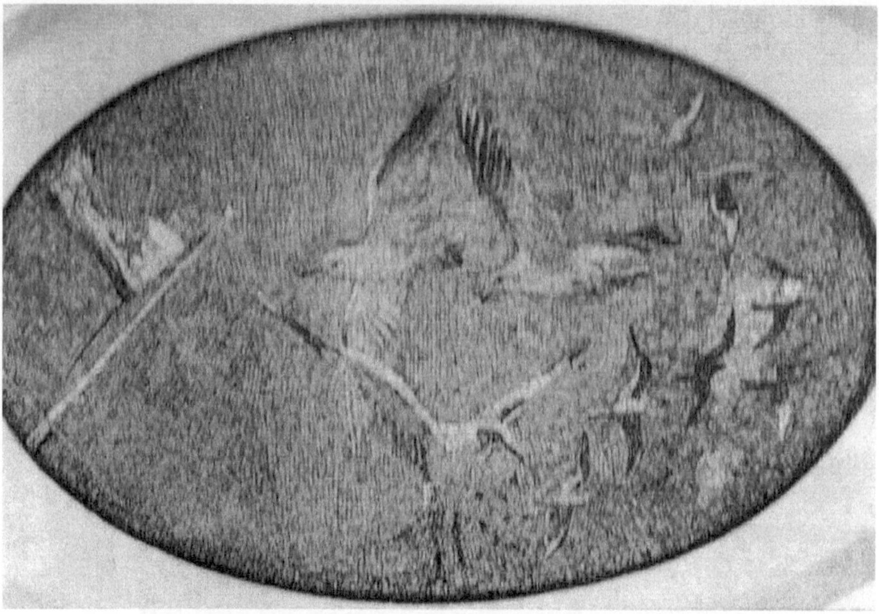

Figure 2.16 Birds in ceiling mosaic at Maiakovskaia metro station (1938). Photograph courtesy of Aleksei Bril'skii

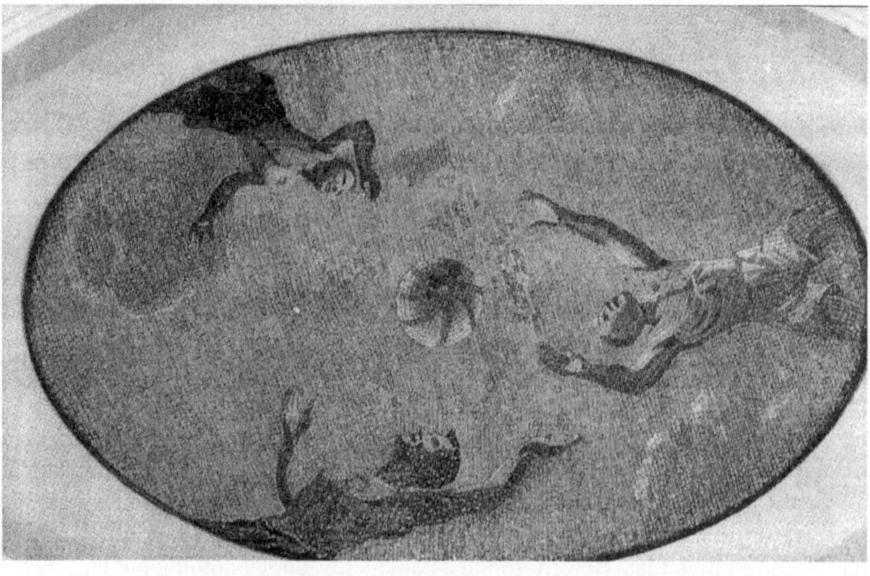

Figure 2.17 Ball players in ceiling mosaic at Maiakovskaia metro station (1938). Photograph courtesy of Aleksei Bril'skii

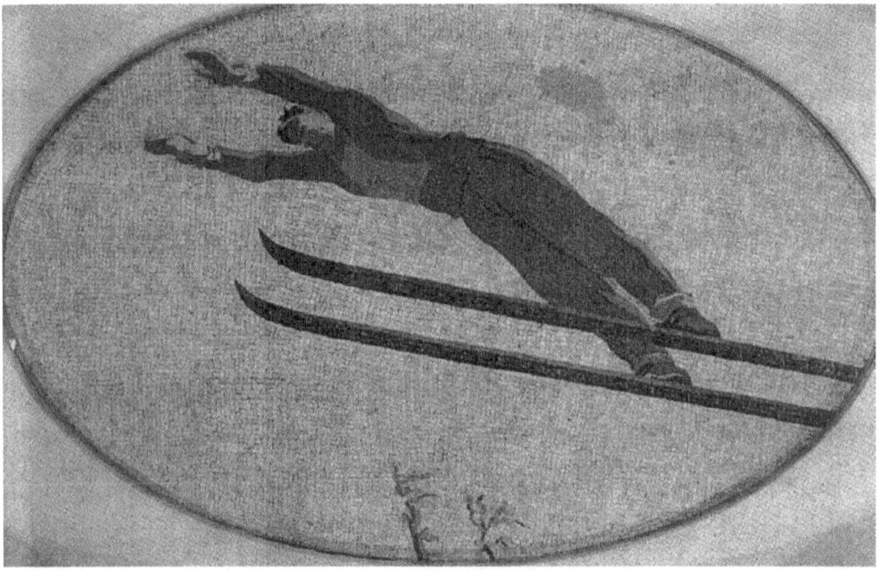

Figure 2.18 Skier in ceiling mosaic at Maiakovskaia metro station (1938). Photograph courtesy of Aleksei Bril'skii

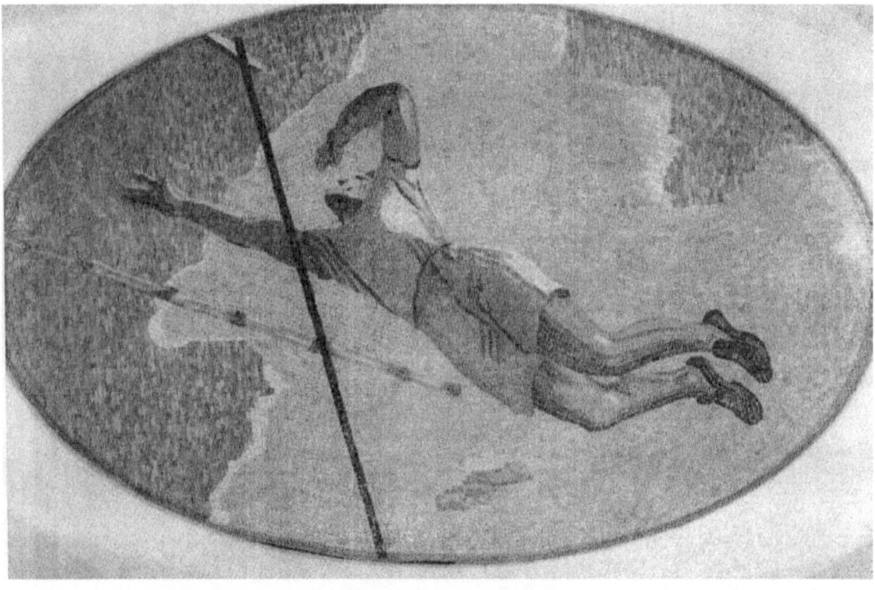

Figure 2.19 High jumper in ceiling mosaic at Maiakovskaia metro station (1938). Photograph courtesy of Aleksei Bril'skii

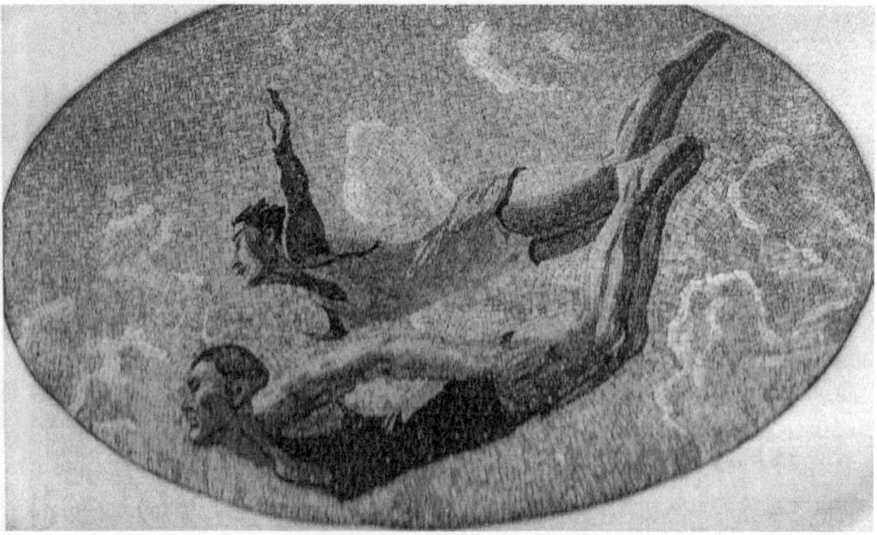

Figure 2.20 Divers in ceiling mosaic at Maiakovskaia metro station (1938). Photograph courtesy of Aleksei Bril'skii

level. The athlete/airplane equation, nonetheless, is clearest in the striking image of the two divers, side by side in mid-dive, though one is slightly higher than the other as they sail through the air (Fig. 2.20). Wearing a red swimsuit, body extended and arms outstretched, the loftier female figure is isomorphic with the airplane, unlike the diver beside her (in a darker red), whose arms are aligned with the length of her (?) body. They approximate two aircraft flying in unconventional formation.[58]

Though Deineka presents both in profile, sporting short hair and swimsuits identical in cut, his specification of gender for the figure close to the bottom of the panel is equivocal enough to have caused confusion. On the one hand, the curve of the right breast and the overall contours of the body, as well as the swimwear, signal womanhood ("women who seem more like airplanes or birds," Friedman 2000: 60); on the other, the roughly outlined facial features and almost military haircut tag the diver as male (O'Mahony 2006: 113). As an artist whose extensive pictorial gallery of athletes was remarkably even-handed, perhaps Deineka here wished to androgynize the pair of celestial divers, though in his other female images, naked or near-naked women tend to be broad-hipped, with generous breasts, solid thighs, and long hair—physical markers differentiating them unmistakably from their male counterparts, with bodies hardly optimal for divers. Whatever Deineka's intention, the athletes in the Maiakovskaia sky represent the pinnacle of human endeavor, the attainment of perfection that yields a transcendent beauty and elevates them to the divine.

Viewed by one scholar as "an extremely lofty example of *Gesamtkunstwerke*" (De Magistris 2012: 239)[59] and by another as a space "imbued with a sense of lightness and elevation and the defiance of gravity" (Friedman 2000: 61),

Maiakovskaia station vouchsafed the requisite conditions for Deineka to showcase his artistic skills, simultaneously providing an outlet for his infatuation with athletes and the expansion of his repertoire of aviation visuals.[60] Five years later three of his seven mosaic panels at Novokuznetskaia (1943)—out of the sixteen originally designed for Paveletskaia station—literally and metaphorically wrested workers from earthly limitations by similarly installing them in the firmament as they miraculously reconfigured the nation's capital seemingly overnight (Fig. 2.21). These designs are congruent with the two panels at Maiakovskaia portraying industrial progress, just as two of the other Novokuznetskaia panels feature planes, while another almost duplicates the Eden-dwelling apple-pickers at Maiakovskaia. The final panel showcases a parade of banner-waving athletes symmetrically arranged around a female athlete carrying flowers atop a huge shield borne by a quartet of bare-chested specimens of physical male perfection. Not only do viewers need to look upward to perceive these foreshortened ideal figures, but the latter themselves all raise their eyes—toward the infinity traditionally associated with the Supreme Being. Here, too, Deineka's *di soto in sù* treatment reveals the spiritual yearning underlying the modern technology revolutionizing human concepts of space and time.[61] For Soviet citizens traveling on the metro and lifting their eyes to the Deinekian heaven, the glimpses of a pantheon of immaculate beings and states daily served as inspirational models of a potential, ever-better self, committed to the principle of *areté*.

Ascent to the sacred

Above all I love form.

A.A. Deineka, 'Notes to a Biography'[62]

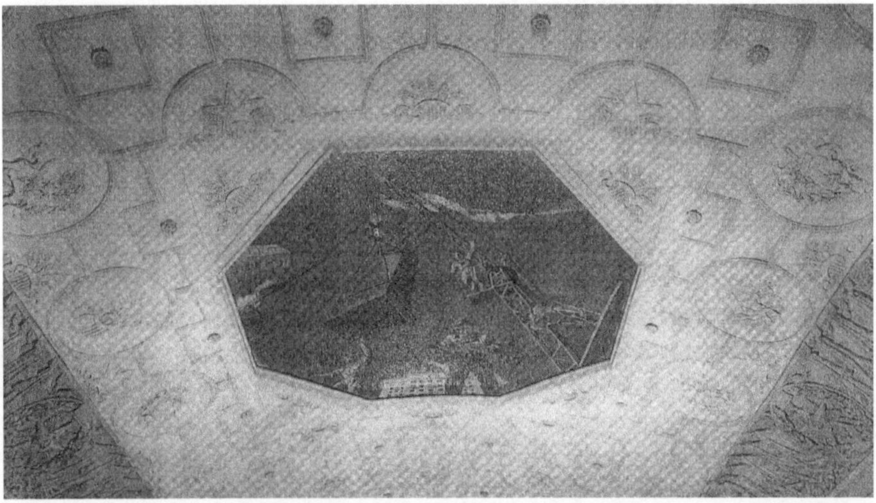

Figure 2.21 Construction in ceiling mosaic at Novokuznetskaia metro station (1943). Photograph courtesy of Aleksei Bril'skii

Commentators indentured to a radically politicized reading of art, exemplified by Boris Groys (2012: *passim*), inevitably locate Deineka's paintings of sport and aviation within the official campaigns promoting both as a means of state control and ideological showmanship. To do so they have to ignore Deineka's attested passion for bodily motion and its representation in art, which persisted as a constant throughout his life, beginning in childhood.[63] He wrote, "I prefer beautiful pictures. I like to see power in a picture—motion, rhythm, light."[64] A profound reverence for aesthetics and a constant search for formal solutions to a host of artistic dilemmas suffuse Deineka's copious writings, which purely political analyses ignore. It is no accident that conservative representatives of the Party's Central Committee and other hardcore conservatives leveled charges of formalism against him in 1948 and 1954 or that reminiscences by Deineka's contemporaries characterize him as a reserved loner. In his ample commentaries on art, Deineka's compulsive recourse to Greek sculpture and the Renaissance rendition of the human body—particularly Michelangelo's 'ideal forms'—as an ultimate standard indicate his shared belief that *areté*, the resolute striving for perfection, breached the boundaries separating the mortal from the divine.

A struggle for some mode of perfection, of course, is hardly unique to Soviet propaganda; indeed, it informs countless religions, thereby rendering ascent to the celestial sphere a richly saturated trope embedded in many cultures' collective unconscious. The ethos of the classical age and the Renaissance codifying the harmony of the physical, mental, and spiritual that produced precisely those artistic works that Deineka and other artists over centuries have prized as aesthetic absolutes united him sooner with those predecessors than with his Soviet contemporaries, though his fascination with athletics and sport intersected with the era's valorization of both. That is precisely the reason that the Deineka of the 1930s—his most original and prolific decade—never wholly 'fit into' Socialist Realism unless one insists on the Proscrustean bed that would ignore those aspects of his art manifestly contravening Socialist Realist requisites. Following Vladimir Sysoev and Susan Reid, Christina Kiaer has justly countered simplistic readings of Deineka's post-1920s works, maintaining that "the transition from Deineka's more overtly Modernist and experimental images of collective labor of the 1920s to his more obviously Socialist Realist depictions of the collective body of the mid-1930s resulted not from his 'bending to the prevailing wind' of official pressure but rather, or at least also, from his changing vision of what constituted appropriate revolutionary art" (Kiaer 2005: 323). A detailed analysis of his paintings during that period, which only Sysoev and, to a lesser extent, Kiaer have essayed, reveals his individual, rather than submissive, concept of sports and aviation. Judging by the available evidence, ideology held little attraction for Deineka, whose artwork and publications corroborate that he never abandoned his love of mobility, which the Soviet programs in sports and industrial progress also happened to champion.[65] His writings testify to his unremitting absorption with aesthetics in both its philosophical and technical aspects—above all, the nature of beauty and his capacity to convey it artistically.

For Deineka athletics and sports constituted a potential route to the sacred category of perfection. Aviation, which he analogized (but never seemed to equate) with athletics and sports, proved more complex, however, and Deineka's response to his experience of flight establishes the nature of that complexity, which his paintings instantiate. Grandiose and impressive on the one hand (the "delight" of his virgin flight), aviation also teemed with dangers and an existential solitude that produced the "horror" he recorded even while appreciating the novel perspective on phenomena and one's place in the world that flight vouchsafed. Combating that horror tested the aviator and parachutist in a modernist battle with unprecedented technological possibilities and perils that also thrilled them. And if *Airborne*, *The Red-Winged Giant*, and *Parachutist over the Sea* transmitted the cosmic alienation and angst inherent in flight, WWII reified its lethal military potential, of which Deineka's *Downed Ace* shows disquieting awareness. Though conventional (if perceptive) views of Deineka absorb him into the Stalinist promotion of sport and aviation, his vision reached higher—into metaphysics, whereby the drive to fulfill one's physical and spiritual potential so as to arrive at a personal perfection raised one to the Olympian sphere that humanity has strived for since the beginning of time. It is the attempt to liberate and materialize the imaginative impulse that so beguiled Nietzsche, for whom, as Bachelard contends, "*Pure air is awareness of the free moment, the moment that opens up a future*" (emphasis in the original; Bachelard 1988: 137). If Nietzsche is, indeed, "the prototype of the *vertical poet*, the *poet of the summits*, the *ascensional poet*" (Bachelard 1988: 127), then Deineka in his lyrical images of athletes, Nikitka, and future pilots is the ascensional artist inspired by the pull of the aerial domain.

Notes

1 [Ia liubliu bol'shie plany i zhiznennost' obraza v veshchakh, poetomu ia vliublen v grecheskuiu skul'pturu, gde garmoniia pravdy zhizni i tsel'nogo videniia prekraskogo identichny.] (Sysoev 1989: 31.)

2 Bachelard 21, emphasis in the original. Translated in 1988, the original, first published in 1943, is titled *L'Air et les songes*. The tenth chapter of Bachelard's study, "The Aerial Tree," anticipated features of Mircea Eliade's "Symbolism of the 'Centre'" in his *Images and Symbols: Studies in Religious Symbolism* (1991).

3 Ann Dumas, "Exhibition in focus: Degas and the Ballet; Royal Academy of Arts," *The Telegraph*, 5 May 2012. Available at http://www.telegraph.co.uk/travel/destinations/europe/uk/london/8826967/Exhibition-in-focus-Degas-and-the-Ballet-Royal-Academy-of-Arts.html (accessed 5 May 2012).

4 See the website *Edgar Degas: The Complete Works* at http://www.edgar-degas.org/ (accessed 5 May 2012).

5 For a chronological outline of Deineka's professional biography, see Sysoev 1989 and *Aleksandr Deineka* . . . 2012: 11–26.

6 Higher Arts and Technical Studios [Vysshie khudozhestvenno-tekhnicheskie masterskie] (1920–26 in Petrograd).

7 According to Deineka, he painted the work in two weeks: "I was afraid to break the rhythm with anything that I added, and therefore didn't put in a number of details which were present in the original sketch. I always saw the figures on the snow as silhouettes." *Art, Life, Time* [Zhizn', iskusstvo, vremia 1967], cited in Sysoyev 1982: np; Sysoev 1989: 103–107.

8 For the avant-garde's addiction to mobility, see Vladimir Paperny, who contrasts the "instability" and constant change of Culture One (1993: 59–60) with the "immobility" of Culture Two, when "upward movement [was] only possible as growth from the ground

while detachment from the ground appear[ed] as heavy as lead weights" (1993: 63). If one accepts such a bipolar division of culture (architecture is Paperny's major, though not sole, concern) and attempts to place Deineka within it, then he likely would be located between the two, albeit closer to Culture One. In fact, his 'middling' position explains the choice of subtitle for the largest (richly illustrated) Anglophone study of Deineka to date, *Aleksandr Deineka [1899–1969]: An Avant-Garde for the Proletariat 2012*.

9 Deineka, "A Keen Sense of the New" [O chuvstve novogo 1961], cited in Sysoyev 1982: 209.

10 It also encourages perspectivism, which, in Nietzsche's sense, would be anathema to monotheistic Soviet ideology.

11 Deineka, "From an Autobiographical Essay" [Iz avtobiograficheskobo ocherka 1947], cited in Sysoyev 1982: page facing plate 97. Burke in *A Philosophical Inquiry into the Origin of Our Ideas of the Sublime and the Beautiful* (1757) detected pleasure in the terror that overcomes one when confronted by the sublime in the form of such phenomena as magnificence, vastness, infinity, and the like. Burke, however, distinguished unequivocally between the sublime and the beautiful, whereas Deineka does not.

12 In this Deineka follows Kazimir Malevich, who viewed flight as the supreme vehicle for transformation of mind and spirit and whose writings on art are saturated with metaphors of flight and ascent. On Malevich's engagement with flight, see Wohl 1994: 158–78.

13 [Nashi gorizonty neobychaino rasshirilis'. My privykli videt' zemliu s samoleta. Slovom, me gliadim ne tol'ko vdal', no vverkh i vniz, chto, konechno, zastavliaet zhudozhnika po-novomu reshat' prostranstvo.] (Deineka, "Iz moei rabochei praktiki" (1961), cited in Sysoev 1989: 2, 53).

14 Friedman also analogizes Deineka's evocation of montage with *aeropittura*, though, true to her political investment, she notes the support of the Italian government for this Futurist initiative, without mentioning Marinetti (Friedman 2000: 58). For an outstanding analysis of Marinetti's artistic contribution to aviation as well as of writers' and other artists' celebration of the phenomenon, see the amply illustrated study by Robert Wohl 1994.

15 Other scholars, such as Friedman 2000, O'Mahony 2006, and Groys 1996, 2012, have dwelled on the political aspects of Deineka's works, whereas my interest lies in the aesthetic and personal investment motivating Deineka's obsession with sports and their capacity for transcendence. Deineka himself reportedly excelled at swimming and boxing (Sysoyev 1982: 9).

16 On the *axis mundi* [world center] see Mircia Eliade, *Images and Symbols*, Princeton: Princeton UP, 1991. 37–54. Jacob's Ladder as described in the Old Testament's Book of Genesis likewise rises to heaven and later would be interpreted as synonymous with the role of Christ as the mediator between heaven and earth.

17 Not only Stalin, but also Mussolini, as Gerald Silk notes, "exploited the venerable fusion of flight and ascent, implying, somewhat metaphorically, that, like the skillful and courageous aviator, he would take Italy to a 'higher level' in all spheres of life—historically, militarily, economically, but also morally and spiritually" (Silk 2003: 256).

18 On this question, see Bowlt 1996, especially 58. Groys's thesis (2012) that for Deineka the athletic body strived for immortality through becoming a machine merely susceptible to endless, depersonalized iterations impresses me as wrongheaded. His dismissal of any relationship between Deineka's concept of the athletic body and that of Greek aesthetics (rather than politics) ignores Deineka's love of classical sculpture and its ideals. Tellingly, Groys's essay teems with generalizations and theory, but avoids specific analysis of Deineka's works (Groys 2012: *passim*).

19 In the Greek original, the full realization of one's potential, for which athletics were deemed essential, so that the body, mind, and soul would be in peak form. According to Bernard Knox and others, *areté* (personified as a goddess) is associated with the Greek

word for pray (araomai/ἀράομαι). See Homer, *The Odyssey*. trans. by Robert Fagles. Introduction and notes by Bernard Knox. London: Penguin Classics, 1996.

20 [Nashe vremia rozhdaet dlia khudozhnika siuzhety v nebe.] Cited in Sysoev 1989: 53.

21 Paintings of Stalin resorted to the proportions, structure, colors, and lighting conventions typical of icons. See A. Gerasimov's *Stalin's Speech at the 16th Congress of the Communist Party* (1933) and his *Hymn to October* (1942), A. Vialov's *Stalin and Voroshilov on the Cruiser* Chervona Ukraina (1933), S. Svarog, *Members of the Politburo of Tsk VKP(b) in Gorky Park* (1936), Iu. Kugach's *Glory to the Great Stalin* (1948), F. Shurpin's *The Morning of Our Fatherland* (1949), A. Laktionov's *The New Apartment* (1952), and many others. The cover of a 1949 issue of the magazine *Ogonek* unashamedly adopted religious iconography by projecting Stalin's portrait in the sky above the Kremlin, with celestial light rays beaming from it down to the population. For this and other images of Stalin, see the website *Stalinka* at http://images.library.pitt.edu/s/stalinka/. See also Evgenia Petrova, "Religious Sources in the Art of Socialist Realism," *Socialist Realisms* . . . . 2012: 161–65.

22 Co-founder and member of the avant-garde OST [Obshchestvo khudozhnikov-stankovistov/Society of Easel Painters 1925–31], Deineka left the group in 1928 and joined Oktiabr' [October], only to transfer to RAPKh [Rossiiskaia assotsiatsiia proletarskikh khudozhnikov/RAPP] in 1931.

23 In addition to studying under Favorskii, Deineka worked alongside Vladimir Maiakovskii, whose poems he illustrated and whose visual priorities and graphic techniques left a visible imprint on Deineka's works of the 1920s. Ten years after Maiakovskii's suicide, Deineka wrote: "He found the verbal form of the present day . . . . He was my teacher; he showed me how to grasp the essence of an event, and, more important still, to find the appropriate imagery to convey this essence." Deineka, "Vladimir Vladimirovich Maiakovskii" (1940) and "From My Work Experience" (1961), cited in Sysoyev 1982: page facing plate 169, which reproduces Deineka's painting *Mayakovsky Working for the Rosta* (1941).

24 *Sport* . . . 1980: page facing plate 48.

25 "From My Work Experience," 1961, cited in Sysoyev 1982: page facing plate 102.

26 "From My Work Experience," cited in Sysoyev 1982: page facing plate 6.

27 "From My Work Experience," cited in Sysoyev 1982: page facing plate 5.

28 Examples in the pre-Soviet era include Kuz'ma Petrov-Vodkin's *Bathing the Red Horse* (1912), and during Soviet times, Arkadii Plastov's *Bathing the Horses* (1938).

29 For the crucial distinction between nude and naked, see the classic studies by Kenneth Clark, *The Nude: A Study in Ideal Form*. Bollingen Series 35.2. New York: Pantheon Books, 1956; and John Berger, *Ways of Seeing*, London: Penguin, 1972.

30 For brief but astute commentary on the making of *Olympia*, see Stefan Steinberg, "Leni Riefenstahl—Propagandist for the Third Reich," *World Socialist Website*, 15 September 2003. Available at http://www.wsws.org/articles/2003/sep2003/rief-s15.shtml (accessed 12 June 2011).

31 Denigrated in the 1920s as a feature of rank capitalism, competition came to the fore during the 1930s in the Soviet Union, which under Stalin favored team sports over athletics [fizkul'tura]. Deineka consistently portrayed both in his artwork.

32 On the various phases of Soviet sport, see Irina Makoveeva, 'Soviet Sport as a Cultural Phenomenon: Body and/or Intellect,' SISC III (2002): 9–32. For sport as battle, see V.T. Ni, *Football Players* [Futbolisty 1971] and Iaroslav Titov, *Football. Attacking* [Futbol. Ataka 1970] and Ice-Hockey. Victory Goal [Khokkei. Pobednaia shaiba 1975], Iurii Pokhodaev, *Khokkeisty* Dinamo [*Dynamo* Ice-Hockey Players 1977].

33 Deineka's athletes and sportsmen seem to compete not with others, but with themselves, in the spirit of *areté*.

34 Sysoev also noticed the similarity (Sysoev 1989: 218).

35 "Ornament and Color" [Ornament i tsvet 1953–5], quoted in Sysoyev 1982: page facing plate 110 (wording modified). The original reads as follows: "pod siluetom

ia ponimaiu ne chistuiu zalivku chernym po belomu, a bol'shuiu formu, kotoraia pogloshchaet v sebe vse neobkhodimye dlia blizkogo vospriiatiia detail. Formu, edinstvuiu po tvorchestoi udachi, kak *Adam* Mikelandzhelo ili *Petr I* Fal'kone. Siluet ne est' absoliutnaia ploskost', a vkliuchaet v sebia ob"em i tsvet. . . . Iasnye, siluetno-zhivopisnye rasstanovki predmetnykh mass blestiashche derzhat vidimost' na bol'shikh rasstoianiiakh, a ikh okonturivanie, kotorye my nadbliudaem v drevnikh freskakh,—odin iz sposobov sokhraneniia bol'shoi formy." Identified as A.Deineka, "Iz moei rabochei praktiki," *Zhizn', iskusstvo, vremia* (Leningrad 1974), ed. V. Sysoev. 52. Cited in Sysoev 1989: 118–19.

36 Sysoyev 1982: page facing plate 156.

37 For Kuptsov's painting, see Fig. 2.7.

38 Of considerably less interest is the 1932 *Bomber* [Bombovoz], flying over an empty field, possibly readying to land.

39 The red volume functions as a displacement of the standard red kerchief worn by the Pioneers.

40 Such is Mike O'Mahony's interpretation of the painting in his excellent study *Sport in the USSR* (2006: 122–4). Here and elsewhere my view of Deineka's paintings about sport and aviation departs from O'Mahony's (and Groys's) in that it favors a metaphysical and aesthetic rather than political model of analysis. My conviction is that Deineka's fundamental allegiances were to principles beyond politics, and his lifelong fascination with movement and the body trumped political orthodoxy, though at various junctures his deeply felt concerns coincided with the political imperatives of the era.

41 In his review of the manuscript of our collection Scott Palmer proposes that the painting is ironic, but I detect no hint of irony in it. Moreover, Deineka's verbal and visual works suggest that irony was not a part of his temperament.

42 Deineka, *Life, Art, Time* (1967), cited in Sysoyev 1982: page facing plate 185.

43 Sysoyev 1982: page facing plate 103.

44 Characteristically, Deineka takes care to observe balance in his composition.

45 Deineka, "Echoes of the Past" [Pereklichka s proshlym 1941], cited in Sysoyev 1982: page facing plate 156.

46 For information about and examples of da Vinci's designs for human flight, see 'Leonardo da Vinci: His Flying Machines' at http://www.angelfire.com/electronic/awakening101/leonardo.html (accessed 2 August 2013). Charles Elachi, director of NASA's Jet Propulsion Laboratory, declared that the principles of flight Leonardo outlined are "basically . . . the foundation of how we do flights." Deborah Zabarenko, 'Leonardo da Vinci's notebook on flight visits the Smithsonian,' *MSN News*, 14 September 2013. Available at http://news.msn.com/pop-culture/leonardo-da-vincis-notebook-on-flight-visits-smithsonian (accessed 14 September 2013).

47 Iskusstvo—eto sovest' cheloveka. From Aleksandr Deineka, *Iz moei rabochei praktiki* (1961), cited in Sysoev 1989: 2, 49.

48 Deineka, "From an Autobiographical Essay" (1947), cited in Sysoyev 1982: page facing plate 192.

49 In aesthetics and design, the sketch for the theater's ceiling dovetails with the visuals for the metro stations: a foreshortened, circular image of the heavens, with three airplanes at the highest level, and three groups of athletes distributed evenly along three points of the perimeter, with the fourth occupied by one young female atop a ladder picking apples and the others, in flowing robes evoking ancient Greece, holding flowers and hailing both airplanes and runners. The upward-directed energy is buttressed by two portals flanking one set of runners, though the four groups face in different directions. Pale blues, flesh tones, and reds dominate the painting, which is unambiguously conceived as a vibrant Eden of physical vitality.

50 Deineka's former teacher, Favorskii, painted the ceiling fresco, showing airplanes in flight celebrating May Day, in the theater of the Central House of Young Pioneers in Moscow (1935–6), and Deineka's first 'ceiling assignment' was for the Red Army

Theater in Moscow (1937), analysis of which lies outside the scope of this chapter owing to considerations of space.

51 See Cooke 1997, De Magistris 2012, Friedman 2000, Groys 2012, O'Mahony 2006, and Wolf 1994.

52 Catherine Cooke's thoughtful and wide-ranging article on the primacy of light in the system of metro stations under Stalin and its indivisibility from the radiant future focuses on the architectural aspect, paying only slight attention to the decorative, albeit she acknowledges the importance of overall aesthetics to Kaganovich, who insisted that the proletariat desired beauty in all structures around them (Cooke 1997: 147).

53 For the background of Dushkin's work on this station and the prizes it received, see (Dushkina 2004: 72–6). The immensely talented Dushkin had undergone a baptism by fire in his collaboration with the aggressive sculptor Mikhail Manizer at the Ploshchad' Revoliutsii [Revolution Square] station, his first assignment, and came to Maiakovskaia determined not to make kindred concessions. These proved unnecessary in the long run, and it was Deineka who had to make adjustments against which he railed (Dushkina 2004: 72–4). For the collaboration of the two men, see Friedman 2000: 53. Dushkin also designed the Palace of the Soviets station, now known as Kropotkinskaia.

54 Though initially he found mosaic far from the easiest medium ("Work with mosaic demands daring and risk" [Sysoev 1989: 2, 55]), he returned to it for the decoration at Novokuznetskaia and later works, such as *Morning* [Utro 1949], *A Good Morning* [Khoroshee utro 1959–60], with its four naked male swimmers (again, the counterpart of the female quartet in *The Bathers* [Kupal'shchitsy 1952]), *Hockey Players* [Khokkeisty 1959–60], and others. During the 1940s and 1950s he also produced a number of sculptures, the majority devoted to the dynamic bodies of athletes.

55 Athletics and sport were the theme of the sculptures by Elena Ianson-Manizer decorating the Dynamo [Dinamo] metro station. On its design and decoration, see Wünsche 2000.

56 On the importance of 'elevation,' see Cooke 1997: 152.

57 To bolster her political reading, Friedman actually found it necessary to identify the bomber planes among the represented aircraft (2000: 57).

58 Intriguingly, the representation of the two is comparable to Leni Riefenstahl's shots of divers in her documentary on the Olympics (*Olimpiad* 1937).

59 For a nuanced and wide-ranging investigation of the relationship between *Gesamtkunstwerke* and the avant-garde see David Roberts, *The Total Work of Art in European Modernism* (Ithaca, NY: Cornell UP, 2011, especially the chapter dealing with the Soviet Union (207–31) and the section explicating Nietzsche's revision of Burke's sublime (201–6).

60 Of his work on the metro design, Deineka wrote, "The dynamics of the figures and their foreshortened silhouettes—bold, new, and convincing—had, with the aid of the composition, to create the sensation of rising up in the air. . . . The slightly gleaming pieces of unpolished smalt create a kind of unity with the polished marble and the columns of stainless steel." Deineka, "From My Work Experience" (1961), cited in Sysoev 1982: page facing plate 161.

61 For a cogently argued and penetrating analysis of how concepts of space evolved over the centuries, and how the philosophy of science negotiated the relationship between celestial and spiritual space, see Margaret Wertheim, *The Pearly Gates of Cyberspace: A History of Space from Dante to the Internet,* New York: W.W. Norton & Co., 1999.

62 [No bol'she vsego ia liublio formu.] Deineka, "Zametki k biografii" (1944), cited in Sysoev, 1989: 2, 20.

63 After WWII the number of works reflecting that near-obsession dwindled, as did his output in general. Despite his retreat into still lifes (fruit, flowers, etc.), however, such late works as *Basketball* [Basketbol 1964] evidence his desire to cast scenes of sport or athletes in sculpture, mosaic, or, in this instance, stained glass. Occasionally he even produced oil paintings of the all-too-familiar boys in sunny Sevastopol, notably *In the South* [Na iuge 1966] just three years before his death.

64 Deineka, "From an Autobiographical Essay" (1947), cited in Sysoyev 1982: page facing plate 194.

65 Hence the introduction of Aviation Day on 18 August, the expansion of the Osoaviakhim [The Defense Society for Assistance to the Aviation and Chemical Industries] into the last mass organization of the decade, and the lionization of flyers such as Valerii Chkalov to the ranks of national heroes (Friedman 2000: 48–9).

Bibliography

Aleksandr Deineka [1899–1969]: An Avant-Garde for the Proletariat. (2012) Madrid: Fundación Juan March.

Bachelard, G. (1998) *Air and Dreams: An Essay on the Imagination of Movement.* Trans. Edith and Frederick Farrell. Dallas: The Dallas Institute Publications (Bachelard Translation Series).

Bowlt, J.E. and O. Matich. (Eds) (1996) *Laboratory of Dreams: The Russian Avant-Garde and the Cultural Experiment.* Stanford: Stanford University Press.

Bowlt, J.E. (1996) "Body Beautiful: The Artistic Search for the Perfect Physique." In *Laboratory of Dreams: The Russian Avant-Garde and Cultural Experiment.* Eds. J.E. Bowlt and O. Matich. Stanford: Stanford University Press. 37–58.

Bown, M.C. and B. Taylor. (Eds) (1993) *Art of the Soviets: Painting, Sculpture and Architecture in a One-Party State, 1917–1992.* Manchester and New York: Manchester University Press.

Bown, M.C. (1993) "Aleksandr Gerasimov." In *Art of the Soviets: Painting, Sculpture and Architecture in a One-Party State, 1917–1992.* Eds. Bown and Taylor. Manchester and New York: Manchester University Press. 121–39.

Clark, T. (1997) *Art and Propaganda in the Twentieth Century: The Political Image in the Age of Mass Culture.* London: Calmann and King.

Cooke, C. (1997) "Beauty as a Route to 'the Radiant Future': Responses of Soviet Architecture." *Journal of Design History,* Vol. 10, No. 2, *Design, Stalin and the Thaw*: 137–60.

Corn, J.J. (1983) *The Winged Gospel: America's Love Affair with the Airplane, 1900–1950.* New York: Oxford University Press.

De Magistris, A. (2012) "Underground Explorations in the Synthesis of the Arts: Deineka in Moscow's Metro." *Aleksandr Deineka [1899–1969]: An Avant-Garde for the Proletariat.* Madrid: Fundación Juan March. 236–45.

Dushkina, N.O. (Ed.) (2004) *Zhizn' arkhitektora Dushkina 1904–1977.* Moscow: "A-Fond."

Efimova, A. and L. Manovich. (Eds) (1993) *Tekstura: Russian Essays on Visual Culture.* Chicago, London: University of Chicago Press.

El Cosmos de la Vanguardia Rusa: Arte y Exploración Espacial 1900–1930. (2010) Santander: Fundación Botín.

Friedman, J. (2000) "Soviet Mastery of the Skies at the Mayakovsky Metro Station." *Studies in the Decorative Arts*, Vol. 7, No. 2 (Spring–Summer): 48–64.

Gerovitch, S. (2011) "'Why Are We Telling Lies?' The Creation of Space History Myths." *The Russian Review*, Vol. 70 (July): 460–84.

Groys, B. (1996) "The Birth of Socialist Realism from the Spirit of the Avant-Garde." In *Laboratory of Dreams: The Russian Avant-Garde and Cultural Experiment.* Stanford: Stanford University Press. 193–218.

—. (2012) "Aleksandr Deineka: The Eternal Return of the Athletic Body." *Aleksandr Deineka [1899–1969]: An Avant-Garde for the Proletariat*. Madrid: Fundación Juan March. 76–83.

—. "Underground as Utopia." (2012) *Aleksandr Deineka [1899–1969]: An Avant-Garde for the Proletariat*. Madrid: Fundación Juan March. 249–52.

Gutkin, I. (1999) *The Cultural Origins of the Socialist Realist Aesthetic 1890–1934*. Evanston, IL: Northwestern University Press.

Harte, T. (2009) *Fast Forward: The Aesthetics and Ideology of Speed in Russian Avant-Garde Culture, 1910–1930*. Madison, WI: University of Wisconsin Press.

Hunt, T. (2001) "From Mantegna to Michelangelo: Illusionistic Ceiling Paintings of the Renaissance Pave the Way for Baroque Excess." *UC Davis Prized Writing*. Available at http://prizedwriting.ucdavis.edu/past/2001-2002/from-mantegna-to-michelangelo-illusionistic-ceiling-paintings-of-the-renaissance-pave-the-way-for-baroque-excess.

Iablonskaia, M. (1964) *Aleksandr Aleksandrovich Deineka*. Leningrad: "Khudozhnik RSFSR."

Jartsev, A. (n.d.) "A Stalinist Dream of Art." *The Art-Bin*. Trans. Henrik Nordström. Available at http://www.art-bin.com/art/asovietarteng.html.

Kaufman, R. (1936) "Aleksandr Deineka." *Iskusstvo*, Vol. 3: 84–100.

Kettering, K.L. (2000) "An Introduction to the Design of the Moscow Metro." *Decorative Arts*, Vol. 7, No. 2: 2–20.

—. (2000) "Sverdlov Square Metro Station: 'The Friendship of the Peoples' and the Stalin Constitution." *Decorative Arts*, Vol. 7, No. 2: 21–47.

Kiaer, C. (2005) "Was Socialist Realism Forced Labour? The Case of Aleksandr Deineka in the 1930s." *Oxford Art Journal*, Vol. 28, No. 3: 321–45.

—. (2012) "Aleksandr Deineka: A One-Man Biography of Soviet Art." *Aleksandr Deineka [1899–1969]: An Avant-Garde for the Proletariat*. Madrid: Fundación Juan March. 56–67.

Kibrik, E. (2012) "The Artist of Modernity" [Khudozhnik sovremennosti 1957]. *Aleksandr Deineka [1899–1969]: An Avant-Garde for the Proletariat*. Madrid: Fundación Juan March. 403.

Makoveeva, I. (2002) "Soviet Sports as a Cultural Phenomenon: Body and/or Intellect." *Studies in Slavic Cultures (SISC)*, No. 3: *The Russian Body* (July): 9–32.

Markovits, A.S. and L. Rensmann. (2010) *Gaming the World: How Sports Are Reshaping Global Politics and Culture*. Princeton and Oxford: Princeton University Press.

O'Mahony, M. (2006) *Sport in the USSR: Physical Culture—Visual Culture*. London: Reaktion Books.

—. (2003) "Archaeological Fantasies: Constructing History on the Moscow Metro." *The Modern Language Review*, Vol. 98, No. 1 (January): 138–50.

Palmer, S.W. (2006) *Dictatorship of the Air: Aviation Culture and the Fate of Modern Russia*. Cambridge: Cambridge University Press.

—. (2005) "Icarus, East: The Symbolic Contexts of Russian Flight." *SEEJ*, Vol. 49, No. 1: 19–47.

Paperny, V. (1993) "Movement—Immobility." In *Tekstura: Russian Essays on Visual Culture*. Chicago and London: University of Chicago Press. 56–69.

Pisano, D.A. (2003) *The Airplane in American Culture*. Ann Arbor: University of Michigan Press.

Riordan, J. (1991) *Sport, politics and communism*. Manchester and New York: Manchester University Press.

Silk, G. (2003) "'Our Future Is in the Air': Aviation and American Art." *The Airplane in American Culture*. Ed. D.A. Pisano. Ann Arbor: University of Michigan Press. 250–96.

Simpson, P. (2004) "Parading Myths: Imaging New Soviet Woman on *Fizkul'turnik's* Day, July 1944." *The Russian Review*, Vol. 63 (April): 187–211.

Socialist Realisms. Soviet Painting 1920–1970. (2012) Skira.

Sport v sovetskom izobrazitel'nom iskusstve. (1980) Moscow: Sovetskii khudozhnik.

Starks, T. (2008) *The Body Soviet: Propaganda, Hygiene and the Revolutionary State*. Madison, WI: University of Wisconsin Press.

Stites, Richard. (1989) *Revolutionary Dreams: Utopian Vision and Experimental Life in the Russian Revolution.* New York and Oxford: Oxford University Press.

Sysoev, V.P. (Ed.) (1989) *Aleksandr Deineka: Zhizn', iskusstvo, vremia*. 2 Vols. Moscow: "Izobrazitel'noe iskusstvo."

Sysoyev. V.P. (1982) *Alexander Deineka: Paintings, Graphic Works, Sculptures, Mosaics, Excerpts from the Artist's Writings*. Leningrad: Aurora Art Publishers.

Wohl, R. (1994) *A Passion for Wings: Aviation and the Western Imagination 1908–1918*. New Haven and London: Yale University Press.

Wolf, W.K. (1994) *Russia's Revolutionary Underground: The Construction of the Moscow Subway, 1931–1935.* Ph.D. Dissertation. The Ohio State University.

Wünsche, Isabel. (2000) "Homo Sovieticus: The Athletic Motif in the Design of the Dynamo Metro Station." *Decorative Arts*, Vol. 7, No. 2: 65–90.

3 Comic cosmonaut

Space exploration and visual satire in *Krokodil* in The Thaw

John Etty

Introduction

The contemporaneous technological developments that precipitated the Soviet Union's unprecedented achievements in space flight, and the political developments of The Thaw, prompted Soviet artists to explore the possibilities of visual representations of life everywhere, from the terrestrial USSR to the bounds of the known universe. In keeping with the orthodoxies of official culture, the majority of these artistic treatments were serious, conveying the significance of the technological achievements and optimistic certainty about the material and spiritual satisfaction that space exploration would bring. This chapter considers a different type of visual language: humorous, satirical visual responses to spaceflight in *The Crocodile* magazine during The Thaw era.[1] Rather than foregrounding the connection between Soviet ideology and technological feats, as other studies have done,[2] I shall explain how numerous topical discourses were combined in the space imagery of this popular humor magazine, and how *Krokodil*'s visual language allowed the magazine's images to communicate nuanced and skeptical visions of Soviet modernity that betrayed the coincidence of a political-social crisis at a time of technological triumph. My contribution to the discussion of the impact of flight on visual language will be to consider the popular representations of space exploration in *Krokodil* magazine.

From its first production in August 1922, *Krokodil* featured satirical cartoons, articles, and poems, and while other similar magazines were closed for various reasons by the early 1930s, it survived in a form that remained remarkably unchanged until the collapse of the Soviet Union.[3] In its Thaw-era format, *Krokodil* was a sixteen-page, four-color magazine, printed on newsprint, published thirty-six times annually (on the 10th, 20th, and 30th of each month), with a circulation of 700,000 copies. While most readers received their copies through subscriptions, which were highly prized, the large posterly cartoons were especially well suited to public display, in the glass cases in shop windows or in the street, where the magazine was also available for public consumption.

In the years of late Stalinism and the early Thaw era, the magazine's front and back covers, as well as its center pages, almost always featured large, full-color cartoons on topics connected with foreign policy subjects, often directly concerning the United States or her allies, and usually in ideologically correct terms. *Krokodil* generally contained a large number of images. As a survey of the thirty-six issues shows, in 1956 an edition of *Krokodil* contained an average of twenty-five images (or sets of connected images). Containing cartoons and texts on diverse social, economic, and foreign policy subjects, the magazine routinely addressed the USSR's technological achievements. Around twenty-five full-time staff members (including the Editorial Board, appointed by the Central Committee of the Communist Party) were employed on the magazine, divided among five departments, but a total of sixty-five writers and artists contributed regularly to *Krokodil*; and all except those who also sat on the Editorial Board were freelancers (Shabad 1964: 22). Artists received commissions from the editors or submitted unsolicited works, and the members of the Editorial Board selected content for the magazine at well-attended and often raucous meetings.

This chapter studies *Krokodil*'s visual exploration of the relationship between technology and humor in the context of a society that valorized scientific, technical, and industrial endeavor, in a state that celebrated the use of machines to achieve economic goals. As many scholars have shown, there were close connections between Bolshevism and science,[4] but little attention has been paid to Soviet technology and humor. Stalinist culture found nothing comic about technology per se, since technology was both a product and a tool of the New Soviet Man. The Soviet machine was generally hailed along with its human operator. In the 1930s, for instance, heroic pioneer aviators such as Valerii Pavlovich Chkalov (1904–38) and Mikhail Mikhailovich Gromov (1899–1985), and aviatrixes such as Polina Denisovna Osipenko (1907–39), Valentina Stepanovna Grizodubova (1909–93), and Marina Mikhailovna Raskova (1912–43), completed numerous record-breaking flights. Their efforts mirrored numerous similar achievements using technology in other fields, notably those of Aleksei Grigor'evich Stakhanov (1906–77), whose 1935 mining exploits were celebrated by the Soviet state. The modern machinery that enabled these achievements was considered a facilitator, an indicator, and a driver of social and political progress in the USSR, but it was not considered to have any comic value. The only humorous potential for technology was in its inefficiency or decrepitude, as in *Volga-Volga* [Volga-Volga 1938, dir. by Grigorii Vasil'evich Aleksandrov],[5] but in *Krokodil* in the years before The Thaw, technology was never satirized. This fact is explained by the peculiar position of the magazine.

As an organ of the Soviet state, produced under the aegis of the Pravda publishing house and subject to Communist Party of the Soviet Union (CPSU) authority, *Krokodil* produced a unique brand of state-sponsored satire. The USSR's only satirical magazine for most of its existence, *Krokodil* straddled official and unofficial zones, simultaneously enunciating and critiquing normative discourses, as well as setting its own agenda. It was the ambivalence of the magazine's official satiricism that determined the nature and range of the content it published, and partly as a result, this very ambivalence has led some scholars to devalue *Krokodil* as little more than visual illustration of official policy.[6] *Krokodil*'s ambivalence is reflective of broader problems with the definition and function of comic genres throughout the Soviet period in particular, and in Russian culture in general,[7] but it also contains the essence of the magazine's worth as a source of material on Soviet modernity. David Hoffman's definition of modernism—"social interventionism and mass politics" (2003: 7)—describes both the existence and the content of the magazine. As the developments of the post-Stalin period altered the nature of Soviet politics, *Krokodil*'s attitude to technology altered, and, ironically, in an age of supreme technological advancement, after Stalin's death several cartoons appeared in *Krokodil* that satirized the technology. As well as the images of space technology discussed here, various cartoons made humor out of Soviet tendencies to over-mechanize.[8]

To question some of the assumptions about *Krokodil* and Soviet media, and to appreciate the magazine's contribution to critical political discourse and visual language, this chapter examines images on the theme of Soviet space explorations drawn by Ivan Maksimovich Semenov (1906–83), which appeared in *Krokodil* between 1958 and 1964. The magazine contained over one hundred and twenty space-related cartoons during this period—almost half of which were published in the breakthrough years of 1959 and 1961—and Ivan Semenov drew at least fifteen of them.[9] The images selected for discussion here were chosen not only for their depictions of the cosmos, but also on the basis of their prominence in their respective issues, which was itself indicative of Semenov's style and his significance as an artist. These cartoons provide insights into the depiction of space exploration in Soviet satirical visual language. Furthermore, given their appearance in a serialized publication over a six-year period, they invite consideration of the changes in the visual language precipitated by the key technological developments of the space race. Finally, studied in the context of the domestic political reforms of The Thaw, they are instructive regarding the cultural consequences of the coincidence of the Soviet political reforms and the achievement of spaceflight.

"A star of the first magnitude"

Ivan Semenov's biography is an ideal story of someone who benefited from the radical social change brought about by Soviet power. Born in Rostov in 1906, he rose from humble origins after a hard childhood in which he had to sell painted boxes of cigarettes to support his family (Cherepanov 2009: 106). Nevertheless, he began medical training in 1925, before discovering a talent for graphic satire. He studied at Rostov Art College (1926–8) and his work first appeared in *Krokodil* in April 1931. Moving to Moscow, he became a regular contributor to the magazine and other Soviet publications, such as *Komsomol'skaia pravda*, during the 1930s (Ioffe 1956: 16), and even exhibited in the influential 'Artists of the RSFSR in the Last Fifteen Years (1917–1932).' He became a member of the CPSU in 1941, and during the war worked mainly for *Krokodil* and *The Red Fleet* [Krasnyi flot]. Semenov continued to work for *Krokodil* in the years after the war, and most issues of the magazine contained one of his cartoons. Although his work is relatively unknown in the West, he became something of a hero in Soviet official culture. By the late 1950s he had become one of *Krokodil*'s most respected and celebrated artists, admired by many contemporaries: for instance, Boris Efimovich Efimov (1900–2008), comparing him with all Soviet and foreign artists, called him "a star of the first magnitude" [byl zvezdoi pervoi velichiny] (Efimov 1976: 147). Semenov became a member of *Krokodil*'s Editorial Board in December 1958 and was awarded the honorary title of People's Artist of the USSR in 1974. He was also famous as a book illustrator (including children's books), and he co-founded and edited the children's journal *Merry Pictures* [Veselye kartinki] in 1956. It is indicative of the relative artistic freedom of this period that the publication became an inspiration for the first Russian comic artists. Semenov's work appeared recently at the Tretyakov Gallery's celebration of the fifty-fifth anniversary of *Merry Pictures* and was also displayed as part of the 2012 Moscow Book Festival in an exhibition titled 'Classic Soviet Cartoons.'[10]

Semenov's style matured over time, but even his earliest cartoons are recognizable. Though his images were distinctive, they were not atypical. While some of his more celebrated *Krokodil* colleagues may have often used visual metaphors, symbolism, or distortion, Semenov's realistic style relied on simple ink lines and cheerful watercolor technique, which maximized *Krokodil*'s limited printing capabilities by juxtaposing different colors for maximum effect. His cartoons often incorporated references to multiple topical discourses, and he created lively but ordinary-looking scenes, capturing a wide-angle gaze, involving large casts of characters that were often busy and full of action (if not movement). He mastered large-scale, multi-figure images [massovyi risunok] (Cherepanov 2009: 106), which were familiar from the grand style of Socialist Realist artworks, but he was also notable for his use of comics techniques sooner associated with American comics and generally avoided by Soviet artists (Alaniz 2010: 65), such as ribbon paths, motion lines, and speech or thought bubbles. These attributes likewise marked the works of Semenov's colleagues. Iurii Nikolaevich Uzbiakov (1916–82) and Evgenii Alimpievich Vedernikov (1918–2009) also employed multi-character scenes, often using wide-angle perspectives. Vedernikov in particular produced extremely detailed scenes and provided some of his cast with individual character, but neither artist's figures were imbued (as Semenov's inevitably were) with a sense of individuality or quite the same sense of mischief. Konstantin Pavlovich Rotov (1902–59) and Naum Moiseevich Lisorgskii (1910–86) filled their compositions with characters and action, and also employed simple comics techniques.

Semenov's representational strategy depended upon the successful combination of different discourses. As I argue below, the combination of multiple discourses in one image provided readers with the unique opportunity to view one aspect of Soviet ideology through the lens of another, in a way that invited a personal engagement with politics not afforded by other media. Furthermore, this chapter will explain the effects of Semenov's thematic and compositional decisions. *Krokodil*'s visual representations of The Thaw's themes of resetting time, expanding into outer space, and combating the evils of the past simultaneously affirmed and critiqued official treatments of the most topical themes in Soviet politics and culture. Certain themes that dominated Soviet media were reflected in a politically correct manner in *Krokodil* cartoons, but the latter also examined, mixed, and refracted these official discourses in everyday terms that suggested profound, if humorous, skepticism.

Krokodil's cosmos

Given its concern with aspects of Soviet modernity, *Krokodil* magazine is an excellent case study for the examination of images of Soviet space exploration and the ambivalent relationship between technology on the one hand, and satire and visual language on the other. In numerous cases, *Krokodil*'s imagery was serious and devoid of satire. The covers of countless issues during the late 1950s and early 1960s straightforwardly depicted Soviet satellites and rockets or visualized Soviet dominance in the cosmos. Many *Krokodil* space-themed covers depicted Soviet achievements as representative of the culmination of all human technological endeavors, and therefore were to be celebrated by people of all nationalities. On 10 April 1962, for example, almost exactly a year after Yuri Gagarin's space flight, *Krokodil*'s front cover, drawn by Vitalii Nikolaevich Goriaev (1910–82), showed a cartoon female admirer gazing, enchanted, at a photograph of the world's first cosmonaut. The woman's globe headscarf and blue blouse decorated with stars suggest that she is a female representation of the earth and the cosmos. She utters the words, "I've loved you for a year now . . . " [Uzh god, kak ia tebia liubliu . . .], implying that the whole world shares affection for, and a sense of intimacy with, Gagarin.[11] In all cases, the connection between Soviet space superiority and the USSR's political system was drawn at least implicitly, although often through the symbolism of Russian nationhood. In Goriaev's cover image, for instance, the female figure could plainly be meant to represent the Russian motherland [Rodina mat'].

Krokodil also featured cartoons about the Soviet space explorations that depicted the feats as events that both astounded and dismayed observers in capitalist nations. The front and back covers of *Krokodil* on 20 August 1962, drawn by Evgenii Aleksandrovich Shukaev (1932–88), featured a wraparound cosmos, populated by tiny stars, a crescent moon, a ringed planet, and a giant white dove, accompanied by the Vostok 3 and 4 rockets, speeding through azure space. The image, the caption, "Escort of Peace" [Eskort mira), and the poem, "The Flying Party Group" [Letaiushchaia partgruppa], which fills one-third of the back cover, envision spaceflight as a supra-national human achievement while at the same time specifying numerous Soviet identity markers. *Krokodil* cartoon images such as those created by Boris Efimov frequently featured Western capitalists or US generals staring aghast at cosmic indicators of Soviet modernity or attempting feebly to launch their own crude missiles into space. Only one *Krokodil* cartoon depicted a US spacecraft in the cosmos: in an untitled cartoon by E. Valter (from Tallinn), one Santa Claus, aboard a small, misfiring rocket-sleigh labeled 'US,' calls out "Hey, buddy, give me a tow: I'm running late!" [Ei, druzhishche, voz'mi na buksir: opazdyvaiu!], as another Santa Claus overtakes him, flying a significantly more powerful-looking spacecraft that proudly wears a red Soviet star.[12]

Other *Krokodil* images—often much smaller, single-frame cartoons—by numerous other artists depicted space exploration as a childhood concern, with children launching themselves into space in miniature rockets, or displaying their enthusiasm for space by their innocent questions or mistaken answers.[13] Nevertheless, in many cases, even during The Thaw, *Krokodil* found humor in contrasting Soviet modernism with inferior technologies. A shift occurred after 1945, away from satirizing the backwardness of the Imperial Russian past in favor of ridiculing the retrograde West. Although the magazine's content and meaning might have been orthodox, its visual language was never as serious or reverential as treatments found in other media, and it was the magazine's combination of socio-political subjects with humor and whimsicality that distinguished it. Indeed, it was often the juxtaposition of more than one discourse that created the incongruities upon which the magazine's humor was based, and Semenov's images frequently combined, very skillfully, multiple topical references in a visual language that was both familiar and atypical.

Semenov's images of the Soviet exploration of the cosmos are in many ways characteristic of the magazine's style. *Krokodil*'s cartoons depict space as already populated: Semenov's cosmos is home to gods, devils, and angels, as well as signs of the zodiac; it contains Martians as well as humanoid creatures; and it hosts visitors from earth, including the USSR's most famous cosmonauts. These inhabitants mark the cosmos as a zone of interaction between real and imagined beings. The literal message of cartoons such as Semenov's *First Furrow in Virgin Space* [Pervaia borozda na kosmicheskoi tseline 1961] (Fig. 3.1) was that Soviet science was capable of delivering Soviet citizens into the cosmos; and, it was implied, they were both at home and entitled to dwell there. No overt reference to geopolitical Cold War struggle was included, and no American satellites or spacecraft

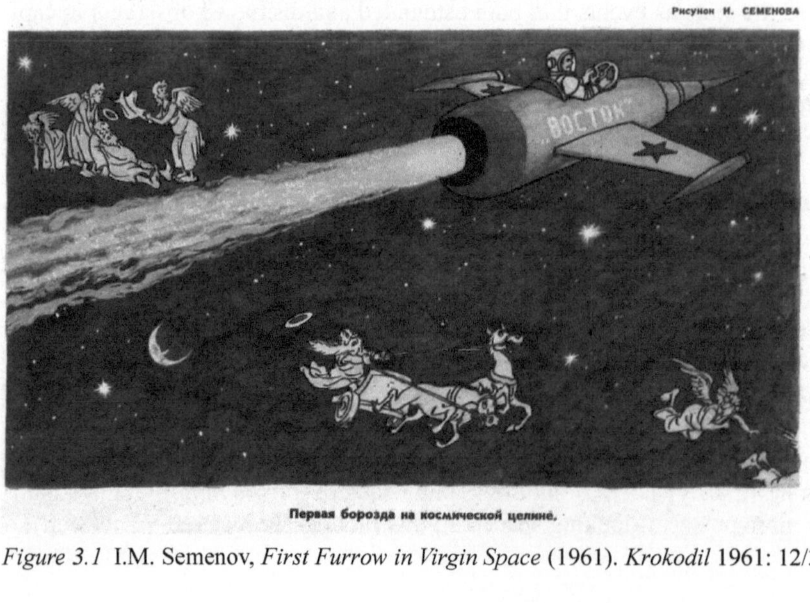

Figure 3.1 I.M. Semenov, *First Furrow in Virgin Space* (1961). *Krokodil* 1961: 12/3

were pictured in any of Semenov's images. Instead, the cosmos imagined by Semenov featured Soviet markers of identity, which served to reinforce and normalize the discourse on space exploration current in other media, as well as implying that the cosmos was naturally predisposed to colonization by Soviet citizens. The cosmos, therefore, was imagined as an arena where the technological and political supremacy that the Soviets argued they deserved went unchallenged.

In Fig. 3.1, a Soviet spacecraft speeds through the cosmos, leaving a vivid exhaust trail, which spreads and separates into multiple individual furrows, marking the path of the missile's flight. As the cartoon's caption indicates, the spacecraft also functions as a tractor, plowing a fiery yellow furrow across the cosmos. Semenov here combines references to two topical Soviet achievements: Vostok-1, the mission that successfully delivered Gagarin into orbit on 12 April 1961, and the Virgin Lands scheme, a key policy element of Khrushchev's leadership during The Thaw. Semenov draws several parallels between these concurrent Soviet initiatives. Conventionally depicted as empty and barren, the cosmos here is meant to remind us of the previously unproductive lands recently cultivated by Khrushchev's agricultural policy, which originated in September 1953 and involved bringing over 40 million hectares in Kazakhstan and Western Siberia under cultivation by 1960 (Durgin 1962: 255). In Semenov's cosmos, the power and thrust of the Vostok-1 is equated with the energy and enthusiasm of the volunteer farmers who flooded into the countryside in 1954. Furthermore, the plowed furrow across the cosmos represents the irreversible effects of Soviet excursions into these previously untouched regions. As *Krokodil*'s readers would have known, the Virgin Lands scheme proved tremendously successful in its immediate aim of providing vast quantities of cheap grain in the very short term. The bumper harvest of 1956 provided over 63 million tons of grain from the Virgin Lands areas, a total representing an increase of 180% on the 1949–53 harvests and a 90% increase on the 1954–5 yields in the same areas (Durgin 1962: 261). By implication, the visual language of the cartoon refers to Khrushchev's triumph over those political opponents who opposed his scheme, while also incorporating a third discourse in its inclusion of a bewildered divine community, who are scattered by Soviet expansion into the cosmos. The placement of white celestial stars in almost perfect alignment with the red stars on the spacecraft's wings seems to naturalize the Soviet presence in the cosmos, and implies that the pale and ghostly divine beings are rightly displaced.

Rather than leaving his cosmos empty and expansive as in Socialist Realist art, Semenov fills his space with figures and planets. His cosmos is huge, but filling up. The idea of populating space (which was associated with a broader fascination with human exploration of the cosmos, including an interest in astrological-occult aspects) had a long history in Russo-Soviet culture. Powers of flight (aerial, celestial, and divine) were important theoretical and aesthetic strands in Russian Orthodox Christian theology, science fiction, Soviet modernist discourse, Socialist Realism, and Russian Cosmism.[14]

Echoing pre-Soviet and modernist ideologies of flight and space exploration, to some extent Thaw-era news media treatments communicated using the same visual language as the early Socialist Realist representations of the aviation exploits of

Stalin's Falcons during the 1930s. Celebratory, hagiographic images of Chkalov being welcomed by Stalin after returning home from an aerial adventure in 1936 were painted by artists such as Samuil Iakovlevich Adlivankin (1897–1966) (*The First Stalinist Route* [Pervyi stalinskii marshrut 1939]) and Petr Tarasovich Mal'tsev (1907–74) (*Greeting of the Courageous Crew* [Privetstvie muzhestvennym ekipazham 1936]). Their paintings depicted the heroic characters up close while their aircraft formed the background, and the artists made no attempt to represent the achievements themselves. Alternatively, some artists, including Aleksandr Aleksandrovich Deineka (1899–1969), chose to depict the pilots' achievements. Images such as *In the Air* [V vozdukhe 1933] and *Above the Snow* [Nad snegom 1938] envisioned the aircraft, mid-flight, but made no direct reference to the crews or their heroism.[15] Cultural responses to these events reflected the dominant discourses of the age: the development of the technological expertise necessary to expand the realms of human endeavor and the bravery of the individuals who performed the feats. The first artists to consider these themes often chose to deal explicitly with only one, and the separation of dominant discourses in visual artworks in the 1930s was perhaps a response to the imperatives of the commissioning and exhibition process that developed in the 1930s. It also was indicative of the difficulty of depicting an aerial technological achievement and individual heroism in the same image.

Cartoons of The Thaw period offered artists the possibility of depicting reality in an unrealistic manner. Unrealistic representations of spacecraft, therefore, were justified, as well as unavoidable at a time when official secrecy kept citizens ignorant of the true appearance of spacecraft. Here, Semenov's cosmonaut sits inside his rocket (face partially visible through his helmet visor) and guides the rocket with a steering wheel. This space rocket-tractor (which also resembles a conventional aircraft) thus reminds us of contemporary and historic technological developments, national triumphs, political policies, and personal heroism—individual and collective. In Semenov's visual language for communicating his vision of the cosmos, the images form patterns of graphic representation based on the color and depth of the background, the presence of planets that are represented by star-shapes, the existence of space creatures of different kinds, and the presence of Soviet spacecraft. Movement is symbolized in Semenov's visual language by motion lines, but in all cases the speed of the motion is incidental to the main message of the cartoons. In the case of Fig. 3.1, for instance, the motion lines serve a triple function. While they literally represent the exhaust trail of the spacecraft's rocket engines and refer metaphorically to the "first furrow" across virgin space, they also are typical of the kind of comics techniques he employed. In numerous earlier images in *Krokodil*, aircraft and birds in flight were depicted without any significant motion lines.[16] Although Semenov's earlier images used motion lines and other comics techniques, he did not apply them to aircraft as a matter of course.[17] In these images of the cosmos, then, Semenov's use of motion lines communicates something of the violent and disruptive, though temporary, effect of Soviet ventures into space, which is contrasted with the vast, deep permanence of the latter. It also reminds us about the pace of Soviet technical development, the vigor of the space program, and Thaw-era politics. Indeed, the period after Stalin's death was characterized by a sense of renewal and, of course, the thaw metaphor, with its temporal connotations of a liminal phase between winter and spring, and its cultural implications of relaxation, freedom, and regeneration, contributed to the sense of rejuvenation through Khrushchev's political style and through his various policy initiatives, which ensured what Nancy Condee calls "the realignment of conflict along a temporal axis–the old and the new" (2000: 167).

Condee's explanation of Khrushchev's contribution to Soviet politics also aptly describes the shift in representations of Soviet modernity that occurred during The Thaw, in what was a return to the type of strategies employed in the early years of the Soviet state. Both were defined by an opposition primarily to the past; the markers of Soviet modernity in the images in *Krokodil* generally envisioned clear distinctions between events either side of 1956, whereas the aesthetic of Stalinist modernity depicted a society opposed to the West and capitalism. In both cases, technological subjects were used to visualize the qualities that made Soviet communism superior. Semenov's cartoons employed unusual comics techniques that conveyed a sense of the novelty and speed of space exploration. Crucially, Semenov's cartoons derive their humor from the incorporation of incongruities, often achieved through references to multiple topical discourses in the same images. In this way, Semenov inverts the utopianism current in official Soviet discourses, to reveal a much more skeptical vision of the power of space exploration. The images discussed in this chapter sideline the importance of Soviet power in space. Heroic Soviet cosmonauts, where they appear, are peripheral. Space, it is implied, is already full of human problems, and Soviet incursions do not solve them. This vision is at odds with the Stalinist imperative of controlling and subordinating nature to the will of Soviet man. Furthermore, Semenov's cartoons are not simply a vision of space, but a cosmic reflection of Earth. Their suggestion that space exploration has an impact in the cosmos is undermined by their own references to markers of Soviet reality, and in fact they predict a disappointment with the outcomes of the Soviet space program. Asif Siddiqi argues that Soviet enthusiasm for the cosmos and space exploration began to wane by the late 1960s (2011: 294), and The Thaw *Krokodil* images pre-mediate and inform some of the criticisms of the Soviet space program made during the following decade. These criticisms and the loss of cosmic enthusiasm were functions of the end of The Thaw era and the beginning of an era of ideological and material disappointments.

The *Krokodil* delusion

The pictorial space of Semenov's work is characterized by its fullness, but also by its lack. His cosmos is full of life, but it is not the exclusive domain of any creature; nor, apparently, is it subject to any particular form of human or natural law. *Krokodil*'s cosmos does not appear to be ruled by the same scientific laws of Marxism-Leninism that, Soviet ideology insisted, governed human history on Earth. Instead, *Krokodil*'s cosmos was home to numerous gods, both pagan and Christian. Khrushchevian official discourse declared religion a remnant of the past, and attacked religion with Leninist zeal. Khrushchev's anti-religious crack-down false-started in 1954, resumed in 1959, and peaked in 1961. Similar in many ways to pre-war anti-religious propaganda, Khrushchev-era atheism was unique in that it reflected its space-age context and targeted schools, aiming to create a new worldview that was "radically utopian" in its teaching of science (Froggatt 2006: 264). In Fig. 3.1, one of Semenov's heavenly creatures is reminiscent of Dazh'bog, the Slavic god of the sun, who daily drove his golden chariot (drawn by a pair of fire-breathing white horses) across the sky (Kennedy 1998: 61–2). While Dazh'bog's power to give light, fire, and life seems to have been superseded by Gagarin's rocket, and while he and the other divine beings appear horrified at the sight of Vostok-1, Semenov's image subverts the standard Soviet vision of the cosmos by populating it with gods and angels. On the narrative level, the image communicates the message that scientific Soviet modernity is scattering the remnants of traditional Slavic beliefs and superstitions by contrasting evidence of the reality of Soviet achievements with a joke about the unreality of religious figures. Ironically, then, technological advances enabling the exploration of space brought Soviet man closer than ever to religion.

Conventionally, *Krokodil*'s cartoons, which reflected aspects of the anti-religion campaigns, called upon religious believers to abandon their faiths, since the triumph of scientific exploration (Gagarin's unsuccessful search for God while in space was famously and repeatedly cited)[18] had apparently disproved the existence of God. Similarly, Soviet news media emphasized the emptiness of the cosmos (Smolkin-Rothrock 2011: 165–6). Semenov's imagery explores the notion of empty space by countering Soviet orthodoxy with speculation that heavenly beings inhabit the cosmos. In this case, these angels resemble ancient and/or Kantian philosophers, and as such they appear to represent the European metaphysical-philosophical strand. The rocket, by contrast, symbolizes dialectical Marxism, and the effect of juxtaposing these symbols of different philosophical and historical epochs in the same image is to reinforce the apparent contradictions of divine spaceflight.

In *Heavenly Cosmodrome* [Nebesnyi kosmodrom 1964] (Fig. 3.2) Semenov invites his viewer to ponder the incongruity of the depiction of angels and cosmonauts in the same image. The juxtaposition of Soviet cosmonauts with angels and gods prompts us to question which type of cosmic flight seems more plausible. The inclusion of modern technical equipment and the confusion it has clearly created among the heavenly beings, along with the heavy sarcasm of the accompanying explanation and the numerous jokes at the expense of the religious characters incorporated into the cartoon, poke fun at the notion that divine flight is comparable to space exploration and, more broadly, critique the logic of belief in the existence of God.

In this image, Gagarin arrives to provide a consultancy service to the angels and gods who are struggling to grasp the scientific concepts involved in launching a space rocket. Two angels sit astride the wings of a tiny, red, arrowhead-shaped rocket, which approaches a celestial rocket base as a missile is prepared for launch, and invite Gagarin to go directly to consult with the chief ["Iurii Alekseevich,

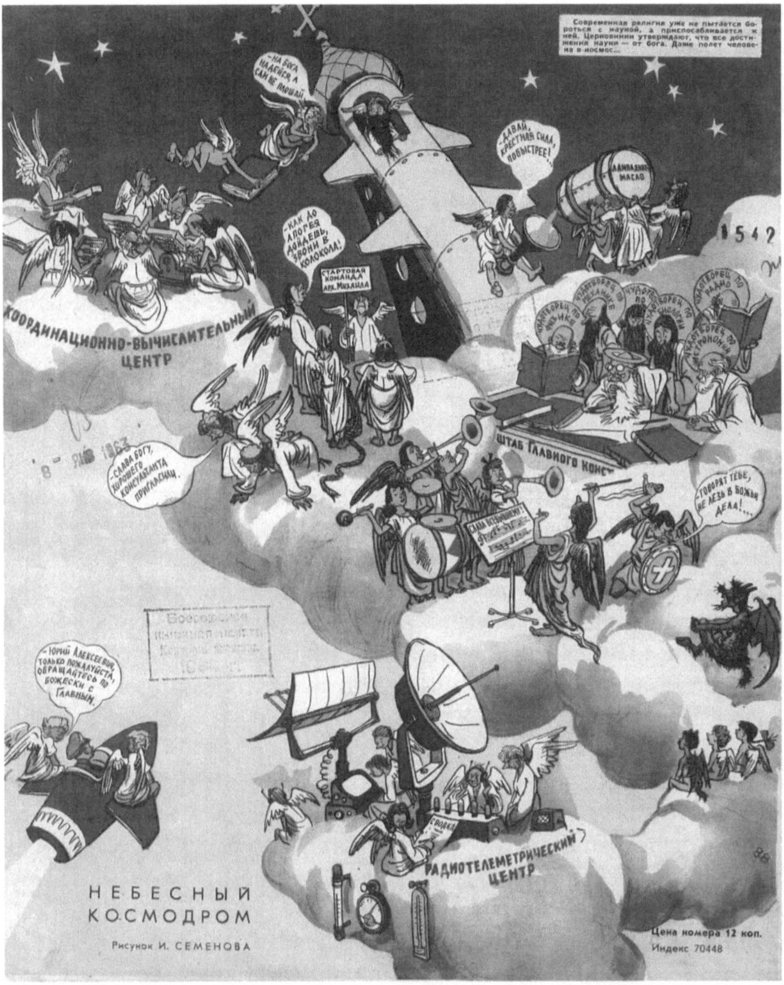

Figure 3.2 I.M. Semenov, *Heavenly Cosmodrome* (1964). *Krokodil* 1964: 1/16

tol'ko pozhaluista, obrashchaites' po bozheski s glavnym"]. Meanwhile, another pair of angels, leaning over the edge of their cloud, offer thanks to God that a good consultant has been requested ["Slava bogu, khoroshego konsul'tanta priglasili"]. Dressed in his military uniform, rather than in his space suit and helmet, Gagarin this time has the role of providing some historical grounding and reminding us that the USSR had achieved spaceflight, while these godly beings have failed to do so. From behind a desk apparently marked 'Headquarters of the Chief Constructor' (the sign is partially obscured by a cloud) glowers a godlike figure. Meanwhile, as the church-tower-cum-space-rocket (already in position on its launch ramp) is fueled with lamp oil, the launch crews shout final instructions to the angelic, if hirsute, cosmonaut, Archangel Michael. All the features of a terrestrial cosmodrome are present (the Radiotelemetry Center, the Technology Coordination Center, and all the attendant technical experts), but in this case all the personnel comprise angels and saints. This heterogeneous scene is explained by an inset at the top of the cartoon, which states that theologians have claimed that all scientific achievements, including manned space flight, derive from God. Semenov's blasphemous whimsy, therefore, imagines what God's cosmodrome might look like. His image ridicules the assumption upon which it is based by including Gagarin and by having two winged devils watch amusedly alongside other angel observers.

This cartoon places markers of Soviet modernity alongside markers of religious faith, with the assumption that the viewer will draw the politically correct conclusion. Here, however, because of the comically unscientific signs in evidence, the strong implication is that a godly space program would falter. The divine rocket remains attached to its launch ramp and the saintly experts pore over their textbooks with anxious expressions. The technicians bend over their technology as if attempting to master it (four abacuses can be seen), while the radio technicians are clearly listening rather than broadcasting. Moreover, the angelic cosmonaut is being delivered a heavy tome (presumably a Bible) and a heavenly choir sings "Glory Be to God on High!" The scene is set in the cosmos, but Gagarin's recruitment as consultant and the appearance of a book by Konstantin Tsiolkovsky on God's table suggest that this mission might be earthbound (a notion that would complete the circularity of the irony of this image), but it remains unclear what the point of this mission might be.

This cartoon's visual language of church domes, angels, and saints draws heavily on Orthodox iconography and symbolism, and while the overt irony of the scene reinforces the normative discourses on science and atheism, it is an unstated (and perhaps unintentional) irony that it relies so heavily upon viewers' familiarity with the aesthetics of the remnants of the past it seeks to destroy. Semenov's scene is diverse and contains multiple miniature episodes, with no indication of the sequence in which they are to be read. The implication, then, is that these events are simultaneous. The band of unfilled yellow background separating Gagarin from the divine beings that fill two-thirds of the image emphasizes his incongruity. His foreignness is further emphasized by his spacecraft, which (in contrast to the immobile religious rocket) has successfully delivered him, and by his status as a special visitor. Semenov leaves the issue of Gagarin's relationship to God slightly ambiguous. Is Gagarin visiting the gods as a consulting expert or has he, in fact, as a result of his courage and impetus to explore the cosmos, been co-opted into the pantheon of divine cosmonauts?

***Krokodil*'s multi-scenic narratives and borderless spaces**

Semenov's images employ a technique of multi-scenic compositions that produce complex images overabundant in rich content and often confusingly constructed. These images contain multiple actions and mini-narratives, and combine multiple topical discourses. Since they often include no artistic directions on how to read the image correctly, they imply that the various episodes pictured take place more or less coterminously and share the same spaces. Although Semenov often used some basic comics techniques, his refusal to sequence his narratives or to separate the action of his images into discrete panels outlined by frames or separated by gutters diverged from the logic of American-style comics. As a result, Semenov's images combined mini-narratives that might otherwise have been separated, which resulted in images that reveled in their perpetual motion and unending actions in an ambiguous merging of space and time. This aesthetic approach is reminiscent of Western European history paintings of the eighteenth and nineteenth centuries, and of the multi-figural canvases of certain canonical Socialist Realist painters during the 1930s and 1940s, such as Vasilii Prokof'evich Efanov's *Outstanding People of the Land of the Soviets* [Vydaiushchiesia liudi strany sovetov 1939] and Iurii Petrovich Kugach's *Glory to the Great Stalin* [Slava velikomu Stalinu 1948]. In other ways, however, it revealed fundamental ambivalences that were central to many aspects of *Krokodil* and more generally to Soviet culture.

Emma Widdis, in her study on the broader economic and cultural mapping project that was undertaken inside (and above) the geographical boundaries of the USSR during the 1930s, has shown how the symbolism of exploration and mapping communicated messages about conquest and domination (2003). In the post-war period, Semenov cartoons, which include references to the planets of Earth's solar system and other features of the Milky Way, may be read as imagined maps of the cosmos. Mapping the cosmos represented the logical extension of the expansion impetus and symbolized the extension of Soviet civilization into outer space. Semenov's cartoons celebrated and simultaneously critiqued this cosmic mapping project, communicating his doubts about the potential of Soviet space exploration by his peculiar visions of merged discourses and by the space he constructed. His images are distinguished not only by their composition and large scale, but also by their lack of a frame, which created borderless and timeless zones. Echoing tendencies toward gigantism inherent in other Soviet enterprises, including Socialist Realist artworks, these *Krokodil* cartoons fill entire pages of the magazine, and stretch to the edges of the paper. In conventional comics and cartoons, the frame performs numerous important functions, which have been the subject of many studies by comics scholars. Thierry Groensteen, for example, argues that the frame has six functions, all of which affect both the contents of the panel and the reader's cognition of the comic (2007: 39–57). The most important aspect of the frame is its power of "closure" (McCloud 1994: 63–5). The absence of a frame, to delimit the area of the image and thereby define the time and space of the action depicted, has important consequences for the meaning of Semenov's images. The undefined time and space represented takes on an infinite quality. Scott McCloud refers to "bleeds," "when a panel runs off the page," with the effect that "time is no longer contained by the familiar icon of the closed panel, but instead hemorrhages and escapes into timeless space" (1994: 103). The expansiveness of Semenov's cartoons communicates messages about the scale of the cosmos and the nature of the task of exploring it. Widdis comments that during the period of The Thaw, "doubt reentered the discourse of Soviet space, unsettling the certainties of the ideological map" (2003: 194). Two factors created doubt: as Widdis suggests, the trauma of de-Stalinization caused political destabilization; but, as I contend, equally responsible for cultural and spatial disorientation was Soviet space exploration, which created the need for new means and new representations for the exploration of such an unmappable terrain. *Krokodil*'s aesthetic thus offered reassuring and politically correct narratives via unsettling techniques that created the sense that the only truly reliable space was private.

Writing more generally about the purpose of a frame, Boris Uspenskii argues that the device is indicative of a shift in perspective, from an internal to an external point of view (1976: 46n39). In *Krokodil*, the frame (or the implied presence of a frame) defined the perspective of the viewer, and therefore delineated the object of the magazine's satire from the subjective gaze of the magazine's reader. With the exception of front and rear cover images, image-bleed along more than two sides of a large cartoon in *Krokodil* was quite rare. Cartoons were often placed close to the edge of the page, and often were not framed by a border line: images that were larger in scale than one page rarely featured no border at all. Though Semenov's images were often closed by conventional frames, he also experimented by merging the form and aesthetic function of the periphery of his cartoons. For example, in *Krokodil Concert* [Krokodil'skii kontsert] (1954: 12/8–9) Semenov framed a circus scene with stage curtains, and in *Where Have You Brought Us?* [Kuda ty zavel nas?] (1954: 9/7), the overhead view of a room being decorated is framed by foreshortened walls. In the images of the cosmos, Semenov's decision to allow his images to escape over the edge of the page refers to the vastness of his subject matter, but its effect is to immerse and disorient the viewer.[19]

In *On the Ocean of the Air* [Na vozdushnom okeane 1958] (Fig. 3.3), Semenov fills the entire image with multiple characters participating in various scenes. This is a comical vision of the cosmos as a busy urban landscape, with diverse everyday educational, commercial, recreational, and other activities, occurring proximally but unconnectedly. The image contains no guidance on how to understand or interpret the action or, in fact, the order in which to consume it. Although the sizes of figures vary, Semenov does not provide any close views of particular characters. Furthermore, he uses an overhead perspective that allows the viewer a

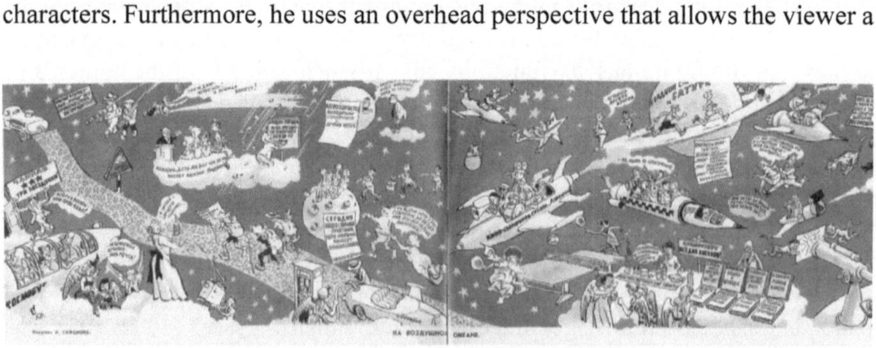

Figure 3.3 I.M. Semenov, *On the Ocean of the Air* (1958). *Krokodil* 1958: 36/6–7

godlike gaze and seems entirely appropriate for his subject. The viewer assumes the perspective of a cosmonaut, and is immersed in the image in a way that creates a sense of suspension in outer space. Despite their naïveté, these images—in their profusion of actions and in their warped geography—are difficult to access.

Semenov's images lack horizons and conventional straight lines of perspective, which are so familiar from Soviet realistic forms. Indeed, the structure of all the images considered here rests upon diagonal lines of composition that cross each other at acute angles. This compositional method, which is dynamic and energetic, reminds us of the posters of the Five-Year Plan era and in particular the photomontage poster work of the Artists' Brigade. This means that the image lacks central axes or focal points and appears chaotic. In the context of the narratives, these omissions, as well as the appearance of convex indicators of spatial depth (the runners on the circular Saturn-ring track in Fig. 3.3 or the Martians appearing from over the bowed line of the horizon in Fig. 3.4), are entirely justifiable. Their effect, however, is to create a distortion in the viewer's perception of spatial depth and distance. In the center of *On the Ocean of the Air*, a half-clothed man sleepwalks toward a sign in front of a cratered planet upon which sit several other sleeping figures. The sleepwalkers apparently are attending a lecture, the punning title of which is advertised by the sign,[20] but the inconsistencies in size of the sleepwalkers, the sign, and the planet create a visual confusion for the viewer. The scale of certain images implies depth, as does the Milky Way road, which is being poured diagonally across the image by a milk tanker. Semenov's inclusion of multiple mini-narratives, each containing little jokes, as well as his compositional choices, creates an image that is participatory and democratic in that it invites a high degree of active interpretation by the viewer.

Semenov's godlike perspective implicitly refers to Soviet spaceflight. Human space habitation (Fig. 3.3) and missions to Mars (Fig. 3.4)—semi-humorous comments about the rhetoric and aims of the Soviet space program—are indicative of the technological and scientific developments of the twentieth century. Yet Semenov chooses to depict references to these modern discourses in terms that are strikingly reminiscent of much older aesthetic traditions. In Russia, both the Orthodox icon and the popular print [lubochnye kartinki] adopted expansive methods when representing space and time. Both forms were therefore used to depict long and short narratives, as well as objects, according to mimetic and non-mimetic rules (Alaniz 2010: 15). Russian Orthodox icons, which were flat, multi-plane images of narratives that remained suspended in time because of their lack of visual techniques for narrative progression, share many characteristics with Semenov's space images. His debt to this native heritage is at least as great as to the Western European artistic tradition of genre painting associated with Pieter Bruegel the Elder (1525–69). Bruegel the Elder's peasant and landscape paintings, which were notable for combining singular subjects taken from religion and classical antiquity with everyday pastoral scenes, and for their profusion of activity and distortion of perspective, also influenced the European tradition of the 'teeming picture book' (Wimmelbilderbuch in German). These children's books have received little scholarly attention, but are conventionally understood as large-scale picture books (aimed at children, but also enjoyable for adults) that are predominantly textless and contain multiple, richly colored, and detailed panoramas[21] featuring "pluriscenic" landscapes containing situations and characters familiar to their readers (Rémi 2011: 117–18). Their development has been traced to the work of the German artist Ali Mitgutsch (1935–), but Semenov's works in *Krokodil* and *Merry Pictures* predate Mitgutsch's.[22]

Like the Wimmelbilderbucher, which contain "an abundance of characters and an inexhaustibly rich source of material, but also a certain degree of disorder and chaos" (Rémi 2011: 116), Semenov's cartoons are democratic. Because he does not provide any close views of particular characters, viewers have no indication of how the action affects individual characters. The result is an absence of empathetic connections. No strand of the narrative is prioritized, and among the various activities in the image no hierarchy is apparent. This strategy differs from the compositional decisions conventionally ascribed to Socialist Realist artists working in genres with a higher status. Countless Soviet visual artworks' spatial configurations place icons of Soviet power at the center, with centripetal and centrifugal forces clearly identifiable. Jan Plamper argues that the products of the Cult of Personality placed Stalin at the symbolic and literal center of artworks (2012: 89). In orthodox Soviet visual culture, then, the center of the image connoted a center of political power. By contrast, Semenov offered no such prioritization of the pictorial space, thereby implying that life in space will be as democratic as his composition. No physical space will be made sacred, and no individuals will be privileged over others. Since he depicts a cosmos lacking representatives of the Soviet government, his vision implies that the human exploration and settlement of the cosmos presents the opportunity to escape from political power. In that sense his visual language about the cosmos subverts the orthodoxies of Soviet representations.

Krokodil's Picturing Time

In the top-left corner of Semenov's untitled 1962 image (Fig. 3.4), the Mars-1 spacecraft travels horizontally, parallel with the top of the image, above the moon, Saturn, and other planets of the solar system. Mars-1 is connected to Earth by long white ribbon lines, which indicate the rocket's speed. Semenov constructed the temporal aspects of this image by suspending different actions occurring at different speeds in the same image. Perspectives, distances, and relative times are distorted, with the effect that the viewer seems to pause, experiencing in slow motion a suspended moment that merges past, present, and future time. As Martians[23] prepare celebrations for the arrival of the Soviet mission, a Martian orchestra plays *We Are for Peace!* [My za mir! 1948], while the Martian choir sings *Motherland (This Russian Land)* [Rodina (Eto russkaia zemlia) 1949]. Other Martians hold misspelled Russian signs, all of which suggests that these space-dwellers are very excited about their visitors from Earth. Their attempts to learn Russian seem comical, but hint at a desire to emulate the Soviets. Although the welcome celebrations being prepared suggest that the Soviet cosmonauts will

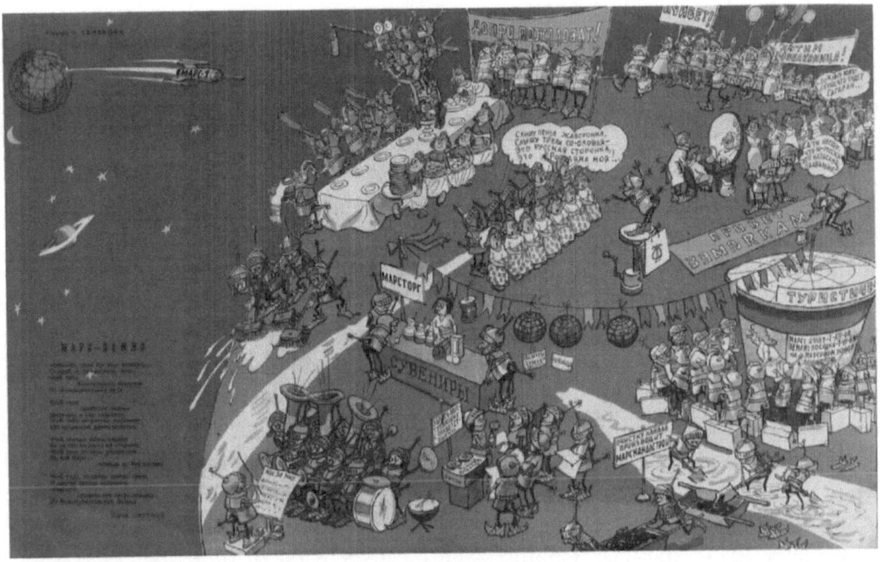

Figure 3.4 I.M. Semenov, Untitled 1962. *Krokodil* 1962: 33/8–9

be treated like dignitaries, the souvenir stall displaying its wares in the center of the image also indicates that these visitors are tourists. Among the gifts on sale, alongside "'Earth' lampshades" [Abazhury 'Zemlia'] and Sputnik models, are Martian matryoshka dolls, reflecting the anticipation that both Martians and cosmonauts will purchase souvenirs of this occasion.[24] The image may also refer ironically to the expansion of Soviet influence into the world's less economically developed regions during this time. Notably, on a smaller scale, the mid-1920s repeatedly equated an airplane's arrival in the countryside with the promise of economic advance and therefore the anticipated salvation of the retrograde rural population—a phenomenon discussed by Scott Palmer in his *Dictatorship of the Air*.[25]

The concept of time and its progression is communicated in this image via Semenov's selective use and rejection of comics techniques. As José Alaniz has shown, Soviet ideology developed an antipathy toward comics culture in general and comics-like visual language in particular (2010: 33). On several occasions during The Thaw, *Krokodil* published images that criticized US comics, and derided American society for tolerating such a negative influence on its youth.[26] Consequently, although exceptions were published, *Krokodil* and other publications tended to feature single, framed cartoon images with captions rather than strips of images with word balloons and other comics techniques. Semenov was notable among *Krokodil* artists for his use of certain visual techniques more commonly associated with American-style comic strips, and although he used structural comics techniques (such as different frames to convey sequentiality in

Merry Pictures), here he employs motion lines and speech bubbles to describe the passage of different instants in time.

Semenov collapses temporalities in this image. Past, present, and future merge in one cartoon, in a manner that implies a comment upon the Soviet Mars-1 program and Soviet space exploration in general, with the effect that the viewer's own temporality is distorted. The image juxtaposes the modern technology of a Mars spacecraft and depicts a television camera right beside a dish of grapes reminiscent of an ancient symposium. These anachronisms imply that Martian civilization remains backward by comparison with the USSR, which recasts the utopianism of Soviet science-fiction stories concerning Mars (*Aelita* (1923) by Aleksei Nikolaevich Tolstoi (1883–1945), for example) as a feature of the present-day Soviet Union, rather than of ambitious Futurism. In Semenov's depiction of the past, present, and future in the same image, the spacecraft may be seen traveling but still in the act of leaving the Earth on the left, while the Martians are depicted in preparation and in celebration of an arrival that cannot yet happen. Our perspective, in fact, allows us to see both ends of a journey to Mars that, in reality, was a voyage never intended to land on Mars; it lasted more than seven months and even then was not successfully completed (Harvey and Zakutnaya 2011: 265–66). Compression or suspension of time was a persistent theme in Soviet modernism. The dominant Soviet political-cultural treatment of temporality was acceleration, but the most important landmark events in Soviet history—the revolutionary year of 1917, the confirmation of Stalin's leadership, the beginning of the era of the Five-Year Plans, and the period of de-Stalinization—were all characterized as zero points, after which time was reset. Indeed, a CPSU Central Committee decree imposed GMT on 1 June 1919, literally resetting the clocks. The abandonment of circular time in favor of the linear, associated with the development of modernism, coincided in the USSR with the mechanization of agriculture and the expansion of central political control in the late 1920s.

The leveling of various moments of differing durations is the key to the construction of meaning in this image. Speed lines convey the distance traveled and the time taken by the Soviet spacecraft, while water splashes illustrate the brush sweeps of the Martians. Apparently simultaneously, the Martian choir sings an entire verse of its song and the brass band plays its refrain, while in the top right, other Martians make comments about the prospects of meeting Gagarin. The excited activities observable in the image, such as the digging of Martian canals, for example, parallel Soviet enterprises that aimed to compress time and industrialize at unprecedented speed. Ostensibly, the parallel events on Mars serve to validate the broad aims of the Mars-1 mission in particular, and the Soviet political project itself. Indeed, the Martians appear to have adopted a collective lifestyle and communal distribution of work tasks that suggest a belief in Soviet-style communism. Semenov equates Martian time with Stalinist time, since their aims and pre-industrial methods have so much in common, but his subject and his techniques for leveling time imply that the technological progress of modern Soviet science has eclipsed even the achievements of Stalinism. Through his visual language, therefore, Semenov implies that the space-era USSR has transcended even Stalinism.

Giving no indication of the correct sequence of events, Semenov relies upon the viewer to place these events in any logical order, at the same time as he implies that they are coterminous. Despite the appearance, these actions could not occur within the same moment in reality, and therefore the events each contribute to the temporal construction of the image. The placement of the spacecraft in the top left of the image leads the reader to presume a linear progression of actions, from left to right across the picture. In fact, no such instruction is delivered, and no such sequentiality is implied by the action within the image. Semenov's comics techniques provide the only indications of movement and the passage of time. Otherwise, owing to the absence of any rhythmic or structuring function in the image, time appears suspended. The extent to which Semenov's imagery represents the triumph of "historical" or "suprahistorical" forces is unclear. Mikhail Epstein suggests that Russia perennially chose a "suprahistorical" path, favoring spatial progress instead (2003:278). The suprahistorical depiction of time was enshrined in Socialist Realism, the USSR's official artistic doctrine. Defined by Andrei Zhdanov (Secretary of the Party's Central Committee) at the first Congress of Soviet Writers in Moscow, 17 August to 2 September 1934, Socialist Realism practiced the manipulation of representations of time in pictorial space: Soviet artists had "to depict reality in its revolutionary development" and produce artworks in which it was possible "to glimpse our tomorrow" (Bowlt 1989: 293). Thinkers from Einstein to Bergson and artists from Picasso to McCloud have considered the extent to which the viewer's relationship to the passage of time (or the non-passage of time) and to the events depicted may be represented visually. Cubism adopted a scientific approach to the simultaneous spatial representation of different perspectives.[27] Comics art stretches or compresses time, and generally delineates it within frames that merge space and time in varying proportions (McCloud 1994: 94–103).

Crucially, Semenov does not border his image by a comics-style frame, with the effect that the viewer is positioned in an indeterminate zone inside the area of the action. Weightlessness appears in all of Semenov's cosmos images, for obvious reasons, but in Fig. 3.4, though Mars exerts its gravitational pull, the timelessness of the image is conveyed through the leveling of time. Semenov depicts all of his figures from the same perspective, grounding viewers and stabilizing their vision. Our perspective places us too close to the surface of Mars to be able to assimilate all of the details at the same time, and the lack of any fixed spatial reference points means that we remain disorientated. The recognition of the spatial and temporal locations of the action is central to the appreciation of all of Semenov's images. The artist invites us to accept the premise that somewhere in the cosmos these events are taking place at roughly the same time as the reader's own experience, and the image's temporality is thus attached directly to the viewer's. In the case of Fig. 3.4, the events in question relate to the launch of the Mars-1 on 1 November 1962. This unsuccessful mission to fly past Mars (the topical discourse serving to contextualize the cartoon in this case) adds references that reassured the contemporary viewer.

In Semenov's images of cosmic exploration, the distortion of time betrays skepticism about the value of Soviet space exploration. The viewer enjoys the dramatic irony suggested by the fact that Mars-1 is traveling to Mars only to discover a society very similar to the USSR's. Through his inclusion of a souvenir stand and a tourist ticket kiosk, Semenov reduces the voyage essentially to tourism, or a temporary vacation. Even the Martians' public celebrations are reminiscent of the terrestrial Soviet versions. The excitement of the Martians about meeting the Soviet explorers and the fact that so many Martians are queuing at the ticket office, ready to buy tickets to visit Earth, indicate that Martian science has not developed a technology capable of transporting them around the cosmos. Visiting Mars, Semenov implies, is therefore akin to visiting the Soviet Union before the development of space technology, and the value of traveling to Mars is therefore questionable. By suggesting that traveling across vast distances is akin to time travel, Semenov further experiments with temporality in his images, compressing time and collapsing distance to the point where the third and fourth dimension appear to become indistinguishable.

Krokodil: the cosmic carnival

As I have argued, Semenov's skeptical view of metanarratives (religious, ideological, and scientific) is communicated in his critique of modernity, which also appears in the cosmos cartoons identified here. Symbols of technological development and Soviet modernity are juxtaposed with markers of classical and ancient Slavic antiquities in Figs. 3.1, 3.2, and 3.5. Semenov places spacecraft—the quintessential indicator of Soviet scientific progress—alongside pagan Slavic gods (Fig. 3.1), Christian deities (Fig. 3.2), Martians in Roman armor (Fig. 3.4), and Roman gods (Fig. 3.5). In *Planets: Our Ranks Have Swollen!* [Planety: Nashego polku pribylo! 1959] (Fig. 3.5), Semenov shows a stylized Soviet rocket ridden like a sledge by a youthful Soviet figure representing the year 1959. This vigorous representative of Soviet technological progress arrives at the center of an incomplete celestial sphere, to the joyful welcome of eight planets and their personified astrological namesakes. Their outstretched hands and beaming smiles are echoed by the faces on the stars in the background of the image and the cartoon's caption (as well as Uranus's punning exclamation "Hurrah!" [Ura!]). Semenov has each of the planets represented by figures carrying their traditional signs. The rightful place of the Soviet rocket among its celestial siblings is reinforced by the repeated use of five-pointed stars (in red, on the Soviet figure's hat [ushanka] and on the nose of the rocket; in blue, in the background of the image) and the presence of two sickles (in red on the rocket; in blue, waved by Saturn). The cartoon implies that, just as astronomers who have discovered new planets have suggested that they be included in the astrological system or that new signs of the zodiac should be incorporated, so the achievement of sending satellites into space equates to the creation of a new celestial body. In other words, as of 1959, Soviet sputniks are permanent fixtures in the solar system. This cartoon, which appeared in the first issue of *Krokodil* to be published that year, celebrates the New Year and the birth of Soviet space exploration simultaneously.

The image of a Soviet youth (rosy-cheeked, with scarf trailing in an invisible wind as if speeding through space, though the rocket is apparently hovering)

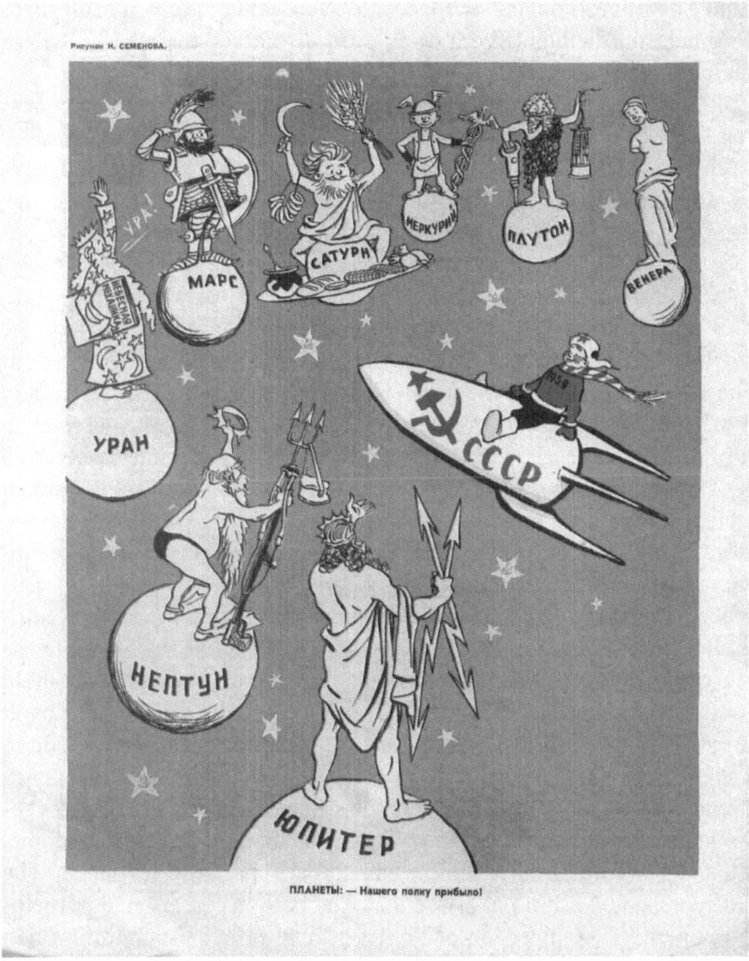

Figure 3.5 I.M. Semenov, *Planets: Our Ranks Have Swollen!* (1959). *Krokodil* 1959: 1/4

astride a terrestrial projectile is incongruous. The jarring juxtaposition of terrestrial Soviet symbols alongside those of astrology is achieved partly by Semenov's striking use of bright white and red against pastel shades of blue. The bright white rocket, compared with the shadowy blue Roman symbols of antiquity, is indicative of the enlightenment that Soviet science was supposed to bring. Indeed, the effects of modernity seem already to have been felt among these Roman gods, since Neptune's trident has been transformed into a spear-gun, Pluto is equipped with a pneumatic pick and a miner's lamp, Uranus carries a textbook titled *Celestial Mechanics* [Nebesnaia mekhanika], and Mars' arsenal now includes a pistol. Nevertheless, Semenov's decision to contrast Soviet space technology with the mythology of Roman astrology invites speculation about the nature of Soviet modernity. Viewers may well wonder whether this young cosmonaut's appearance in the celestial realm represents hubris, and they may question the implication that Soviet cosmonauts or spacecraft were taking their rightful place alongside other celestial beings (Fig. 3.5) or even displacing them (Fig. 3.1) or disproving their existence (Fig. 3.2).

Mikhail Bakhtin (1895–1975) famously described "the intrinsic connectedness of temporal and spatial relationships" as the "chronotope (literally 'space time')" (1981: 84). Where the chronotope appears in artistic works, "spatial and temporal indicators are fused into one carefully thought-out, concrete whole. Time, as it were, thickens, takes on flesh, becomes artistically visible; likewise, space becomes charged and responsive to the movements of time, plot and history" (1981: 84). As Bakhtin notes, the term "space time" has particular relevance to astrophysics and the exploration of the cosmos, and the application of the chronotope to the visual culture of Soviet space also involves a degree of literal interpretation. Not only does Semenov fill his spaces with characters and actions, but he depicts multiple concurrent scenes and mini-narratives that thicken the temporality of the image. Semenov's multi-scenicity may be interpreted as a visualization of polyphonic narrativity, in which multiple points of view interact dialogically. His compositions may also be understood in terms of the chronotope of the threshold, which, Bakhtin observes, "can be combined with the motif of the encounter, but its most fundamental instance is as the chronotope of crisis and break in a life" (1981: 248). The threshold chronotope is always metaphorical and symbolic, and it refers to another crisis, which is outside the novel. In the case of Semenov's cartoons of space exploration, the chronotope of the threshold extends beyond the images and indicates the crisis of modernity in the Soviet Union precipitated by the political upheavals of de-Stalinization and the socio-cultural effects of space exploration.

Soviet official culture generally viewed time as linear: the modern Soviet present sat between a rotten and corrupt past, which was already largely overcome, and a radiant and beautiful future, which was already partially achieved. The Soviet present represented a liminal phase that was both necessary and perpetually drawing to an end. The direction of Soviet official political-cultural motion was unstoppably forward. The collapsing of temporalities into the present, therefore, countered ideological emphases, yet it was a phenomenon envisioned in the pages of every issue of *Krokodil*. The journal's visual language routinely juxtaposed images and symbols of the past, present, and future for satirical effect. This was perhaps a response to the sense of distortion of the passage of time in everyday life, as people adjusted to the permanent postponement of the arrival of the glorious future that Soviet politicians promised and grew accustomed to the absence of a livable present. In Semenov's images of the exploration of the cosmos, time is flattened, and we are presented with the chronotope of a threshold on which the Soviet Union stands. The coincidence of The Thaw and the success of Soviet space exploration exploded the temporal and spatial boundaries of Soviet civilization, bringing together the Soviet present, technological prospects for the future, and the splendors of the history of the Roman Empire. Soviet achievements in space seemed to unite vast expanses of time, surveyed in both directions, and Semenov's cartoons transcended these boundaries.

Semenov's works contemplate the notion that Soviet ideology, including those elements concerning technological development, over time inevitably would bring material benefits. Moreover, they imply that the hyperbole of Soviet official media falsely analogized Soviet culture with divinities and the contributions of the world's ancient civilizations. Semenov suggests that the discourse on Soviet space exploration represented a metanarrative equivalent to those of previous societies, and his implicit rejection of those older metanarratives invites the reader to critique the validity of the USSR's modernist discourses. Thus, by placing the Soviet vision of modernity alongside older metanarratives, Semenov proposes a new vision of a different modernity, in which all grand narratives are equally valid and equally suspect.

Conclusions

As this chapter shows, in the late 1950s and early 1960s the imagery of *Krokodil* reflected and responded to the technological achievements that enabled the Soviet exploration of the cosmos. I have examined the effect of flight on the visual language of popular representations of spaceflight in the works of Ivan Semenov, whose multi-figural compositions borrowed visual conventions from several different traditions, including the Russian Orthodox icon and Socialist Realist imagery, and treated politically correct subject matter in a manner that implied a parody of official modern Soviet culture. The publication of these images in a serialized, satirical magazine altered their meaning and form because they were intended as humorous commentary. Semenov's frameless and multi-scenic images, which combined numerous topical discourses in the same set of mini-narratives, produced a disorientating and confusing effect for the viewer. Though he constructed meaning via certain limited comics techniques, such as ribbon paths, motion lines, and speech bubbles, his images lacked instruction on how to read them. They combined familiar symbols and references to numerous topical discourses in an unconventional manner. Imagining a cosmos that was simultaneously full and empty of spatial, temporal, and political indicators, he effectively created images that were reassuringly humorous in their acknowledgement of the USSR's state of permanent change.

Khrushchev's Thaw redefined cultural and political codes, and resituated the Soviet Union in its temporal relationship to its own history and future. During The Thaw, when politico-aesthetic discourses were shaken by the rigors of de-Stalinization, Khrushchev's leadership brought the references to difference (from the Imperial Russian past or the modern capitalist West) to bear on the Soviet Union, so that the USSR became temporally and politically contrasted with itself. Technology's ability to affect time and the duration of events (and in particular travel through physical space) was a theme that held cultural significance in its own right in the 1920s and 1930s, but according to the rhetoric of those earlier decades time and distance would only be shortened: past and present grew further apart, while present and future were drawn closer together by technology. As space flight appeared to demonstrate, in accordance with Khrushchev's politics, time during The Thaw became fluid: politics delineated the past, while technology rendered the future tangible. Different temporalities at once seemed more similar to and more different from the present, and that is why Semenov's cartoons leveled all temporalities and visualized them side by side, reflecting the sense of crisis created by The Thaw. Soviet self-visualizations came to reflect the crisis at home, as the USSR became the locus of the spatial, temporal, and political indicators of difference. Ironically, this crisis of self-image coincided with the literal projection of Soviet technological achievements into space after the launch of Sputnik in 1957. As the Soviet technological gaze shifted upward, some of the aspects of the Thaw-era Soviet crisis were projected into the cosmos, where, according to Soviet scientific utopianism, social problems would be eliminated. Religious belief, alcoholism, technological backwardness, and other corrupt remnants of the past were cast into space, where Soviet cosmonauts eradicated them. Semenov presented a nuanced and skeptical vision of a Soviet cosmos. When he depicted space exploration and its technological markers bringing Soviet modernity into proximity with classical civilizations, he referred implicitly to time travel, but he suggested a criticism of the value of spaceflight. As this chapter has suggested, the conflicted official satiricism of *Krokodil* and the irony of the USSR's moment of technological triumph, which occurred during a phase of profound reassessment, find expression in Semenov's visual linguistic techniques.

Notes

1 This chapter will hereafter refer to the magazine as *Krokodil*, as it is commonly known. Cartoon images from the magazine will be referenced according to the following convention: year: issue number/page number. For example: *Krokodil* 1962: 10/1.
2 See, for example, Bailes 1978, Higham *et al.* 1998, McCannon 1998, and Palmer 2006.
3 *Krokodil* magazine began as the illustrated Sunday supplement to *The Worker's Newspaper* [Rabochaia gazeta]. It was published in this form for the first time on 4 June 1922, but appeared as *Krokodil* No. 1 (13) from 27 August 1922 (Stykhalin and Kremenskaya 1963, 176–9). The magazine ceased publication after the dissolution of the Soviet Union, although it has since been revived (unsuccessfully) more than once.
4 See, for example, Bailes 1978, Graham 1993, and Andrews 2003.
5 In this film the technological backwardness of the town of Melkoretchensk is contrasted with the modernity of Moscow.
6 See, for instance, Richard Stites's explicit criticism of the magazine (2010), and the more implicit treatments in Fitzpatrick 1999; Kelly 2007; LaPierre 2012.
7 See Oushakine 2011; Ryan-Hayes 2006; Yurchak 2005: 249–50.
8 In the front cover image of *Krokodil* 1959: 34 by Sergei Aleksandrovich Kuzmin (1916–?) a huge and unfeasibly complicated-looking machine is being worked on by numerous engineers, inspected by several designers, and admired by one official, who asks what such a machine was built for, only to be told that it was designed to harvest dill! A similar cartoon on the front cover of *Krokodil* 1962: 33 by Iurii Nikolaevich Fedorov (?–1976) features a confused designer, scratching his head and admitting that his impressive new machine is built for harvesting coconuts (and is therefore useless in the Soviet Union).
9 The authorship of some of the smaller images in *Krokodil* was not indicated.
10 "Merry Pictures: The First Russian Comics" ran at the Tretyakov Gallery, Moscow, 15 December 2010 until 20 February 2011. "Classic Soviet Cartoons" http://www.forbes.com/sites/kenrapoza/2016/04/27/; "look-who-funds-ukraines-anti-putin-internet-tv"/ #1ecad5a84e4c ran at the Moscow Central Artists' House between 9 and 12 June 2012.

11 Similar messages about the global significance of Soviet space exploration were communicated by cartoons such as *Without a Word about the Satellite* [Bez slov o sputnike] (1958: 2/9), by a Czech artist, which visualized the change inspired by the flight of a Soviet satellite: the imperialist who stood over the toiling workers in one frame is made to flee when the oppressed gain a sense of pride and power from seeing Sputnik pass overhead.
12 See *Krokodil* 1958: 36/12.
13 See, for instance, *Children of 1965* [Deti 1965 goda] by Egor Glebovich Gorokhov (1927–?) (1959: 4/5), *Without Words* [Bez slov] by Anatolii Tsvetkov (1923–76) (1961: 12/9), and the untitled cover image by Ivan Semenov (1963: 18/1).
14 Also see the Introduction and Chapter 5.
15 See Chapter 2, on Deineka.
16 See, for instance, *Air Raid Warning* [Vozdushnaia trevoga] by Vitalii Nikolaevich Goriaev (1910–82) (*Krokodil* 1952: 33/8), *Pests of the Fields* [Vrediteli polei] by Iulii Abramovich Ganf (1898–1973) (*Krokodil* 1952: 25/16), and *In Formation* [Pristroilis'] by B. Efimov (*Krokodil* 1954: 15/10). See, also, the very faint motion lines present in *The Aircraft [Came in] Prepared to Land* [Samolet poshel na posadku] by Boris Mikhailovich Leo (1906–75) (*Krokodil* 1952: 7/4) and *In the Skies* [V podnebes'e] by Aleksandr Vladimirovich Bazhenov (1904–?) (*Krokodil* 1952: 30/15).
17 See Semenov's *Hot Day at the North Pole* [Zharkii den' na severom poliuse] (*Krokodil* 1954: 25/8–9).
18 This story about Gagarin's report is disputed (Interfax 2006).
19 On the vastness of space and its effect on the viewer, see Chapter 2.
20 The sign reads: "Today—General meeting of sleepwalkers. After the meeting, the concert: Luna Sonata." The joke revolves around the pun on the Russian word 'lunatiki,' which means both 'sleepwalkers' and 'lunatics,' and has the same root as the Latin word for 'moon.'
21 On the topic of historical panoramas, see Chapter 10.
22 This style has more recently been popularized by Martin Handford's series *Where's Waldo?*
23 Semenov has chosen to depict his Martians wearing suits of armor, like the Roman god Mars. This refers to the poem in the bottom left of the image, and follows the visual joke devised by the American animator Chuck Jones, whose character Marvin the Martian also wore armor from his very first appearance, in 1948.
24 As Susan Reid has shown (2002), tastes in and acquisition of consumer goods became a vital concern in the domestic Soviet response to The Thaw and, more generally, to the Cold War.
25 See particularly Chapters 1 and 2.
26 See, for example, the image by Naum Moiseevich Lisogorskii (1910–86) in *Krokodil* 1962: 4/8, which (under the caption: "Door to the criminal world" [Dver' v prestupnyi mir]) depicts a young innocent opening the door-like front page of a comic, unaware of the violent and murderous goings-on inside.
27 "Duration thus assumes the illusory form of a homogenous medium, and the connecting link between these two terms, space and duration, is simultaneity, which might be defined as the intersection of time and space" (Bergson 1959: 110).

Bibliography

Alaniz, J. (2010) *Komics: Comic Art in Russia*. Jackson, MS: Mississippi University Press.

Andrews, J.T. (2003) *Science for the Masses: The Bolshevik State, Popular Science, and the Popular Imagination in Soviet Russia, 1917–1934*. College Station, TX: Texas A&M UP.

Andrews, J.T. and A.A. Siddiqi (Eds.) (2011) *Into the Cosmos: Space Exploration and Soviet Culture*. Pittsburgh, PA: Pittsburgh University Press.

Bailes, K. (1978) *Technology and Society under Lenin and Stalin*. Princeton, NJ: Princeton University Press.

Bakhtin, M.M. (1981) "Forms of Time and of the Chronotope in the Novel." *The Dialogic Imagination: Four Essays*. M. Holquist, Ed. Austin, TX: Texas University Press. 84–258.

Bergson, H. (1959) *Time and Free Will*. London: Taylor & Francis.

Bittner, S.V. (2008) *The Many Lives of Khrushchev's Thaw: Experience and Memory in Moscow's Arbat*. Ithaca, NY: Cornell University Press.

Bowlt, J.E. (Ed.). (1988) *Russian Art of the Avant-garde: Theory and Criticism, 1902–34*. London: Thames & Hudson Ltd.

Cherepano, Iu. (2009) *Reportazh iz "Krokodila."* Moscow: Kontakt-Kul'tura.

Condee, N. (2000) "Cultural Codes of The Thaw." *Nikita Khrushchev*. P.W. Taubman, P.A. Gleason, and S. Khrushchev, Eds. New Haven, CT: Yale University Press. 160–76.

Durgin, F.A. Jr. (1962) "The Virgin Lands Programme 1954–1960." *Soviet Studies* 13.3: 255–80.

Efimov, B. (1976) *Shkol'nikam: O karikature i karikaturistakh: Kniga dlia uchashchikhsia*. Moscow: Prosveshchenie.

Epstein, M. (2003) "Russo-Soviet Topoi." *The Landscape of Stalinism*. E. Dobrenko and E.Naiman, Eds. Seattle, WA: Washington University Press. 277–306.

Fitzpatrick, S. (1999) *Everyday Stalinism: Ordinary Life in Extraordinary Times: Soviet Russia in the 1930s*. Oxford: Oxford University Press.

Froggatt, M. (2006) "Renouncing Dogma, Teaching Utopia: Science in Schools Under Khrushchev." *The Dilemmas of Destalinisation: A Social and Cultural History of Reform in the Khrushchev Era*. Jones, P., Ed. New York: Routledge. 250–66.

Graham, L.R. (1993) *Science in Russia and the Soviet Union: A Short History*. Cambridge: Cambridge University Press.

Groensteen, T. (2007) *The System of Comics*. Jackson, MS: Mississippi University Press.

Harvey, B. and O. Zakutnyaya. (2011) *Russian Space Probes: Scientific Discoveries and Future Missions*. New York: Springer-Praxis.

Higham, R.D.S. *et al.* (1998) *Russian Aviation and Air Power in the Twentieth Century*. London: Frank Cass.

Hoffman, D.L. (2003) *Stalinist Values: The Cultural Norms of Soviet Modernity*. Ithaca, NY: Cornell University Press.

Ioffe, M.A. (1956) *Ivan Maksimovich Semenov*. Moscow: Sovetskii khudozhnik.

Interfax. (2006) *Ia gorzhus' obvineniiami v tom, chto vvel Iuria Gagarina v pravoslavie*. Available at <http://www.interfax-religion.ru/orthodoxy/?act=interview&div=73&domain=1> (accessed 26 April 2013).

Kelly, C. (2007) *Children's World: Growing University Pressin Russia, 1890–1991*. New Haven, CT: Yale University Press.

Kennedy, M.D. (1998) *Encyclopedia of Russian and Slavic Myth and Legend*. Santa Barbara, CA: ABC-CLIO.

LaPierre, B. (2012) *Hooligans in Khrushchev's Russia: Defining, Policing, and Producing Deviance during the Thaw*. Madison, WI: Wisconsin University Press.

Maurer, E. *et al.* (Eds.). (2011) *Soviet Space Culture: Cosmic Enthusiasm in Socialist Societies*. Houndmills, Hampshire: Palgrave Macmillan.

McCannon, J. (1998) *Red Arctic: Polar Exploration and the Myth of the North in the Soviet Union, 1932–1939*. New York: Oxford University Press.

McCloud, S. (1994) *Understanding Comics: The Invisible Art*. New York: HarperCollins.

Oushakine, S.A. (2011) "Laughter under Socialism: Exposing the Ocular in Soviet Jocularity." *Slavic Review*, 70.2: 247–55.

Palmer, S. W. (2006) *Dictatorship of the Air: Aviation Culture and the Fate of Modern Russia*. Cambridge: Cambridge University Press.

Plamper, J. (2012) *The Stalin Cult: A Study in the Alchemy of Power*. New Haven, CT: Yale University Press.

Reid, S. (2002) "Cold War in the Kitchen: Gender and the De-Stalinization of Consumer Taste in the Soviet Union under Khrushchev." *Slavic Review*, 61.2: 211–52.

Rémi, C. (2011) "Reading as Playing: The Cognitive Challenge of the Wimmelbook." *Emergent Literacy: Children's Books from 0 to 3*. Kümmerling-Meibauer, B., Ed. Amsterdam: John Benjamins Publishing. 115–40.

Ryan-Hayes, K.L. (2006) "Introduction." *Contemporary Russian Satire: A Genre Study*. K.L. Ryan-Hayes, Ed. Cambridge: Cambridge University Press. 1–10.

Shabad, T. (1964) "Behind the Smile on Crocodile." *The New York Times*. June 7: 22.

Siddiqi, A.A. (2011) "From Cosmic Enthusiasm to Nostalgia for the Future: A Tale of Soviet Space Culture." *Soviet Space Culture: Cosmic Enthusiasm in Socialist Societies*. Maurer, E. *et al.*, Eds. Houndmills, Hampshire: Palgrave Macmillan. 283–306.

Smolkin-Rothrock, V. (2011) "The Contested Skies: The Battle of Science and Religion in the Soviet Planetarium." *Soviet Space Culture: Cosmic Enthusiasm in Socialist Societies*. Maurer, E. *et al.*, Eds. Houndmills, Hampshire: Palgrave Macmillan. 57–78.

Stites, Richard. (1992) *Revolutionary Dreams: Utopian Vision and Experimental Life in the Russian Revolution*. New York: Oxford University Press.

—. (2010) "Heaven and Hell: Propaganda Constructs the World." *Passion and Perception: Essays on Russian Culture*. D. Goldfrank, Ed. Washington DC: New Academia Publishing.

Stykalin, C. and I. Kremenskaia. (1963) *Sovetskaia satiricheskaia pechat', 1917–1963*. Moscow: Pravda.

Uspenskii, B.A. (1976) *The Semiotics of the Russian Icon*. Lisse: Peter de Ridder Press.

Young, G.M. (2012) *The Russian Cosmists: The Esoteric Futurism of Nikolai Fedorov and His Followers*. New York: Oxford University Press.

Yurchak, A. (2005) *Everything Was Forever, Until It Was No More: The Last Soviet Generation*. Princeton, NJ: Princeton University Press.

4 Flying city or housing freed from gravity

Ideas of space travel and internationalism in G.T. Krutikov's *City of the Future*

Aleksandra Idzior

Introduction

In 1922, as Vladimir Tatlin, Liubov' Popova, Aleksandr Rodchenko, and other Russian Constructivists eager to support the new state through their art were gaining critical attention, Georgii Tikhonovich Krutikov (1899–1958) moved from his birthplace of Voronezh to Moscow. For Krutikov the newly reappointed capital city—*the* socialist metropolis after the October Revolution—constituted not only living space, but also space for the artistic imagination, a provocative launching pad for what would become his representation of a utopian future megalopolis. Six years later, Krutikov presented his diploma project, titled *City of the Future* [Gorod budushchego],[1] executed in the studio led by Nikolai Aleksandrovich Ladovskii at VKhUTEMAS/VKhUTEIN.[2] Although Krutikov referred to his work as a proposal for a future city, it subsequently became known as the design for a "flying city" [letaiushchii gorod].[3] Actually, the only flying element in Krutikov's vision was an independent "cabin" [iacheika], a unit that served a twofold function: that of a vehicle and that of a partial dwelling. Though the inhabitants rather than the city itself would fly, Krutikov designed the entire spatial structure of the *City of the Future* with aerial transportation in mind.

Appreciating the power of the imaginary and fascinated by the vision conceived by Krutikov, I examine his projected city in the context of discourses related to the phenomenon of space exploration and the 'frontier.' My analysis reevaluates Krutikov's proposal, assessing the politics of representation involved in its production.

Krutikov's diploma project

Krutikov's diploma project consists of images accompanied by textual commentary, with the visual material divided into two parts. The first is devoted to his analysis of existing architecture as affected by various means of transportation. In this section all the visual material is spread over sixteen panels on which Krutikov collaged images that include cut-and-paste photographs interspersed with drawings of plans, diagrams, and illustrations. By contrast, the second part offers a model of the future city and is laid out on four panels, with images evoking the new city conveyed through Krutikov's own set of drawings, plans, and diagrams, and only occasional use of mass-produced photographs. His objective was to demonstrate the crucial elements of the envisioned future city, such as housing complexes hovering above the earth, and a mobile cabin for transporting the occupants to their dwellings. Whereas the visual material is quite extensive, the textual segment is relatively succinct, with Krutikov's comments presented in point form.

According to Krutikov, his overall goal was to establish a "series of new questions in architecture."[4] These included "the expansion of the architect's horizon in accordance with a number of scientific disciplines"[5] (such as mathematics, technology, or knowledge of space in nature), and the introduction of mobile planning that would foster an architecture in outer space. Sharing the Constructivist passion for linking art and architecture with the new social order, Krutikov argued that architecture and the built environment could and should become active agents in the socialist transformation of the economic, political, and social basis of society—a fundamental transformation that was in full swing and to which architecture had much to contribute.

Technology, movement, velocity, and architecture

The analytical part of Krutikov's project is centered on the existing state of affairs in architecture, as impacted by various means of transportation and speed. These issues are presented in twelve groups. Starting with the problematic related to the "theory of architecture of a mobile form," and ending with considerations devoted to "dreams, fantasies, beginnings, caricature, achievements," the architect managed to include quite diverse topics. They ranged from the "form-evolution of car, train, boat, dirigible and airplane," through the "foreign mobile country homes," up to the "progress of power engineering," and the "conquest of a new space."[6] Krutikov reasonably believed that architecture was directly affected by the evolving means of transportation and ever-increasing speed as exampled by flight, among other means.

The dynamism of the polymorphous images re-organized by Krutikov and their syncopated rhythm render the pulse of the modern city, the evolution of which involves precisely the intensification of speed, which in turn correlates to the progress in means of transportation. Krutikov's interest in velocity corresponds to the widespread cult of speed that in turn was part of the aesthetic of modernism in the early part of the twentieth century, as demonstrated, for example, by the Futurists on the one hand,[7] and Le Corbusier on the other.[8]

In Russia, even before the Revolution, artists embraced Futurism out of a commitment to avant-garde art by finding a source of inspiration in the city and industry. Airplanes, automobiles, trains, and speed were the emblems of modernity (Harte 2009). Knowledge of analytical Cubism and Futurism reached Russia almost simultaneously, around 1911–12. The Russian artist challenging Symbolist art before WWI responded to the burgeoning urban centers and industrialization by promoting values consistent with Futurism and combining it with the style of Analytic Cubism. Because knowledge of Analytic Cubism and Futurism arrived in Moscow almost at the same time, and because they were superficially similar, the Russian artists—Mikhail Fedorovich Larionov, Natal'ia Sergeevna Goncharova, Kazimir Severinovich Malevich, Vladimir Yevgrafovich Tatlin, and poets Vladimir Vladimirovich Maiakovskii and Velimir (Viktor Vladimirovich) Khlebnikov—tended to link them. Thus, the style that evolved from them in the period 1911–14 is known as Cubo-Futurism, although it had less to do with Cubist than with Futurist values. In 1912, while speaking in St. Petersburg, Filippo Tommaso Marinetti stimulated Goncharova and Larionov into developing a Russian version of Futurism known as Rayonism. In 1913 Malevich applied the term Cubo-Futurism to works by Goncharova such as *Cats* [Koshki 1913], and to his own painting, *The Knife Grinder: Principle of Flickering* [Tochil'shchik: Printsip mel'kaniia 1912]. The reception of Cubism and Futurism occurred at a moment of friction between the pro-Russian, pro-rural, and pro-agrarian Slavophiles, centered in Moscow, on the one hand, and, on the other, the pro-Western, pro-urban, and pro-industrial European stance dominant in St. Petersburg. The Cubo-Futurists aligned themselves strongly with the latter movement. Overall the Italian Futurism and the Russian Rayonism, whatever their divergences, both represented the new reality of urban growth, metropolitan culture, and humankind's dependence on machines. Ultimately, they shared an interest in dynamism, mechanical movement, and speed, light, and energy.[9]

There is a parallel between the time when Futurist discourse entered Russia and Krutikov's preoccupation with the contemporary urban environment characterized by speed. Namely, when asked during his defense examination how much time he had spent on developing his ideas about the future city, Krutikov answered that it took him fifteen years to develop this concept (Khan-Magomedov 1987: 30–5). If indeed Krutikov spent so much time deliberating his project, it means that its gestation had started a few years before WWI and the October Revolution, in 1913—the time of Futurism's growing popularity in Russia. It was also in 1913 that a series of eight popular postcards was issued representing images called 'Moscow of the Future' [Moskva budushchego],[10] which projected a science-fiction vision, with fantastic images of city streets filled with dirigibles, airplanes, and cars. Considering that the publication of these postcards coincided with the increasing knowledge of Futurism in Russia, their appearance on the market may be interpreted as Russia's growing desire for modernization and technological progress. It is entirely possible that the colorful and whimsical prognosis of the future metropolis evoked by the postcards would make a memorable impression on the then fourteen-year-old Krutikov. Thus this uncanny imagery, representative of calls for industrial development in still-Tsarist Russia, might be regarded as inspiring Krutikov's interest in means of transportation and their effects on the city of tomorrow. These images vividly announcing the city yet to come, which while certainly tantalizing the imagination with the glamour of technological advancement, represented at the same time the darker underpinnings of urban modernization, which were congestion, noise, pollution, and chaos.

Perhaps Krutikov considered the Revolution of 1917 and the explosive changes it triggered as an historic moment that constituted a catalyst for speeding up the delivery of the bright and industrialized life envisioned on the postcards.

Consequently, as the postcards embodied desires under the old regime, the new conditions brought about by the Bolshevik government opened up possibilities to act on them, at the same time safeguarding against the disorder of the past system. Many Soviet intellectuals during the 1920s were searching for novel concepts for a newly established society, and several of them reached rather similar conclusions. One of the dominant elements emphasized in this discourse was, predictably, movement; hence the link between Krutikov's design of future houses and the urban visions of Khlebnikov that recur throughout the poet's *oeuvre* (Stobbe 1986). In *The City of the Future* [Gorod budushchego 1920], Khlebnikov, for example, envisions domiciles that foreshadowed those of Krutikov. According to the poet, the houses of tomorrow would consist of a framework of steel bays to accommodate individual glass living units that could be docked and undocked. This would allow residents to move from one building and from one city to another, simply by disconnecting and moving their apartments by rail, by steamship, or by air (the last by being equipped with a self-propelling engine). The veritable visualization of Khlebnikov's images as rendered by Krutikov on paper a decade later indicates Krutikov's strong tendency to reach out for the vocabulary that permeated Russian culture shortly after the October Revolution. In fact, throughout his project, he critiqued the past to improve it, not to destroy it. However, what is striking in his choices is that the visual material he gathered demonstrates advances, though with shortcomings, that represent solely foreign, not Russian, milieux.

The omission of national examples of architecture not only amplifies the international dimensions of Krutikov's project, but may also indicate his resistance to Stalin's attitude toward the national architectural heritage announced in 1928. Constituting a part of the 'Cultural Revolution,' this hostile approach to the Russian heritage manifested itself in the escalation of attacks, from verbal bashing, expressed shortly after the Revolution, to the systematic destruction of the historical monuments in Moscow at the end of the 1920s.[11] In the section of his project in which Krutikov scrutinized historical achievements in architecture and transportation, and which served as the rationale for the construction of the new city, Krutikov introduced iconographic references to technology and the implied role of the architect/artist as engineer. By juxtaposing images of built structures with pictures of various means of transportation, he followed the modernist tradition in perceiving metropolitan life as a sublime manifestation of technology and engineering work, of which aviation was a key aspect.

Krutikov's work created an overall effect by exaggerating, altering, and fragmenting reality. He mixed visual materials, juxtaposing image with image, photography with drawing, and elevation with plan, his calculated arrangement ultimately conveying a sense of control and constraint. Yet this variable iconic material creates a collision of form and metaphor in invoking aesthetic connotations of Cubism (and particularly Dada), but with a content that is optimistically futuristic. The technique of collage associated with synthetic Cubism was a well-known and an often used mode in Russia among the Cubo-Futurists even before the Revolution. Similarly, during the 1920s, Dada collages and photomontages were not only highly popular but also masterfully handled by Gustav Gustavovich

Klutsis, Aleksandr Mikhailovich Rodchenko, and El Lissitzky [Lazar' Markovich Lisitskii]. While the Dadaists manipulated images to convey chaos and crumbling order, Krutikov applied the same manipulation to opposite ends. He used images of existing structures and evolving means of transportation as building blocks to construct a new image of an improved city. Because current models were inefficient and confusing, they needed to undergo this justifiable confrontation and revision that would lead to the optimal, new urban form. The motivating desire was not so much to negate as to radically improve.

Borrowing modernist devices, Krutikov throughout his work incorporated popular imagery as disseminated by postcards and magazine illustrations. In doing so he followed avant-garde traditions by engaging and identifying himself with what Thomas Crow calls "marginal, 'non-artistic' forms of expressivity and display" (Crow 1983). All this heterogeneous material drawn from many published sources, such as books, magazines, and journals, was both Soviet and foreign. Since the introduction of New Economic Policy [Novaia Ekonomicheskaia Politika/NEP] in 1921, the Soviet market had been saturated with foreign publications.[12] Krutikov numbered among the many who took advantage of illustrated magazines, which provided a rich source of 'raw' visual material featured through crowded, innovative layouts. In short, he turned to both 'low' and 'high' cultural sources for his project.

Photography and architects' drawing

In organizing all the images he had assembled, Krutikov separated his own vision of the optimal future city from the material illustrating changes in the means of transportation and their relation to architecture. The former he executed in drawings, while the latter are rendered predominantly through reproduced photographs and illustrations.[13] Photography indeed was a well-established and widely used technique for creating images in Bolshevik Russia when Krutikov was working on his project. During the 1920s, the use of photographic images was grounded in such popular products as magazines, postcards, illustrations, and advertisements. Bolshevik artists understood photography as the most efficient and objective of all media, and the mass-produced magazines containing photographs were regarded as the epitome of the recombination, recycling, and reproduction of visual language.

Photography in montage and collage functioned as a mode between a modernist critique of the representational conventions and an emerging awareness of the demand to construct novel visual representation for a new mass audience. At first, the gathered and mixed photographic pictures revealed the power to produce a sense of estrangement through the process of extracting the reproduction of an object from its common context by rearrangement, to situate it, artificially, on a different plane of reality. In this way, the distorted images linked to the technique of collage broke the chain of habitual associations. Montage amplified the displacement of time and space within the work, vis-à-vis those of real life. This disordering of images allowed a new recognition of the relationship between art and life. The new objectivity sought not to domesticate or beautify reality, but to document the visible world by new, at times shocking, but ever enlightening juxtapositions.

Krutikov did not take the photographs used in his project, preferring to rely on clippings from a variety of magazines that conveyed the sense of being ready-made and of spontaneity. This medium suggested a documentary quality that was helpful in revealing the mechanics and materiality of the world.[14] The person taking and reproducing photographic images with the use of a camera and chemical processes could be perceived less as an artist and more as a constructor or technician. Also, the variety of images selected by the Soviet architect from a wide range of items might function to undermine the myth of individual authorship. Accordingly, Krutikov became a constructor, forcing the audience to reconsider the significance of the interplay of visual codes consolidated by society and re-introduced by him in the project of re-visioning.

By creating this polarized arrangement separating his own works from the images of others, Krutikov distinguished his input into furthering urban planning and housing from the inefficiencies and shortcomings of the existing state of architecture and means of transportation. Despite this split, by incorporating heterogeneous visual material produced and reproduced by various people, Krutikov conveyed in his project a collective identity—a notion promoted by the newly established Soviet state. These juxtaposed images, their eclectic subject matter, the multiplicity of their techniques and media, and finally the lack of attribution (there is an overwhelming absence of names) constitute a grand visual collage of mixed but nonetheless common experiences. For Krutikov, as for European avant-garde artists, this inscription of commonality was a conscious choice, entailing the disappearance of the author's hand behind the anonymous tools of the graphic designer and the director of images. These tactics were all the more appealing during a period when artists searched for new modes by which to convey contemporaneity and to break down barriers between high art and everyday life.

Krutikov's approach resonated also with Productivist sentiments and ideas expressed by its theoretician, Aleksei Kapitanovich Gastev. In 1919, Gastev proclaimed that collectivity, and not individuality, was the hallmark of the new system, stating that, "a new working-class collectivism . . . is manifested not only in relations between persons but in relations of whole groups of people to whole groups of mechanism" (Gastev 1919: 44–5). After all, the primary objective in Soviet Russia was the dissemination of a promise of social transformation and a collective culture.

Space exploration

Krutikov's main desire was to free architecture from the laws of gravity. His ideas were based not on the current technological abilities of Soviet Russia, but rather on his conviction that the new socialist state, by catching up with the advanced technologies of the Western world, would soon take on the leading role as promoter of a cutting-age science that would advance the technological solutions to support his proposal.

The cult of technology saturated the arts and the political discourse of the age, and Russia's fascination with aeronautics was part and parcel of this phenomenon. Rocketry and space travel had a similar appeal. Long before the Revolution,

Konstantin Edvardovich Tsiolkovskii (1857–1935) had explored its possibilities and made discoveries that were later incorporated into practical research by the first generation of Soviet rocket scientists. Tsiolkovskii's work first became widely known in scientific circles in 1911, and he became a public figure in the 1920s (Holquist 1996: 100–17; Stites 1989: 167–71). In 1925, Moscow University organized a forum on the subject of 'Flight to Other Worlds.'

That year, the association responsible for spreading information about aeronautics published Maiakovskii's science-fiction poem *Flying Proletarian* [Letaiushchii proletarii].[15] The original first edition bore a revealing cover designed by Grigorii A. Berdshatskii. It represented the titular flying proletarian through the anthropomorphic arrangement of objects associated with the technological accomplishments of the emergent socialist state: the human head is outlined with a circular formation of airplanes; the eyes comprise enlarged silhouettes of two airplanes in profile; the nose is adopted from the silhouette of the Shabolovka Radio Tower.[16] A large red sphere represents the mouth, which seems to be wide open, as if shouting the words of the poet who is proclaiming the global reign of the proletariat, in the process of taking over the skies. This round, geometrical shape actually plays a double role by suggesting, on one hand, the orifice, and on the other, the planet earth. Hovering in the blue firmament as it revolves in the sun's orbit, the red globe seems to enact the Soviet version of the Copernican thesis, while being 'crowned' by the technological insignia of progress, therefore demonstrating the superiority of the globe taken over by communism—a realization of the poet's prediction. Ideas related to the conquest of space thrived within Moscow's artistic circles during Krutikov's studies there, for the possibilities opened up by such discoveries dovetailed with the dramatic innovations promised by the new state.

Krutikov's project contained the name of only a single individual—a scientist and space explorer who, unexpectedly, was not Russian. In the explanatory notes to panel number fourteen, called the "Conquest of New Space," Krutikov revealed the name of the author whose construction he represented by inserting a drawing of an airplane labeled "a project of an airplane by M. Val'ie that [in the stratosphere] is able to reach the speed of 2000 meters per second."[17] An Austrian rocket scientist working in Germany, Max Valier participated in the First World Exhibition of Interplanetary Machines and Mechanisms held in Moscow in 1927.[18] Evidently, Krutikov was familiar with Valier's activity, because on plate number fifteen, still under the rubric""Conquest of a New Space," we find another image of Valier's model. This time, in the lower-right corner, there is an illustration showing a hermetically sealed 'ether' plane, identified by Krutikov as a "rocket into interplanetary space." The repeated selection of Valier's prototypes and the fact that the architect credited the scientist through specifying his name suggest Krutikov's fascination with the inventor who targeted outer space in vehicles breaking speed records. It is also possible that Krutikov's emphasis on Valier had political overtones: perhaps his choice of Valier's projectiles into the earth's orbit reprised the hopes that the Bolsheviks had invested in Germany during the very early years of the Revolution—hopes that this industrially developed and then politically radical country would help technologically backward Russia with the global spread of international communism. This did not happen, but ten years later, Krutikov seemed to believe that German science could be useful in Russia's advancement of outer space exploration, this time in terms of scientific/ technological experiments.[19]

Krutikov expressed his 'homage' to Valier through the arrangement of images on panel number fourteen, of which the lower corner contains the figure of Gulliver from Jonathan Swift's *Gulliver's Travels* (1726), watching the flying island of Laputa. When extended, Gulliver's diagonal and upward trajectory, an original part of the composition, includes another drawing of a rocket and the image of Valier's airplane. Krutikov seemed to indicate a progression of flying objects, according to which the "space vehicles" evolve, ascending toward the upper left, where Krutikov positioned the Valier airplane. We may assume that the culmination of this evolving chain would be Krutikov's "flying city." Despite the prominent place allocated to Valier's inventions, we should remember that all the material presented by Krutikov on the sixteen panels demonstrate progressive achievements in architecture and mobility, containing, however—and this is a crucial point—shortcomings. The drawbacks of all those previous attempts at mobility are emphasized by the drawing of Gulliver's figure pointing to Laputa (Swift 1995).[20] Krutikov's own, original project was intended to overcome these shortcomings.

Krutikov's engagement with space in the realm of architecture actually preceded his diploma project. In 1926, along with El Lissitzky and the engineer Pavel Vinogradov, he had participated in the transport section of AsNovA—the Association of New Architects—to devise an architectural formulation for Tsiolkovskii's dirigible (Kazus 1999: 197). A year later, in 1927, Krutikov stated, "The very idea of movement has a great potential for development."[21] While working under the guidance of the architect-educator Nikolai Ladovskii (1881–1941), he also wrote a research paper, "On the Path to a Mobile Architecture: Its Social, Technical and Formal Foundations."[22] In effect, his graduation project crowned his previous explorations of mobile architecture.

Krutikov's basic thesis was that, with the development of civilization, human beings would necessarily acquire a strong desire to free the planet of most of its structures. This in turn could be done, he felt, by providing buildings with the capability of hovering above the earth and using nuclear energy at a certain point in the foreseeable future. This utilitarian view of limited living space on Earth and consequently the call for restructuring society and its natural environment recalls ideas proposed by Nikolai Fedorovich Fedorov, according to whom one had to devise other sites for habitation in outer space (Siddiqi 2010: 106).

Visual representations of the machine in Soviet culture were intended to evoke an ethos of efficiency and productivity, particularly after the Civil War, when the development of the technological means of production became the most urgent national priority. Around 1928, however, as a reaction to the rhetoric of the Stalin's Five-Year Plan, there was a shift when "the aura of the god-machine was eclipsed by the aura of the god-man" (Clark 1985: 101). The theme of socialist construction was transformed from a focus on huge agricultural and industrial projects, demonstrated as triumphs of technology and social organization, to an emphasis on individual heroism in conquering the elements. Over time, this individual hero would be replaced by the figure of Stalin.[23] Krutikov's project may be viewed as an interpretation and exploration of a machine as represented by various means of transportation. The technologically improved city, with its hovering dwellings, which was supposed to be reached via a sort of omnibus vehicle (serving plural functions), was thus envisioned as an advancement at an urban scale along the path to communism. If there was a hero in Krutikov's vision, it was his hovering city—literally a shining star in the firmament.[24]

"Flying city"

Krutikov's city of the future consisted of two main elements, one positioned vertically, hovering in the stratosphere and intended for residential purposes, the other fixed to the ground horizontally and meant for industrial use. Krutikov cleared the planet's surface of houses by lifting them into space, and allocated this freed land for work, tourism, and leisure. The industrial and recreational component was planned outwards from a central point along a spiral, a figure that was associated in Russia during the 1920s with the notion of freedom and liberation. According to Nikolay Nikolaevich Punin, "Societies with antagonistic classes have been fighting for the possession of the land, the line of their movement is horizontal; the spiral represents the movement of liberated humanity. A spiral is the ideal expression of liberation" (Punin 1988: 344–6). The residential structures were suspended in space, in a bowl-like shape that was dipped, with its narrow end pointing toward the industrial area along an axis rising vertically from the center of the latter. The dwelling complexes were arranged in tiers along the surface of this parabola.

The "cell/cabin" constituted a crucial component of the future urban settlement. Krutikov actually allocated to the image of the cabin a dominant position within the whole sequence of his project. The relatively large drawing of the cabin dramatically enters into the picture plane of the first panel that opens the visual part of the *City of the Future*. According to Krutikov, it was meant to be the universal method of transport, accessible to everyone, and used between the ground and the floating buildings. Besides flying, the cabin was capable of conquering distance on land, in or under water, thus embodying the perfect model of expanded mobility and convertibility.[25] Krutikov treated this cabin as both an individual vehicle and a short-term mobile home, which would provide people with the requisite degree of comfort in travel and at stops outside the 'flying city.' There is a striking similarity between Krutikov's 'cabin' [iacheika] and Tsiolkovskii's spaceship designs dated 1903, 1914, and 1915. The drawings of both vehicles demonstrate comparable streamlined bodies that aimed to conquer the resistance of air and to increase their velocity.

Krutikov designed three types of houses in the air. All of the dwellings were communal—the type of living quarters expected to induce collective behavioral patterns. The first variant of the 'communal house' [dom-kommuna] that Krutikov advanced in his project was a labor commune consisting of eight vertical, five-story housing blocks connected by lift shafts with a lower, circular, communal building. The second type was a vertical block with, as Krutikov called it, a hotel-type accommodation, including a honeycomb parking system for cabins, a central portion with living space, and a top part reserved for communal use. The third variant was a more compact version of the first type, in which all the vertical dwellings were consolidated into a single eight-floor cylindrical block, with the communal space gathered into a sphere. Already in 1920, Krutikov's supervisor, Nikolai Ladovskii, had exhibited a design on the theme 'architectural manifestation of a communal house.' Its dynamic structure and bent axis seemed to indicate the desire to overcome gravity, while recalling the silhouette of a rocket missile on its launching pad, similar to Max Valier's prototype used by Krutikov in his project on panel number fifteen, the "Conquest of New Space."

As mentioned above, Krutikov's investigation has the strongest affinity to the ideas on dynamic spatial form devised by the Rationalist movement in Soviet Russia represented by members of AsNovA. The fact that he based his whole proposal for the new city on the premise of mobile architecture and questions of aero- and astro-nautics might actually be traced to the concepts promoted by this association. It was AsNovA that advised the architects to consider in their work "earthly installation of aerial paths of communication" or "avia construction."[26] There are also strong similarities between Krutikov's approach and the ideas of El Lissitzky, also associated with AsNovA. Their commonality lies, first of all, in a strong dissatisfaction with existing skyscrapers, and second, in a shared perception of the correspondence between architecture and human mobility. Krutikov's criticism of skyscrapers reverberated with Lissitzky's scrutiny of American high-rises and their negative effects upon the city.[27] Among the commentaries to the twelfth panel titled 'Evolution of Constructions—from the Cave to Housing in the Air,' one is directed toward an image showing a set-back skyscraper of dominating height. According to Krutikov, there was a "skyscraper crisis" that he wished to resolve. He wrote: "Due to the fact that the skyscraper [neboskreb] takes over a huge amount of space on the ground with its foundations, and because it contains insignificant sections on top, there is a need to escape this waste by dividing the skyscraper into elements that would be separately hung in space."[28] To demonstrate the problems, as he (mis)perceived them, Krutikov attached a schematic drawing representing the tapering silhouette of a high-rise. What follows, then, is the architect's response to this alleged skyscraper crisis, in the form of a remedy, a technologically advanced floating architecture.[29]

In criticizing the construction of a skyscraper for its supposedly uneconomic, elongated, pyramidal shape, Krutikov repeated Lissitzky's negative evaluation of American high-rises upon a city. Lissitzky's antidote to the alleged problem caused by skyscrapers can be seen as the middle point between the status quo and Krutikov's suggestion. Between 1923 and 1926, Lissitzky designed eight cloud-irons to be raised around Moscow's center. His aim was to intensify the city by taking up a relatively small ground area for a vertical support, and by elevating street-level activities onto the horizontally positioned structures above economized and diminished foundations. The overall tectonics in Lissitzky's project created a sense of a rather unstable equilibrium of the balanced horizontal structure above street level, making the architecture appear *as if floating* weightlessly, though still attached to fundaments on the ground. Krutikov pushed Lissitzky's concepts further, proposing dwellings not only removed from the earth's surface, but afloat in the upper level of the stratosphere.

What also allied Krutikov's ideas with Lissitzky's was the above-mentioned correspondence they elaborated between architecture and means of transportation. While Krutikov favored an evolutionary progression in the means of transportation and its impact on humans and architecture, Lissitzky envisioned a system of evolutionary changes among man, mobility, and architecture. He advocated a complete transcendence of earth-embedded building foundations with the advent of a "flying man," declaring, "We are faced with floating and hanging. I want to help and discover this form."[30] A few years later it was precisely Krutikov who pushed Lissitzky's ideas forward and upward.

Indeed, between 1928 and 1929, several other students besides Krutikov in Ladovskii's atelier implemented the principle of a city hovering above the earth for their graduation projects. Viktor Kalmykov imagined a suspended city-ring at the equator, which he called "Saturn" and which would hover in the air with the aid of rigid constructions (Kalmykov 2011). The ring would remain in a fixed position with respect to the earth by revolving at the same speed as the globe. In a kindred spirit, Isaak Iusefovich developed the theme of a floating USSR Hall of Congresses that could be moved and moored to towers set up in the cities where the congresses were being held. These towers were given the function of vertical access, but they were also both residential and public structures. In Soviet architects' projects of those years, the dirigible became a symbol for the global connection of world civilization.

Overall, Krutikov's tone throughout his project is earnest, presumably because of the assumed urgency of his objective and because the issues at stake are of great importance. He explicates his vision of the city with a sense of purposeful determination; his project aspires to be more than another fantastic divagation on the future. While Krutikov professes the new city as his goal, he agitates for the cause of a socialist "great experiment,"[31] implicitly assuming that only the Soviet state, with its socialist ideology, would fulfill the promise of superior technology yielding a new rearrangement of social and spatial interaction. In the Soviet regime's intention to transform human consciousness and modify behavior, housing was expected to play a crucial role. For architects and for party and government leaders, architecture became one of the tools of what at that time was called "the restructuring of everyday life" [perestroika byta], or the reconstruction of the way of life. The "new socialist man" was supposed to regard work as voluntary participation in the common task, which would translate into buildings that would constitute Soviet citizens' new environment, encompassing factories, clubs, and communal houses; Krutikov's project emphasized dwellings.

The concern at that time was that capitalism, which the Revolution supposedly had stamped out, was still evoked in the remnants of bourgeois housing. Therefore it was necessary to eliminate not only the capitalist system of production itself, but all its remnants, such as the old housing and the old way of life. Perceiving in men and women the reflection of their living conditions, architects aimed to transform human nature through the agency of the environment. Krutikov's attempt was to create a new physical setting for the new society. In 1928, the same year in which Krutikov presented his ideas, the Constructivist architect Moisei Iakovlevich Ginzburg in a speech called for the creation of "social condensers" [sotsial'nye kondensatory]. He regarded buildings and even entire cities as tools of social change. Like electrical condensers that alter the nature of the current, the proposed "social condensers" were to transform the self-centered individual of capitalist society into a new, socialist person. For Krutikov, the building blocks for the new city of the liberated new society consisted of buildings as a means of transportation or variants of "machines for living" levitating in space. What is explicit about these houses suspended in air is the conviction that they can be launched only under the auspices of the Soviets, and achieved through the accelerated advances of communism. In short, Krutikov harnessed his dream-like vision, which he made palpable throughout his representation of the future city, to promote agendas beyond the architectural realm.

Krutikov's "flying city" and the frontier

To further unpack Krutikov's project of levitating structures in space it is useful to situate it in the socio-political framework of what I call the 'frontier' or border factor and internationalism. My use of the term 'frontier' in the context of post-Revolutionary Russia involves two historical moments loaded with contrasting ideologies. The first occurred shortly after the October Revolution, when transgression of the geographical, physical frontier between Soviet Russia and Europe (as well as the rest of the world) was deemed a sine qua non for the global spread of communism. As the slogan phrased it, the proletariat of the world was supposed to unite. The initial Bolshevik intention was not to build up a planned industrial economy in a national state, but to stimulate revolution in the already industrialized Western countries. Internationalism was perceived as a foundation of socialism and a necessary tool to overcome the capitalist economy on a world scale. To catalyze world revolution, the Third International was established in Moscow in 1919, with the objective of coordinating the efforts of all communist parties in their struggle to spread the Revolution.

The second situation followed the decision in 1926 to build socialism, not on an international scale, as previously advocated, but within the geographical borders of the Soviet Union. This policy as part of Stalin's 'innovations' emphasized the division between the USSR and the rest of the world; it reinforced the notion of a physical frontier and ideological barriers separating that country from the 'others' [chuzhie]. Krutikov's project of the future city, though presented in 1928, was imbued with the utopian philosophy that characterized the early 1920s, especially ideas that championed the overcoming of existing national boundaries—ideas decidedly contrary to officially promoted policies in 1928.

Artists as mediators between official ambitions and popular notions nurtured ideas relating to the unified globe under the tutelage of the Comintern. These reveal an interesting common phenomenon, which was the designing of dynamic, often sky-piercing structures that remained part of unrealized, paper architecture. Vladimir Tatlin's *Monument to the Third International* [Pamiatnik III Internatsionala] of 1919 was revered for its ability to relocate humans into an ever-higher domain; as the critic Nikolai Nikolaevich Punin put it, "The life of the new humanity rises above ground" (Punin 1920: 344–46). In Krutikov's vision, the elevated city was not site specific. As the drawing of its radiant but vaguely paraboloid urban shape demonstrates, the city was situated high above the earth's surface. This is the only reference that points to the city's position vis-à-vis the globe. Suspended in the air, this fantastic rendering of the residential quarters of the future metropolis evokes a celestial spectacle, with a luminous comet-like body announcing the dawn of a new era.[32] Although ethereal, its presence solidly occupies a generic but implicitly international—indeed, universal—common space. The notion of internationalism is, I argue, embodied in the form of Krutikov's future city. The free-floating complexes of the houses and the flying cabin were projected not only as conquering earth's gravity, but also as occupying unspecified, globally shared outer space.

Cosmic allusions were also conveyed directly in an abstract language. Lissitzky's *Beat the Whites with the Red Wedge* [Klinom krasnym bei belykh 1920] may be read as agitating not for the propaganda of the Comintern, but for a universal class conflict. Using the Suprematist vocabulary of Malevich and incorporating his own architectural training, Lissitzky represented a military assault by the Red Army penetrating the vestiges of encircled Whites. The charge of the sharp red wedge, symbolizing the dynamism of the Bolshevik military force, comes down from the left side of the composition, represented by a pristine pictorial field that renders the bright outer space occupied by the Red Army. In the logic of his symbolic vocabulary, Lissitzky completed the depiction of the opponents to the Revolution as a round white blob. The enemy, which is conveyed as feeble and shown as closing up in a circular formation to defend its position, is situated against a background executed in a sinister black color, suggesting a compromised realm occupied by the bourgeoisie. The battle, in sum, is abstracted into a general class opposition.

In the early 1920s, politics and the arts had both presented metaphorically and envisioned literally attempts to transfer the lived and experienced domain into a higher level. Kazimir Malevich claimed, "We can only be aware of space if we break away from the earth, if the fulcrum disappears" (Malevich 1968: 20). Krutikov followed these revolutionary predecessors when he proposed a free-floating city bound not as much to a nation or a state as to the ideology of breaching geopolitical boundaries. Krutikov's convictions reverberate with Lissitzky's comments of 1922 that only the Soviet Union is able to secure the future for the architects with "an international mental horizon, revolutionary activity, and collective thinking" (Lissitzky 1922: 1–4). Over the course of the 1920s, however, the promotion of and fascination with internationalism in the Soviet Union started to wane, culminating in Stalinist preference for an exclusively national transformation.

Krutikov appeared opposed to the constraints of 1928, with its First Five-Year Plan, and escaped from the reality of the present into ideals of the near past. As Walter Benjamin observed, "The utopian images which accompany the emergence of the new, at the same time, reach back to the primal past" (Benjmin 1999: 893). For Krutikov, this "primal past" resembled the "original" moment of the early post-Revolutionary period, when the new order brought about by the Bolsheviks promised to deliver a collective and international proletarian community. By engaging political ideas that were already unpopular, however, Krutikov exposed himself to harsh criticism and to negative reception in the professional press, where his work was branded as an "irresponsible proposal of (a) utopian kind" (Mordvinov 1930: 12–15).

Indeed, in 1928 Krutikov was in a position that Piotr Piotrowski calls "between revolution and reaction" (Piotrowski 1993). His project of the future city demonstrates that Krutikov aligned himself with rhetoric that was highly popular shortly after the October Revolution, but that period's call for internationalism, mixed with an unrestrained burst of imagination over the course of the decade, became highly contested and vehemently criticized as the decade drew to a close. My contention is that Krutikov, by choosing to ally his work with the policies and visual vocabulary of the heroic revolutionary period in Russia, became, however unintentionally, part of an opposition to the official political and artistic doctrine of the late 1920s.

Post scriptum to the "flying city"

Krutikov's "flying city" or, more precisely, his proposal to equip its citizens with flying machines, met rejection and harsh criticism from proponents of the official doctrine, largely because it was utopian. One artist, however, continued exploring concepts that were close to Krutikov's *idée fixe*. Between 1929 and 1931, Vladimir Evgrafovich Tatlin (1885–1953) built an apparatus for flying, which he named *Letatlin*—a verbal punning of his name and the Russian word *letat'*, meaning 'to fly.' Tatlin envisioned his glider as an extension of ideas related to the material culture that he had been practicing for almost two decades. The quite fantastic *Letatlin* may be perceived, to use Boris Ignat'evich Arvatov's phrase, as "a materialized utopia,"[33] and as a symbol of Tatlin's unrelenting faith in the imaginary. The fact that Krutikov's flying city remained only on paper, or that *Letatlin* never really flew, does not detract from the significance of these two projects as embodiments of a liberating desire and the literal expansion of a horizon, simultaneously combining the artistic with declared utilitarian goals.

Although it was as fantastic as Krutikov's hovering city, Tatlin's flying object nevertheless found a large group of supporters, particularly among aviation aficionados, who were convinced of the usefulness of *Letatlin*. Tellingly, the late 1930s witnessed the design, construction, and decoration of the Maiakovskaia subway station in Moscow.[34] Its platform, built with stainless steel, forty meters under the surface of the city, prominently displayed a series of thirty-five mosaics by Aleksandr Aleksandrovich Deineka, all representing images of flight and aircraft, placed within the domes of a high ceiling.[35] In the Moscow subway, built on Stalin's order, the paradigm of Russian culture, with its interplay of oppositional characteristics, was fully utilized.[36] In this case the low and the high, or the realm beneath the earth and the sphere high above it, were conflated in the space in which engineering achievements of underground travel were imbued with the technological prowess that allowed the Soviets to conquer the sky. When commuters descended under the paved streets of Moscow, they were thus shown Russia's mastery over the firmament, prominently displayed above the subway platforms—a veritable everday utopia.

There is, however, another side to utopia—its dystopian counterpart. In 1930, Rodchenko created a photomontage, *War of the Future* [Voina budushchego], in which the dream of constructing a new life is replaced by fearful projections of a newer means of annihilation. On a dark background, Rodchenko placed photographs of New York's skyscrapers under attack. While two dirigibles hovering above the scene strike the city with what looks like laser beams, two barrels of a canon situated below are aiming upwards at the buildings. Thick clouds of smoke float above the streets. Amid this destruction, two human figures in gas masks and safety uniforms emerge in stiff, frontal position as if ascending from the hell below. The 'things that came' with the 1930s caused political, social, and economic tensions, effecting the course of utopian projection to which states, institutions, and individuals had adhered. In this process, utopia revealed its Janus-face potential. In the ninth year of the third decade, frontiers would be redrawn, borders would be attacked from the sky by high-flying bombers, and building projects would focus on digging deep trenches.

Notes

1 Georgii Tikhonovich Krutikov, "Gorod budushchego. Predposylki k rabotie: 'Gorod budushchego,'" and "Gorod budushchego. Evolutsiia arkhitekturnykh printsipov v planirovkie gorodov i organizatsii zhilishcha: Ob"iasnenia k chertezham, risunkam i skhemam raboty," MS, Diploma project. Collection of A. V. Shchusev State Research Museum of Architecture, Moscow, R1a 11200/1–16 and R1a 11196/1–18.

2 Established in 1920, VKhUTEMAS, Vysshie Khudozhestvenno-Tekhnicheskiye Masterskiye (Higher State Art and Technical Studios) was reorganized in 1926 and its name was changed to VKhUTEIN, Vysshii Khudozhestvenno-Tekhnicheskii Institut (Higher State Art and Technical Institute).

3 According to Selim Omarovich Khan-Magomedov, Krutikov himself always called his design "City of the Future." The 'flying city' term was coined by critics and applied to the project soon after Krutikov's diploma defense (Khan-Magomedov 1973: 282–3).

4 Georgii Tikhonovich Krutikov, "Gorod budushchego," *op. cit.*, R1a 11196/1.

5 Krutikov, "Gorod budushchego," Ria11196/1.

6 Georgii Tikhonovich Krutikov, "Gorod budushchego," *op. cit.*, R1a 11196/1.

7 Filippo Marinetti in the 'Foundation and Manifesto of Futurism' of 1909 extolled the virtues of male energy and audacity, while asserting the supreme magnificence of mechanical speed. Under number four of his Manifesto, published in *Le Figaro*, Marinetti wrote: "We affirm that the world's magnificence has been enriched by a new beauty: the beauty of speed. A racing car whose hood is adorned with great pipes, like serpents of explosive breath—a roaring car that seems to ride on grapeshot is more beautiful than the *Victory of Samothrace*." He then went on to idealize the driver of an automobile as being integral to the trajectories of the universe. Marinetti finished his

program with the apotheosis of the ideal context of a Futurist architecture: "Factories hung on clouds by the crooked lines of their smoke; bridges that stride the rivers like giant gymnasts . . . and the sleek flight of planes whose propellers chatter in the wind like banners and seem to cheer like an enthusiastic crowd" (1992: 147–8). His evocative passage was a homage to the triumph of industrialization, the technical and social phenomena as they were then being extended through aviation and electrical power. Marinetti correctly recognized the advent of a new cultural milieu dedicated to anything large-scale and a highly mobile society. In the face of the passé values of Italian Classical culture, it proclaimed the primacy of a mechanized environment that later informed to an equal degree the architectural esthetic of Italian Futurism (Antonio Sant'Elia's *Città Nuova* of 1914).

8 Le Corbusier also took up this idea, cherished by the Futurists, and in his projects proposed an open city form, which was supposed to facilitate locomotion, in accordance with his motto, "A city made for speed is made for success" (Corbusier 1925: 179). This statement was a part of Le Corbusier's rhetoric accompanying his 'Plan Voisin' project for Paris of 1925, a rather paradoxical concept—that the car, having effectively ruined the great city, could now be exploited as an instrument for its recovery.

9 The relationship between Italian Futurism and the Russian Cubo-Futurist movement, however, is complex and controversial. In fact, Marinetti during his visit to Russia in 1914 received a rather cold welcome for his blatant glorification of war.

10 For a discussion of the postcards, see Bielitskii and Glezer 1993: 250–2.

11 Some of the buildings that disappeared in this period were in the Kremlin: the Church of Spas na Boru of 1330, the Chudov Monastery and the *Voznesenskii* Convent, and the Nikolaievskii Palace of 1775–6; Kazan' Cathedral on the Red Square (of 1636); an outstanding example of Moscow Baroque, the Assumption Church on the Pokrovka (of 1696–99); the Cathedral of Christ *Spasitiel'* of 1838–80.

12 I have not been able to recognize the exact sources of the images appropriated by Krutikov. There are, however, references to access to a plethora of Western publications by other Soviet artists. For example, Zakhar Bykov, a student at VKhUTEMAS, while visiting Aleksandr Rodchenko's studio, found the entire floor covered with piles of French and German periodicals, from which the artist was clipping illustrations for his montages (Lavrentev 1995). Rodchenko had firsthand access to foreign material via his own trips abroad or through his friends (Vladimir Mayakovskii, Osip Brik, etc.) who sojourned in Western countries. Lev Kopelev wrote: "I kept my passion for reading . . . *Berliner Tageblatt*, the *I-Z* and the most expensive but most interesting magazine, *Die Woche*. At that time [1926] these publications were sold freely in a store on Vladimirskaia Street which specialized in publications from other cities and from abroad" (Kopelev 1980: 92).

13 For a discussion on the use of photography by architects, see Cohen 1988.

14 Walter Benjamin in his seminal essay 'The Work of Art in the Age of Mechanical Reproduction' argues that the "aura" of the original, unique work of art is lost to reproducibility, and that this, far from being a loss, opens up progressive possibilities. For Benjamin, photography, film records, etc. are more democratic than painting or other traditional art media. He argues that "painting simply is in no position to present an object for simultaneous collective experience" (1968: 234).

15 The lengthy poem runs to 1860 verses. In the foreword, Maiakovskii states that the newspapers give "facts" and "news," while what interests a poet is what would happen in two or one hundred years. Maiakovskii's own story is set in 2125, thus two hundred years from the time he wrote it. The poem is divided into three parts. The first, called 'The War That Is Being Waged Presently,' features a giant air battle between the Soviet proletarian and the American bourgeois air forces. The latter prevails until an uprising of New York workers against their government turns the tide. In the communist future, envisioned by Maiakovskii in the second part, called "Future *byt* everyday life," electricity brings comfort and ease (electric razors, toothbrushes, and such), and labor

is totally mechanized—the worker merely operates a keyboard. Everybody owns an airplane, and Moscow no longer has any streets, just airports. The closing, third part, titled 'Appeal,' identifies Maiakovskii as the agitator of the future, whose goal is to prepare his contemporaries to conquer the sky (Maiakovskii 1955–61, v. 6: 311–61). For a discussion of the motif of flight as a dominanta in Soviet culture of the early post-Revolutionary years, particularly its architecture and literature, see Beaujour 1983; Bowlt 1978.

16 The Shabolovka Radio Tower designed by Vladimir Grigor'evich Shukhov was built in Moscow in 1922. From this tower, Radio Comintern made its first international broadcast in 1922. The new regime hailed radio as the medium for disseminating its ideology and reorganizing society. For the significance of radio under the Bolsheviks, see Gough 1996.

17 Georgii Tikhonovich Krutikov, "Gorod budushchego," *op. cit.*, R1a 11196/1. Panel 14, point 4.

18 For information on this event, see Winter 1983. Valier was a co-founder of the Spaceflight Society, established in 1927. Its monthly magazine, *The Rocket* [Die Rakete], was published until the end of 1929. Valier was killed in 1930 in an accident by an exploding liquid-propellant rocket. On Valier, see Sänger-Bredt and Engel 1974.

19 After WWI, the Treaty of Versailles forbade the Germans to have any sort of military aviation. Despite that, in 1924, the Germans and Soviets agreed to cooperate. In effect, the Soviets allowed the Germans to construct military aircraft and train pilots clandestinely in the airfield at Lipetsk, near Moscow. While the Germans would continue to conduct their research, in exchange, the Soviets would acquire technical knowledge for their own aviation industry. See Higham, Greenwood, and Von Hardesty 1998.

20 Ever since Aristophanes' pioneering play (*The Birds*, first performed in 414 B.C.), other writers had invented living in the air. Swift, with his fanciful flying city, however, was the first author who attempted to explain Laputa's properties in line with the findings of contemporary science. Therefore, his *Gulliver's Travels* may be read as science fiction, although its intention was to satirize men and their institutions. In the case of Gulliver's third travel adventure, during which he is picked up by the flying island of Laputa, Swift was ridiculing the court of Great Britain and England's Royal Society. At the time Swift published his book, Western Europe was experiencing the Enlightenment. Swift, whose intention was to castigate all the human follies, did not hesitate to take on the cult of reason, often alluding to Isaac Newton.

21 Georgii Tikhonovich Krutikov, "Krugloe i polukrugloe zhilische," *Stroitel'naia promyshlennost'* 8 (1927): 617–20.

22 Krutikov, "Arkhitekturnaia nauchno-issledovatel'skaia laboratoriia arkhitekturnogo fakul'teta VKhUTEINa. Rabota laboratorii v 1928–29 uchebnym godu," in *Arkhitektura i VKhUTEIN* 1 (January 1929): 4.

23 On the cult of Stalin and his tactics of self-representation, see Tucker 1974.

24 For the sky as heaven and flight as transcendence, see Chapter 2.

25 Exploration of new means of transportation was popular during the 1920s. Petr Vasil'evich Miturich, for example, a teacher of drawing at VKhUTEMAS between 1923 and 1930, worked on models for undulators [volnoviki]—mechanisms based on wave technology, as an alternative mode of transportation. According to his design, the undulators were adapted for various elements and could fly, be suspended in water, or move over the ground's surface while using principles found in nature (bird, snake, caterpillar). See Lodder 1983: 217–23.

26 Letter from AsNovA to El Lissitzky written on 9 June 1924, in Kazus, *op. cit.* 197.

27 America, wrote Lissitzky, "had created a particular type of high building by transforming the European horizontal corridor into a vertical lift shaft. . . . The spread of this type occurred in a completely anarchistic way, without any concern at all for the larger organization of the city" (Lissitzky 1923: 5).

28 Krutikov, Panel 12: *Evoliutsiia stroitel'stva*, "Gorod budushchego. Ob''iasneniia k chertezham, risunkam i skhemam raboty," *op. cit.*, 1120/9–12, 1832/33-36 kp.

29 As Scott Palmer has noted, the Amercian skyscraper was pioneered in the late nineteenth century to economize space, and did so through radical verticality (in comments on a draft of this chapter). Obviously, dwellings suspended in air would eliminate the issue of space on the ground.
30 Quoted in Stommer 1990: 60–3.
31 The concept and term are borrowed from Camilla Gray's benchmark work, *The Great Experiment: Russian Art 1863–1922* (1962).
32 For the equation between sky and heaven, see the Introduction and Chapter 2.
33 B.A. (Boris Ignat'evich Arvatov), "Oveshchestvlennaia utopia" (The Materialized Utopia), *Lef* 1, no. 1.
34 For an analysis of Moscow's subways station, see Kettering 2000.
35 For a history of the station, its architecture and decoration, and especially the symbolism of Deineka's mosaics, see Friedman 2000 and Chapter 2 of this volume. For an analysis of this particular project, see Chapter 2.
36 I am referring here to the dichotomy that Iurii Mikhailovich Lotman and Boris Andreevich Uspenskii have mapped out within Russian culture, and Vladimir Paperny has traced during the Soviet period. See Lotman and Uspenskii 1985; and Paperny 1996.

Bibliography

Barkhin, Mikhail Grigor'evich. (1979) *Arkhitektura i gorod. Problemy razvitiia sovetskogo zodchestva*. Moscow: Nauka.

Benjamin, Walter. (1999) *The Arcades Project*. Trans. Howard Eiland and Kevin McLaughlin, ed. Rolf Tiedemann. Cambridge, MA: Belknap Press.

Belitskii, I. M., and G. N. Gl'ezer. (1993) *Moskva neznakomaia*. Moscow: Stroiizdat.

Bowlt, John and Olga Matich. (1996) *Laboratory of Dreams. The Russian Avant-Garde and Cultural Experiment*. Stanford, CA: Stanford University Press.

Clark, Katerina. (1985) *The Soviet Novel: History as Ritual*. Chicago: University of Chicago Press.

Crow, Thomas. (1983) "Modernism and Mass Culture in the Visual Arts." In *Modernism and Modernity*. Benjamin H. D. Buchloh, Serge Guilbaut, and David Solkin, eds. Halifax: Press of the Nova Scotia College of Art and Design. 215–64.

Durant III, C. Frederick, and George S. James, eds. (1974) *First Steps Toward Space*. Washington, DC: Smithsonian Institution Press.

Harte, Tim. (2009) *Fast Forward: The Aesthetics and Ideology of Speed in Russian Avant-Garde Culture, 1910–1930*. Madison, WI: University of Wisconsin Press.

Higham, Robin, and Jacob W. Kipp, eds. (1978) *Soviet Aviation and Air Power: A Historical View*. London: Brassey's and Boulder, CO: Westview Press.

Kalmykov, Viktor. (1973) "Goroda v vozdukhe." *Arkhitektura CCCP* 6: 58–60.

Khan-Magomedov and Selim Omarovich. (1983) *Pioneers of Soviet Architecture: The Search for New Solutions in the 1920s and 1930s*. Trans. Alexander Lieven. New York: Rizzoli.

—. (1973) "Proekt 'letaiushchego goroda." *Dekorativnoe iskusstvo* 1: 30–5.

Kopelev, Lev. (1980) *The Education of a True Believer*. Trans. Gary Kern. New York: Harper and Row.

Krutikov, Georgii Tikhonovich. (1929) "Arkhitekturnaia nauchno-issledovatel'skaia laboratoriia arkhitekturnogo fakul'teta VKhUTEINa. Rabota laboratorii v 1928–29 uchebnom godu." *Arkhitektura i VKhUTEIN* 1 (January): 4.

—. "Gorod budushchego. Predposylki k rabote: 'Gorod budushchego,'" "Gorod budushchego. Evolutsiia arkhitekturnykh printsipov v planirovkie gorodov i organizatsii zhilishcha: Ob''iasneniia k chertezham, risunkam i skhemam raboty." MS, Diploma project. Collection of A. V. Shchusev State Research Museum of Architecture, Moscow, R1a 11200/1–16 and R1a 11196/1–18.

—. (1927) "Krugloe i polukrugloe zhilishche." *Stroitel'naia promyshlennost'* 8: 617–20.

Lavrentev, Alexander. (1995) *Alexander Rodchenko: Photography 1924–1954*. Cologne: Könneman.

Lissitzky, El. (1922) "Amerikanizm v evropeivskoi arkhitekture." *Veshch'-Gegenstand-Objet* no. 1–2: 1–4.

Lissitzky-Küppers, Sophie, ed. (1968) *El Lissitzky: Life, Letters, Texts*. Trans. Helene Aldwinckle and Mary Whittall. London: Thames and Hudson.

Lodder, Christina. (1983) *Russian Constructivism*. New Haven, CT and London: Yale University Press.

Mordvinov, Arkadii Grigor'evich. (1930) "Leonidovshchina i ego vred." *Iskusstvo v massy* 12 (December): 12–15.

Piotrowski, Piotr. (1993) *Artysta między rewolucją a reakcją. Studium z zakresu etycznej historii sztuki awangardy rosyjskiej.* Poznań: Uniwersytet im. Adama Mickiewicza.

Siddiqi, Asif A. (2010) *The Red Rocket's Glare: Spaceflight and Soviet Imagination, 1857–1957*. New York: Cambridge University Press.

Stites, Richard. (1989) *Revolutionary Dreams: Utopian Vision and Experimental Life in the Russian Revolution*. New York, Oxford: Oxford University Press.

Stommer, Rainer. (1990) "The Dream of the Flying Cities: On El Lissitzky's Sky-hook." *Daidalos* 37 (September 15): 60–3.

Swift, Jonathan. (1995) *Gulliver's Travels*, Christopher Fox, ed. Boston and New York: Bedford Books and St. Martin's Press [1726].

Tucker, Robert C. (1974) "The Rise of Stalin's Personality Cult." *American Historical Review* 84, no. 2 (April): 347–66.

Winter, Frank H. (1983) *Prelude to the Space Age: The Rocket Societies: 1924–1940*. Washington, DC: Smithsonian Institution Press.

Zhadova, Larissa Alekseevna, ed. (1988) *Tatlin*. London: Thames and Hudson.

5 Neo-cosmism, empire, and contemporary Russian art

Aleksei Belyaev-Gintovt

Maria Engström

Introduction

In this chapter, I examine the conceptual legacy of Russian and Soviet Cosmism in the work of the Moscow artist Aleksei Belyaev-Gintovt (b. 1965), who over the course of many years has been working consistently to resurrect the Soviet "imperial sublime," which is directly linked, both aesthetically and ideologically, to the mythology surrounding aviation and space exploration. In the early post-Soviet period, the heroism of aviation and space exploration in the Soviet era lent itself to deconstruction as material for the study of totalitarian propaganda, or it was transformed into a commodity—one of the most marketable elements of "Soviet retro" (Goralik 2007).[1] At the beginning of the 1990s space became an icon within youth club sub-culture.[2] In Russian culture today, however, an opposing tendency has emerged, namely nostalgia for the heroic 'common cause,' for the lost synthesis of the individual and society. As a result, the notion of space as a 'national idea' finds support among conservative intellectuals, writers, and artists. In the mythology of contemporary Russian conservatism, a new imperial utopia is directly related to the philosophy of Russian and Soviet Cosmism. The best-known advocates for the Soviet modernization project today are Aleksandr Prokhanov (b. 1938), the writer and political figure Maksim Kalashnikov (b. 1960), and the actor and Orthodox priest Ivan Okhlobystin (b. 1966), who in his provocative manifesto 'Doctrine 77' (Doktrina 77: 2001) singled out space as the foundation of Russian identity. The increasing interest in the theme of space in Russia during recent years is not coincidental; it fits in with the government's rhetoric concerning the development of advanced technologies, which is evident, for example, in the extensive media coverage of the Mars-500 project and in a string of events marking the fiftieth anniversary of the Soviet space program.[3] Even remnants of the Communist Party are advocating the symbolic inheritance of that program. Consider, for example, the popularity in the Russian blogosphere of the congratulatory message posted on national Cosmonauts' Day, which fell on the anniversary of the founding of the organization Communists of Petersburg and the Leningrad Region:

> There will come a time when the first delegation of interplanetary aliens, who were discovered as a result of Gagarin's distant flight in 1961, will bow their quaint green heads and place wreaths at Lenin's tomb.
>
> (Kommunisty Peterburga i Leningradskoi oblasti 2007)[4]

One of the most interesting and well-thought-out interpretations of the legacy of the philosophy of Cosmism and of the Soviet sublime in contemporary Russian art is found in the works of Belyaev-Gintovt, who combines a countercultural understanding of the Soviet legacy with a direct expression of it in the style of the above-cited passage by the communists.

The contextualized analysis of Belyaev-Gintovt's works that I am proposing begins with an interrogation of the history of Russian art that focuses, in part, on the research of Boris Groys (1993, 2003), and, first and foremost, on his scholarly works dealing with bio-political utopias in new Russian art (2006, 2007). I also engage the work of Katerina Clark (2009, 2011), as well as that of Alexander Yurchak dedicated to the language of the late-Soviet underground (2005, 2009).

Art of the second modernism

One of the defining features of Russian art in the 2000s is an excessive emphasis on the Soviet period in Russian history and a new take on the recent Soviet past, especially on the events of the 1930s–50s as a heroic era, forever and irrevocably cut off from the contemporary world.[5] This new 'epic' point of view on the events of recent Soviet history is expressed, on the one hand, in the rejection of a politically engaged, personal approach to this tragic period, and on the other, in the mythologization of the Soviet era, which is increasingly represented as a radical alternative to contemporary Russian society—the latter characterized by the total dominance of popular culture and of corporate and state capital. The state of Russian art today is often described as "a new patriotism," "a second modernism," "a new conservatism," and "progressive nostalgia" (Miziano 2007). As Harry Lehmann puts it in the article "The Art of Reflexive Modernism" [Iskusstvo refleksivnoi modernosti 2010],

> the search for an understanding of the second modernism assumes not a reexamation of the concept of postmodernism, but its radicalization to a certain extent, and, specifically, the linking of tradition and modernity, as well as a return to a system of expression with its own language: columns, pediments, and facades in architecture, figurativeness in art, melody, rhythm, and harmony in new music.
>
> (Lehmann 2010: 99)

Lehmann also emphasizes the fact that this second modernism has overcome the irony and distance of the first, which is also characteristic of the postmodern, underscoring the fact that in a society or culture of reflexive modernism, "tradition and modernity are linked without a hint of irony, and for this reason one can look at the achievements of traditional and of modern art with equal impartiality, giving free rein to both one's own understanding and the new seriousness" (2010: 99).

Although the theme of empire, as well as the sacralization and the aestheticization of politics, has been central to the artists working within the style of the new modernism, it should be noted that the discourse on empire in contemporary

Russia is not just about empire as such. It is as much about *imperium*, or the sacred nature of power.[6] In many of the critical works dedicated to the analysis of art of the second modernism, one encounters the following concepts, which are often used as synonyms: "reactionary avant-gardism," "right-wing postmodernism," "aesthetic postmodernism, "punk conservatism," and "the aesthetic avant-garde" (Popov 2004; Bobrikov 2004). I refer to this art as *imperial pop-art* or *imper-art.* Representatives of imper-art in contemporary Russia include Belyaev-Gintovt, Gleb Kosorukov (b. 1970), Andrei Molodkin (b. 1966), Dmitrii Mishenin (b. 1972), Mikhail Zaikin (b. 1956), Mikhail Rozanov (b. 1973), the architectural group Iofan's Children, as well as a host of other individual artists and groups. They all represent imper-art, in which aesthetic delight in the power of utopia is combined with a severe and cold sense of estrangement.[7] In its relationship to power, imper-art is close to traditional ancient cultures in which power holds a unique ontological status, and the necessary condition for the assumption of power is the presence of will, desire, the ability to take action, strength, and energy. This is why imper-art, like sots-art, admires its object—hierarchical totalitarian systems. Here we see all the fundamental devices of a postmodern poetics (estrangement, double readings, the combination of the marginal and the magisterial, etc.), except the axiology has now shifted. Unlike works of politically engaged totalitarian art or of the no less politically engaged works of their critics representing pop and sots-art, imper-art is preternaturally calm, surveying Russia's past and future from a great height, with an almost otherworldly point of view.

The second important feature of art of the second modernism, according to Lehmann, is its desire for an active role in society and for ideas, or "richness of content" (Lehmann 2010, 104–8). These characteristics also apply to the works of Belyaev-Gintovt, which advocate a new seriousness and neoclassicism, and promote an active role for the artist in society (Belyaev-Gintovt 2009). From the beginning of the 2000s, Belyaev-Gintovt has been strongly attracted to the socio-political movement known as Eurasianism and currently serves as the official artist of the Eurasian Youth Movement. We should note that post-Soviet Eurasianism is not so much a political as it is a metapolitical cultural project; it is a utopian vision of government, where power is located in the hands of free, risk-taking poets, philosophers, musicians, and artists. The founder and chief ideologist of this movement is the well-known philosopher and public figure Aleksandr Gel'evich Dugin (b. 1962). The movement was born out of Moscow traditionalism (Iurii Mamleev's Iuzhinskii Circle) and has preserved its countercultural imprint to the present day.[8] The goal of Eurasianism is the creation of a Eurasianist empire; that is, an alternative to the American project of cultural globalization. It is indicative that Dugin characterizes his movement as precisely "right-wing" or "radical" postmodernism (Dugin 2007: 66–71), though his critics unapologetically refer to him as a fascist.

Interest in Belyaev-Gintovt peaked in December 2008, when, amid considerable controversy,[9] the artist received the Kandinsky Prize, Russia's most prestigious non-governmental award in the field of contemporary art. His *Daughterland*

[Rodina-Doch'] won Project of the Year. Belyaev-Gintovt began to write primarily about his political views (Degot' 2008), but none of the critics noticed the connection of these writings to the *Daughterland* cycle or to his earlier projects involving the intellectual legacy of Russian and Soviet Cosmism and the archeo-modernism that characterized that philosophical movement, as well as Soviet aviation.[10] This is all the more surprising given that Belyaev-Gintovt's first famous work, a collaboration with Kiril Preobrazhenskii (b. 1970) titled 'U-87' (1993), first exhibited in the Ridzhina gallery, was a homage to the fighter pilot Joseph Beuys (1921–86), featuring a life-size representation of his airplane held together by *valenki*, or 'leg wrappings' (Fig. 5.1). (This plane was nicknamed *lapotnik* or *lapot'*, meaning 'bast shoe maker' or 'bast shoe,' respectively.[11])

In the late 1990s the artist began working under the traditionalist slogan "Time, backward!" and abandoned Beuys for Soviet modernism in the form of the levitating athletes of Aleksandr Deineka (1899–1969),[12] preserving in the process both the monumentality of his installations and canvases and the traditionalism of his choice of materials.[13] Images of military, aviation, and space technology are present in virtually all projects by Belyaev-Gintovt, always combined with archaisms in the form of plowed earth, felt strips, or a hammer, and always accompanied by gold and neoclassical architecture. The artist found models for this archeo-modernism primarily in early Soviet cinema, such as Alexander Dovzhenko's (1894–1956) film *Aerocity* [Aerograd 1935], in which modernization—building the city of the future—relied on chtonic powers; that is, on forest partisans and

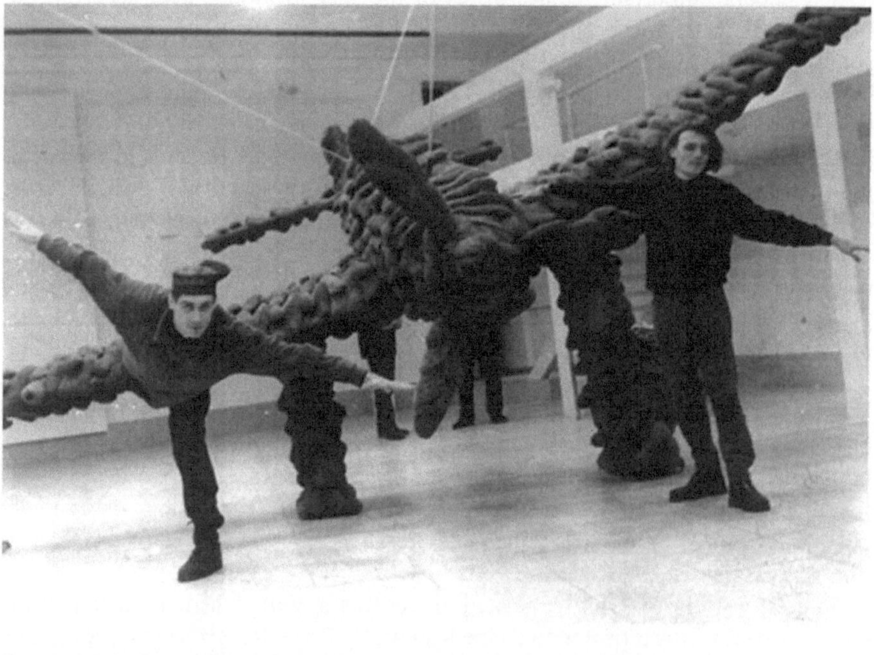

Figure 5.1 *U-87*. A. Belyaev-Gintovt and K. Preobrazhenskii (1993)

uneducated peasants in flap-eared caps and leg wrappings. The contradictory combination of archaic, folkloric traditions with the latest technological achievements that Belyaev-Gintovt sought in his work was a distinctive feature of Soviet culture of the 1930s. Consider those works of faux folklore on the theme of aviation and the presentation of pilots as epic heroes, as "bright falcons."[14] Remember that the synthesis of mythology, the latest technology, and monumentalism characterized Russian aviation from the moment of its birth. Symbolic names were used in Russian aeronautics, names taken from the heroic folktale tradition of the *bylina* [epic]. Some of the first Russian pre-revolutionary "giant planes" [samolety-giganty] were I.I. Sikorskii's (1889–1972) *Russian Warrior* (1913) and *Il'ia Muromets* (1913), and V.A. Slesarev's (1884–1921) bomber, *Sviatogor*, which, incidentally, was never flown.[15]

The resurrection of the utopian imagination and the concept of the political sacred in the twenty-first century have been interpreted by scholars as evidence of a retreat of rationalism and of a return to the dominance of emotionalism, of sensory perception, and of the demand for new forms of communalism capable of being transmitted to the masses through physical and visual practices (see Filippov 2008). Neoeurasianism and imper-art appeal precisely to these recent tendencies, linking the political with the emotional, demonstrating the magic of power and order, covering their ideocratic models in alluring artistic images understood by each and every viewer. In the post-Soviet context, aviation and Space are the most recognizable cultural metaphors, representing the synthesis of individual heroism and collective effort. The monumental canvases of Belyaev-Gintovt address the sensitive citizen bored by the modern world and tired of theories. In this, the artist follows the traditions of the Russian avant-garde, in particular the work of Nataliia Goncharova, who attempted to define the clear, figurative quality of radical art in the following way:

> If religious art and art glorifying the state were always marked by the same majesty, by the same perfect expression of the creative work of the individual, this is because what was considered art never committed the sin of theorizing. The artist firmly understood what he was representing and why he was representing it: thanks to this, the artist's intention was always clear and defined; all that remained to accomplish the work was to find a correspondingly clear and defined form.
>
> (qtd. in Livshits 1989: 363)[16]

Empire and counterculture

The conservative turnaround in Russian art did not begin in the 2000s when critics started talking and writing about the relevance of conservatism, of a new communality, and of empire. It started much earlier, at the beginning of the 1980s in the countercultural sphere; that is, in the underground and marginal intellectual circles of Moscow and Leningrad. At first, hostility to official Soviet culture and to the market liberalism that was then emerging prompted interest among avant-garde artists and musicians in the conservative tradition and in the imperial aesthetic. The allure of glamour and power in post-Soviet kitsch necessitated that something stronger and more captivating be opposed to it; that something become the imperial idea. While in Moscow the idea of empire as the most seductive form of power came from traditionalism, from the principle of the Absolute and the Authoritative (Iurii Mamleev's circle), in St. Petersburg the city itself inspired imperial reveries. Its architecture, palaces, and myths were directly tied to traditions of dandyism and cold imperial style, as in the Neo-academic School of Timur Novikov.[17] The genetic turn toward imper-art represents the actualization of the Petersburg text of Russian culture and can be seen, following Vladimir Toporov, as the "summit of Apollonianism," which since the 1700s has marked Russian culture at the turn of every century (Toporov 2003; Goscilo and Norris 2008). The Moscow traditionalists and the Leningrad-Petersburg Neo-academicians turned to mythology, history, and canonical works of art, the hierarchical structure of which served as a model of the hierarchical structure of the sacred worldview (Lotman 1992). Belyaev-Gintovt participated and continues to participate in these circles, functioning as a link between the Moscow and Petersburg countercultures.

Belyaev-Gintovt first presented his version of the capital of a sacred Eurasian empire in the early 2000s in the project *Novonovosibirsk* [Newnovosibirsk], in collaboration with the artists Andrei Molodkin and Gleb Kosorukov. Together they formed the group F.S.B., or Front Spokoinogo Blagodeistviia [The Front for Peaceful Prosperity].[18] The exhibit ran from 2001–3 in the State Russian Museum and in the New Academy in St. Petersburg, and in Moscow, in the State Architectural Museum Named for A.V. Shchusev. *Novonovosibirsk* includes seven canvases painted by Belyaev-Gintovt and Molodkin, and twelve photographs by Kosorukov. This art project had a profound effect on the artistic community and today it remains the most famous and recognizable cycle of works by these artists. On enormous 8 x 14 foot canvases they portrayed neoclassical sculptures for the new capital of the Eurasian empire, Novonovosibirsk. The city was moved closer to the North Pole, the 'geometric center of Eurasia.' This was an unpopulated and frigid utopia, a city for people of spirit, hyper-warriors, and conquerors of the Arctic. According to the design of these visionary city planners, the capital of this agrarian-cosmic state was organized around twenty-four gigantic monuments, in which functionality does not conflict with aesthetics insofar as Novonovosibirsk was intended to embody the Vitruvian principles of utility, endurance, and beauty. The exhibition consisted of several projects: *Apollo in Power* [Apollon v silakh], a representation of a 328-foot-tall titanium statue of the Apollo Belvedere, with a rocket launcher in his quiver; the cosmodrome *Shambala* in the form of an enormous swan; the *Golden Colossus* [Zolotoi Koloss], a Mir rocket launcher in the shape of a gigantic ear of wheat [kolos]; and a submarine named *Atomic Orthodoxy* [Atomnoe pravoslavie] (Fig. 5.2).

The frightening militarization of the capital of the new Hyperborea is a sham; these are only "weapons of peace" and "climatic weapons." As Marina Oblacheva notes, based on the artists' own comments, the weapon on Apollo's back is designed exclusively for regulating nature:

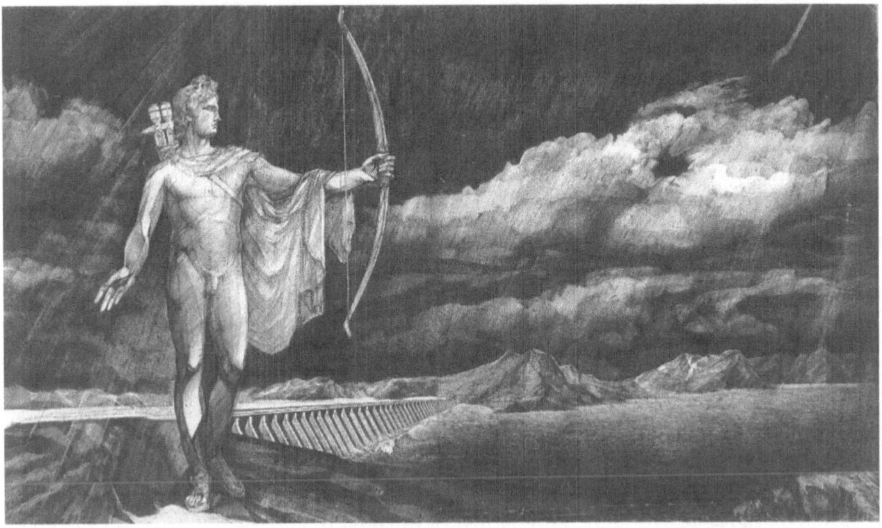

Figure 5.2 Arrows of Apollo, the 'Novonovosibirsk' project. A. Belyaev-Gintovt and A. Molodkin (1999–2002). Ballpoint pen on canvas

> The rockets launched from the "quiver" of the Greek god have a profoundly peaceful purpose—heavy artillery and geodesic spheres allow residents to regulate the meteorological conditions in their megapolis. In general, the ability to influence the climate and the environment as a whole is a pre-requisite for the flourishing of an agrarian-cosmic state, the capital of which must be Novonovosibirsk.
>
> (Oblacheva 2001)[19]

And so, in this project, as in a series of others that will be discussed below, Belyaev-Gintovt visualizes the postulates of Russian Cosmism and the ideas of Nikolai Fedorovich Fedorov (1828–1903) on the regulation of the climate as one aspect of the "common cause."

At the same time, the panoramic vision—the combination of the verticality of enormous statues and the expanse of endless deserts of ice—correctly directs the viewer to images of the "imperial sublime" of the Stalinist period (Clark 2009; Clark 2011: 276–306). Clark emphasizes the role of verticality, height, and spatial expanses in creating the effect of the sublime:

> Thus no matter how immobile its elements may be, a sublime scene is fundamentally dynamic. It is all about affect and elicits an emotional response. The Sublime is a site of drama, derived in part from the very dimensions of the phenomena presented. The term itself, 'sublime,' is in Greek *hypsos*, height and a dramatic verticality—towering cliffs, craggy towers, giant trees—is crucial to the notion of the sublime. . . . But along with a marked verticality

> goes an extreme horizontality, either implicitly or explicitly an affect of this verticality. The heights afford a panoramic vision stretching almost endlessly. Indeed, all the dimensions are hypostasized, including depth (deep caves, steep ravines falling away).
>
> (Clark 2009)

The deathly, icy color of the paintings does not so much bear witness to the lifelessness of the space of this new utopia as it symbolizes the sphere of the extreme, the sphere of pure possibilities (the North, space, night, death). In its colors and thematics, *Novonovosibirsk* references Nicholas Roerich's (1874–1947) pictorial cycle *The Himalayas* (cf. the cosmodrome is called Shambala), as well as the famous painting by Tair Salakhov (b. 1928) *To You, Mankind!* [Tebe, chelovechestvo! 1961]. In other words, the artist consciously inscribes himself within a tradition of Cosmism in Russian and Soviet art (Figs. 5.3 and 5.4).

The sphere of the Absolute is incarnated through the use of Vis ballpoint pens, which reflects the fundamental principle of the Neo-academicians: an orientation toward handicrafts and painstaking manual labor. On the other hand, such a clear disconnect between the scale and pathos of the design and the techniques chosen to execute it underscores the deliberate totalitarianism of the project, testifying to the authors' observation of the canons of "the new seriousness" [novaia ser'eznost'] (Belyaev-Gintovt 2009). The dark-blue, even strokes of the ballpoint pens can be interpreted as the 'trace' of the little man, as either the automatization of the heartsick individual painstakingly rendering his obsessive images or the dream of the boy Kai sketching the magical world of the Snow Queen during a boring lesson.[20] While the name of the new capital itself—Novonovosibirk—is an echo of the neoclassical New Academy movement, it also refers to the New Vasiuki from *The Twelve Chairs* [Dvenadtsat' stul'ev, 1928] by Ilf and Petrov. One can assume that in the *Novonovosibirsk* project, where the frightening otherworldliness of the images combines with jesting and playfulness, Belyaev-Gintovt is answering the demand for a postmodern variant of the imperial sublime.

It should be noted that two representational principles distinguish the imper-art of the Front for Peaceful Prosperity: a play with perspective and scale. This is not

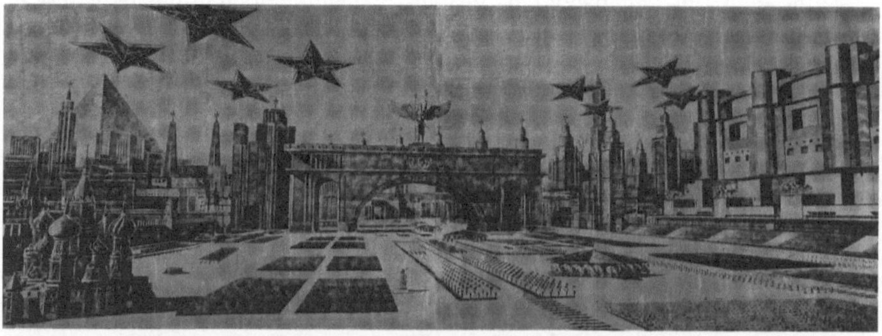

Figure 5.3 Cosmoparade, or the Victory Parade of 2937. A. Belyaev-Gintovt (2010). Gold leaf, red printing ink, handprint on canvas

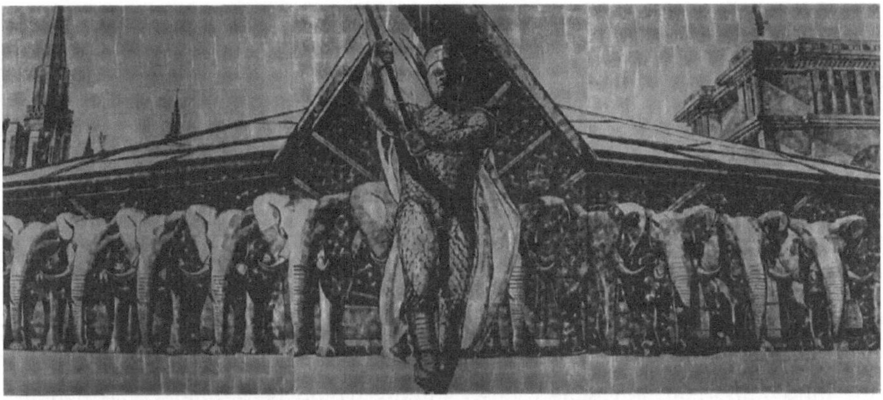

Figure 5.4 *Cosmoparade, or the Victory Parade of 2937*. A. Belyaev-Gintovt (2010). Gold leaf, red printing ink, handprint on canvas

a total return to classicism, but rather mimicry, an eclectic take on this aesthetic through the mediation of totalitarian art on the one hand, and Russian Symbolism on the other. It is also an attempt to revive the Dionysian-Apollonian discourse that unites the two tendencies. Imper-art rejects the horizontal, amorphous, and style-less Moscow Conceptualism and emphasizes the frightening, cold, and superhuman aspects of classicism as the art of Apollo. In the words of Ekaterina Bobrinskaia,

> In this cold style the total illumination, the logical legibility of art disappears. Paradoxically, behind this clear, precise, and simple form that presents no interpretive contrivances, one observes something impenetrable, something that appeals to another dimension in art. The viewer, left alone with this cruel and cold form, must breathe its rarefied air. He experiences an irritating sense of devastation and discomfort. There is no tremor of life in the images of the new classicists. Rather, this form invokes images of death or, more precisely, images of a boundary. But perhaps by means of this coldness, which is so foreign to all the conventions of 'the spirit of modernism,' something forgotten might return or something as yet unknown might enter contemporary culture.
> (Bobrinskaia 2004)[21]

This "forgotten" and unrecognized "spirit of modernism" is the "sublime"; it is the *noumenal*; that is, the mystical tremor that unites fear and joy in the presence of the Absolute.[22] Ice, snow-covered expanses, technology, aggressive masculinity and heroism—these are all concrete manifestations of the noumenal, as relevant for the cultures of the 1930s as it was for the counterculture of the 1980s. Insofar as the sublime in European culture is linked to the masculine principle and the promise of adventure (Clark 2009), it is no surprise that pilots and flight technicians are associated with the category of the sublime. It should be emphasized here that many of the functions and qualities of the Absolute were projected onto aviation as the epitome of modernity and as the most sacred technology of that era (see Chapter 2). And so, the weight and monumentality of Russian/Soviet aviation technology were called upon not so much to strike down the enemy as to fulfill the symbolic functions of divinity—to frighten, to sap the will and fighting spirit of the enemy, inspiring in him both delight and tremors (Palmer 2006: 204–6). These military pilots were the cultural heroes of the period from the 1910s through the 1940s in Russia and the Soviet Union, as well as in Europe and America (the Red Baron Manfred von Richtoven [1892–1919], Charles Lindberg [1902–74], the Normandie-Nieman Squadron, and so on).

On the other hand, a commanding view of the world from a great height, which characterizes the work of Belyaev-Gintovt, distinguishes him from the hero of unheroic times—the dandy. If we were to examine the philosophical and literary passions of the Moscow traditionalists or the Petersburg Neo-academicians, the most important and widely cited authors would be aesthetes and dandies, loners and critics of the snobbery and zeal of the European bourgeoisie, kings, adventures, and soldiers: Ludwig II of Bavaria (1845–86), Konstantin Leont'ev (1831–1891), Oscar Wilde (1854–1900), Nikolai Gumilev (1886–1921), Ernst Jünger (1895–1998), and Julius Evola (1898–1974). Dandyism represents the contradictory union of political engagement and aloofness, self-love and self-parody; it is an image of the disappearance of triumphant masculinity.[23] The dandy stresses his love for the artificial, his dubious relationship to women and to 'warm' nature; he is self-sufficient, vain, and solitary. Longing for the verticality of the Absolute is expressed in imper-art in nostalgic images of the active male principle, a fascination with toned masculine bodies destined for death, with the army, uniforms, and parades (Engström 2012). Portraits of Soviet pilots, soldiers, and sailors occupy a central place in the work of Georgii Gur'ianov, one of the leaders of the Neo-academic movement and one of the most famous Petersburg dandies of the late 1980s and 90s.[24]

Coldness, orderliness, objectivity, and an amoral aestheticism are key concepts in both dandyism and imper-art. Aleksandr Dugin singles out distance as the defining behavioral trait of a hero:

> The hero, by definition, must be 'cold.' If he does not separate himself from those surrounding him, if he does not freeze the warm energies of everyday humanity, he will not rise to the level of achieving the Impossible, that is, the level at which a hero becomes a hero. A hero must leave people behind. But beyond the boundary of social comforts rage the biting winds of objective reality, cruel and inhuman. Earth and stones rise up against the worlds of animals and plants. Aggressive vegetation devours minerals while wild beasts mercilessly stomp down stubborn grasses. Elements that exist outside of society do not know leniency. The world itself is a triumphal banquet of matter that is short and flat and merges with blocks of cosmic ice. The hero is cold because he is objective, because he takes from the world the baton of spontaneous power that is wild and malevolent.
>
> (Dugin 2005: 75)[25]

In this passage, Dugin naturalizes the category of the sublime and brings together the romantic hero who has been lifted above the masses and beyond the insurmountable boundary of the possible with "cosmic ice" and the laws of malevolent nature. Characteristic for both Eurasianists and Neo-academicians, as elite and closed countercultural communities that stand in opposition to the gray Soviet masses and to the new global order, is a Manichaean worldview and an identification with the principles of heroism and sacrifice.[26]

Cosmism and victory over death

In addition to aviation and early Soviet cinematography, other important sources for Belyaev-Gintovt's archaeo-modern imagery include Nikolai Fedorov's philosophy of supramoralism, which also combines a bold futurology with imperial archaism, as well as the ideas of the biocosmic-anarchists, a largely forgotten group of poets and ideologues of immortality in the 1920s (Hagemeister 1997; Siddiqi 2008). Of course, the conquest of space was never a goal in itself for either Fedorov or Konstantin Tsiolkovskii. Instead, it was imagined as the conquest of the environment for the resettlement of our resurrected fathers and mothers; in other words, conquest of space served to eliminate time. The article "Supramoralism, or Universal Synthesis" [Supramoralizm, ili vseobshchii sintez 1902], contained in volume one of Fedorov's *Philosophy of the Common Cause* [Filosofiia obshchego dela], is an action plan. Here Fedorov attempts to present his teachings in a systematic way, with the help of twelve "paschal" questions. Each of these questions exposes the difference between moralism and supramoralism. Supramoralism is, for Fedorov, our "debt to our forefathers," and the "resurrection of our fathers" represents the highest and most natural morality. The "paschal" questions must display the absolute priority of the value of life over all other values and must designate the fundamental task of humanity, which is to build paradise on earth and in space:

> Supramoralism requires a paradise, a Kingdom of God, not one that is otherworldly—but one that is everywhere; it requires the total transformation of earthly reality, and the transformation of everything scattered throughout the heavens, to bring us closer to all that is unknown to us from the otherworld. This paradise, or Kingdom of God, is not within us; it is not only abstract, not just spiritual; it is visible, tangible, fully palpable through sensory organs that have been developed through psycho-physiological regulation (that is, the management of spiritual-physical phenomena), organs that have access not only to herbs and vegetation but to the movement of molecules and atoms of the whole universe, which makes possible resurrection and the total transformation of the whole universe.
>
> (Fedorov 1980: 500)[27]

Supramoralism assumes the absolute priority of labor and historical knowledge, as well as the elimination of the contradictions between "us" and "them," the global and the local, the center and the periphery, insofar as everyone needs to be resurrected—our own forefathers and those of “others” (although, according to Fedorov, there are no “others”).

A group of poet-biocosmists headed by A. Sviatogor saw immortality and interplanetarianism, as well as the struggle against localism in time and space, as the fundamental goal of the October Revolution, although they never acknowledged Fedorov’s work as a source of their ideas:

> Along with the problem of immortality, biocosmism introduces the problem of interplanetarianism. If the first root of evil in the life of individuals and societies lies in death (localization in time), then the second root of this evil is localization in space, that is, the primacy of the domestic hearth, one’s native city, one’s nation and state, and one’s race. Even internationalism, in the final analysis, is just localization in the universe.
>
> (Sviatogor 1922: 8)

Overcoming “localism” in all its manifestations by equating one’s own fathers with those of others in the space of the past and the future is a distinguishing feature of Belyaev-Gintovt’s project *Patria Filia. Rodina-Doch’* [Patria Filia. Daughterland].[28] In the classical interior of Triumph, the gallery where Belyaev-Gintovt’s exhibit was held, one could see, side by side, aflame in black and gold, Tamerlane’s mausoleum in Samarkand, the Hall of Prayer for Good Harvest in the Temple of Heaven in Peking, the Colosseum, the Tsar’s Tower in the Kremlin, the Doge’s Palace in Venice, a Soviet fighter jet, the Parthenon, and the Palace of the Soviets, crowned with a statue of Apollo. All of these iconic and timeless images of the future-past reflect, as I see it, not so much an aggressive neo-imperialist agenda as they do Fedorov’s supramoralism and the overcoming of localism in the ideology of the bio-cosmists. The title of the project is interesting in this regard. Belyaev-Gintovt, as the party visionary for the Neoeurasianist youth movement, represents the future Eurasian Russia-3 in the form of a beautiful young Valkyrie who is ready for battle, armed with a hybrid golden axe and a Kalashnikov automatic rifle. The artist develops the tradition of referring to Russia as a mother or a maiden, but alongside the image of the homeland as mother or of Russia as a young woman or a new bride, which was characteristic of the Symbolists,[29] Belyaev-Gintovt introduces a new image of the homeland as daughter, an image that has yet to take root in official government rhetoric. In my view, Belyaev-Gintovt with his image of the homeland as daughter is developing the geopolitical and ethical potential of Fedorov’s teaching or doctrine, viewing Russia as the daughter of the ancient world empires, such as Rome, China, and the USSR—a daughter who must carry out her supra-moral duty to sew together the fragments of the imperial bodies resurrected by the artist into a single canvas. It should be pointed out here that Fedorov emphasized the gendered aspects of the various stages of resurrection, leaving daughters the task of ‘embodying’ fathers and mothers, lending a somewhat occult and futuristic take on the female tradition of weaving:

> Death, one might say, is a form of anesthesia, during which the most complete corrosion, decomposition, and dispersion of substances takes place. Recombining these scattered particles is a question for cosmo-telluric science and art, and consequently the male task of assembling the collected parts is a physiological and histological problem, while the problem of weaving together, as they say, the fabrics of the human body, the body of our fathers and mothers, is a female task.
>
> (Fedorov 1982: 419)[30]

One of the most important aspects of Fedorov's utopia present in the works of Belyaev-Gintovts is the direct connection between the resurrection of fathers and the idea of empire. For example, in the article "Autocracy" [Samoderzhavie 1906] Fedorov writes of the necessity of a strong centralized government for "the common good" and links the image of the supreme leader to the archaic image of the priestly king:

> Autocracy in its primary sense is dictatorship incited by danger, not from people like us but from the blind power of death, which threatens everyone without exception. Autocracy . . . came about immediately after the death of the first father who had joined everyone together in a single will, a single desire, inspired by loss, by death.
>
> (Fedorov 1906: 375)[31]

It should be noted that the motif of resuscitated empires as a common cause for both sons and daughters is clearly delineated in the mission statement of the Eurasian Union of Youth:

> Our goal is a Eurasian empire. . . . We will have betrayed the heroes of the ancient world, the great empire builders, if we fail to construct something somewhat grander than they did. And only when we have extended our borders from ocean to ocean will the eyes of our dead take on a calm, transparent hue: They—our offspring—did that! Otherwise, the dead will never leave us in peace.
>
> (Manifest ESM)[32]

On the other hand, Belyaev-Gintovt's images of a Eurasian empire of the future always reference the Red Soviet empire, and in his adoption of revolution, the artist is closer to the anarchy of the bio-cosmists than to Fedorov's monarchism.

Cosmoparade, or the Victory Parade of 2937

The connection between the idea of empire and the victory over death and time is especially evident in Belyaev-Gintovt's most recent exhibition, *Cosmoparade, or the Victory Parade of 2937* [Kosmoparad, ili Parad Pobedy 2937], held at the

Triumph Gallery in June 2010. The Cosmoparade of 2937 takes place on Red Square and is a procession marking the apotheosis of the Eurasian empire's resuscitated and unified forces, as well as the victory of the principle of coexistence over that of supersession. The very name of the capital of the Eurasian empire of the future—New Moscow [Novaia Moskva]—is a reference to the 1938 film of that name by Aleksandr Medvedkin (1900–89). One especially notable episode in the film involves the main character's making a living mock-up of the new Soviet capital as it would look according to Stalin's plan for the city's reconstruction. The new Moscow of 2937, in Belyaev-Gintovt's version, has undergone a general reconstruction under the sign of supramoralism and the regulation of nature. Crossing Red Square, shoulder-to-shoulder, are people and zoomorphs in armor made of scales, military elephants, and red flying starships.

Victory Parade is a luxurious and fantastic manifestation of Neo-Sovietism, of the victory of collectivization and the common cause, purified of those banal routine practices that eat away at the beauty of the project. Appearing before the viewer are familiar but at the same time altered images of Moscow, such as pyramids and skyscrapers. Though the Kremlin, Lenin's Mausoleum, and St. Basil's Cathedral make the Red Square recognizable, the rest of the Eurasian metropolis has been rebuilt. The center of the city is occupied by the Ministry of Space, the Ministry of Truth, the Ministry of the Struggle against Chaos, and the Ministry of Love. *Victory Parade 2937* represents Eurasian civilization as an alternative form of globalization. The artist imagines a space that has conquered 'aesthetic pollution'—a space that is free from advertising and the forces of the capitalist market, which is represented by the Cosmodrome in place of the department store GUM and by the Statue of Responsibility, an alternative to New York's Statue of Liberty, which is framed by the arch leading onto Red Square (Fig. 5.5).

The artist portrays the parade in Red Square from two different perspectives: either from the deck of a spaceship (the view downward) or from the viewpoint of a little person dispirited by the height of the statues, the size of the aircraft, and the orderliness of the ranks of soldiers (the view upward). One might conjecture that the main objective in these shifts in foreshortening and point of view is to allow the viewer to visualize the desired unity of the public and the private realms. The viewer is shocked by the vertical and horizontal perspectives; the city expands, and there is an illusion of infinity, for in the new Moscow of 2937 the severe orderliness of forms is linked to the total obliteration of boundaries: the parade combines time, nature, technology, history, and mythology. This mystery play unfolds against a golden backdrop that serves, as it did in both the Roman and Byzantine-Russian traditions, as a symbol of sacred space (Averintsev 1973). Belyaev-Gintovt has acknowledged that the model for this series of paintings was the athletes' parade of 1937, dedicated to the twentieth anniversary of the Bolshevik Revolution and the adoption of the Constitution of the USSR:

> Last year he watched an old film clip of the athletes' parade of 1937 on Red Square. The great choreographer Igor 'Moiseev (1906–2007) had his debut as a producer. "This is an amazing clip!" the artist exclaimed. "It looks majestic,

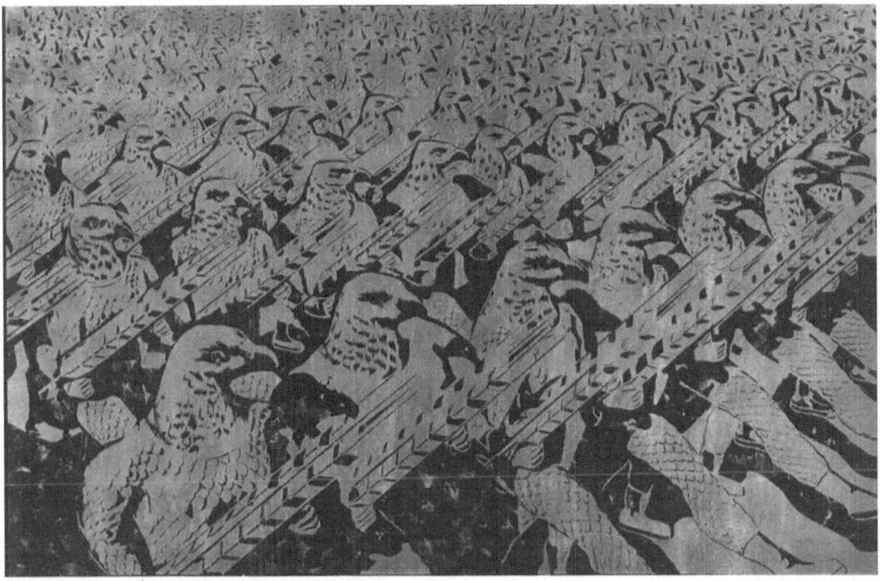

Figure 5.5 Cosmoparade, or the Victory Parade of 2937. A. Belyaev-Gintovt (2010). Gold leaf, red printing ink, handprint on canvas

> triumphant. In general, I've translated this liberating energy of a united people, this great passionate outburst of happiness, into the future. Of course, that was a tragic time, such a tragic time. Such tragic tension, but at the same time I have yet to see before or since a similar outburst of happiness, an outburst of the new man, and an eruption into Space," asserted Aleksei Belyaev-Gintovt.
> (Karpekina 2010)[33]

It is no coincidence that the 1937 parade of athletes, as it was reworked in the Cosmoparade project, teems with air and space craft. Red Square was a place to exhibit not only a new physicality but also new technology, including the latest aircraft. In the massive parades of athletes in the 1930s, which were produced by the great ballet choreographers Moiseev and Kasian Goleizovskii (1892–1970), the marching athletes were always accompanied by dirigibles or hot air balloons. Pilots and parachutists were always in the first rows of marchers, who totaled around 45,000, representing every type of sport at the 1937 parade. This was a triumphant time for Soviet aviation: in that same year on Red Square, aircraft crews led by V.P. Chkalov (1904–38) and M.M. Gromov (1899–1985) celebrated these men, who were the first in history to fly from Europe to America across the North Pole. But even earlier the paraded athletes were accompanied by displays of the achievements of Soviet aviation. For example, in 1929, in preparation for the May festivities on Lobnoe Mesto, the metal ANT-9 airplane, the first Soviet passenger plane, built by A.N. Tupolev (1888–1972), was displayed. The enormous plane ANT-20, named *Maksim Gor'kii*, headed by an entire squadron, opened the 1 May festivities in 1934.

Belyaev-Gintovt links the historical context with the mythical and sacred. In *Victory Parade of 2937*, the artist reproduces not only Soviet athletic and military parades but also their historical precedent—the ancient Roman victory celebration. The latter marked Roman triumphs in major wars and was a ceremony both military and religious—a kind of mystery play in which the people thanked the gods for their victory. As close as it could be to a religious ritual, the spectacle involved the observation of sacrificial rites and oblations.[34] This archaic sacred is present in Belyaev-Gintovt's *Victory Parade*, first and foremost in the archeo-futuristic images of resurrected gods, priests, and warriors from the most highly regulated empires and *katechonic* kingdoms: Egypt, Greater Mongolia, India, Rome, Byzantium, and the Soviet Union.[35] The Soviet parades of athletes were directly associated with the cult of youth, health, beauty, and heroism that was important to all European countries in the interwar period (Keys 2003). In the USSR, however, this youth and beauty were presented to the mausoleum, the party elders, their forefathers, still living or long dead, which intensified the archaic subtext of the victory parade as the prelude to a sacrifice. This is how the aging Gor'kii, fantasizing about a revitalized proletariat, wrote about the parade of athletes on 1 June 1935:

> Every year our parades of athletes become more high-spirited, well-defined, and rich. The step of our youth is more confident and strong, and their eyes gleam more brightly with the happiness of living in a country where the body is so quickly and beautifully developed; and within that body burns ever stronger and more victorious a heroic spirit, which manifests itself daily in their work to enrich our people and to defend our motherland against enemies. Almost daily this spirit sparkles with daring feats for the good of the country. Seeing these thousands of young men and women marching in orderly rows toward our great future, you experience an emotion that could tear your heart in two. You feel sadness, too—that you have no place in the ranks of that mighty army, that you no longer have the strength to march alongside them—and then, when you've reached the mausoleum, to let out a heartfelt "Hurrah!"[36]
>
> (Gor'kii 1949: 96)[37]

The Kremlin and Lenin's mausoleum, often depicted by Belyaev-Gintovt in his early works, occupy a central place in *Victory Parade*. The artist, an architect by training, had this to say about his plan for the new Moscow of 2937:

> This is a gigantic chalice in the center of which stands the Moscow Kremlin. The further away from it, the taller the buildings become. If we look at the plans for this city of the future, we can see three circles with intersecting radii. At the points of intersection skyscrapers rise up. In the first circle there are 40 skyscrapers, in the second, 60, and there should be about 90 at the

> intersection with the third circle. Visible from any point, from any distance in the center of this majestic composition, is the Moscow Kremlin.
>
> (Karpekina 2010)[38]

The centrality of the Kremlin in this project, which is associated with the philosophy of Cosmism, is not coincidental. For Fedorov the Kremlin was a paschal site, the burial vault of our fathers, which, as was the case in traditional societies, stands at the very center of the "village":

> Easter ends only with prodigal sons, who have wiped the dust [prakh] from their feet, that is, when the village is transformed into a city, which is most visibly expressed by moving the remains [prakh] of the dead beyond the borders of the community; in the village, in a true village, not in an infected city, these remains [prakh] always occupy a central place.
>
> (Fedorov 1982: 493)[39]

Fedorov also directly parallels the Kremlin, resurrection, and the coronation of the tsar:

> Having disappeared in the West, . . . Easter remains in Russia, and especially in the Kremlin, before the graves of those gatherers or unifiers of sons for the task of resurrecting fathers, in the *Kremlin, that is, in a fortress meant to defend the ashes of our fathers and to serve as an instrument of defense that will turn into an instrument for reanimating those ashes* when fraternal hostility disappears. A coronation, that is, designating the one standing on the site of the fathers for the general task of resurrecting the fathers, would make no sense if it occurred, not in the Kremlin but in some other place.
>
> (Fedorov 1982: 494; emphasis added)[40]

In this way, the Soviet cult of the leader as a modified cult of the father was directly linked to Fedorov's supramoralism.

With a screenplay by Iurii Olesha, Abram Room's film *A Stern Young Man* [Strogii iunosha 1936]—popular among New Academy artists and a cult classic among lovers of early Soviet cinema—demonstrates the relevance of supramoralism for Soviet culture of the 1930s, as well as for artists of the second modernism (Sal'nikov 2003; Panchenko 2008).[41] One of the major themes of this film, which was banned in the Soviet period, involves the principle of top-down leadership and elitism, which lay at the foundation of Soviet culture and science. The protagonist of the film, Grisha Fokin, creates the *Third GTO Complex*, a moral code for Komsomol members, which includes the fundamental *moral* precept of victory over death. The text is read out loud in the film by the cosmonaut Olga, whose life has been saved by the doctor Iulian Nikolaevich Stepanov:

> There are no people who are the same and there can't be. This is bourgeois equality. The very idea of competition eliminates any notion of equality. Equality is stagnation, while competition is movement. Be equal to the best,

> help those who lag behind, and work to raise everyone up. The best among us are our leaders . . . The best are those who create ideas, knowledge, technology, music . . . These are great minds . . . Those who do battle with nature are *victors over death.*
>
> (emphasis added)[42]

Although Belyaev-Gintovt's project imagines the capital of a *Eurasian* empire of the future, in its details it has less to do with Dugin's Neoeurasianism than Prokhanov's Neocosmism.[43] In his 2002 novel *Mr. Geksogen* [Gospodin Geksogen], Prokhanov connects the mission of the Soviet project with that of all empires: the resurrection of the dead and victory over death:

> The "Red Idea" . . . consists of one thing and one thing only: overcoming death. The ancient Egyptians, who believed in the resurrection of Osiris, were "red." The pantheists of India, who believe in the incarnation of the soul and the illusory nature of life, were "red." Christ, who "trampled down death by death and bestowed life to those in the tombs," was also "red." Nikolai Fedorov, who preached resurrection from the dead and called upon mankind to unite and to resurrect their dead ancestors and settle them on the other planets of the galaxy with the help of Tsiolkovskii's rocket, was "red."
>
> (Prokhanov 2002: 309)[44]

Belyaev-Gintovt's new Moscow is aflame and aglow with the Red Idea, and at the head of this mystery play marches a resurrected Lenin, covered in scales (Fig. 5.6).[45]

'Material' evidence in support of my contention regarding the importance of Fedorov's views to Belyaev-Gintovt can be found in the artist's technique. In 2006 at the Iakut Gallery in Moscow he had an exhibition titled *Soil* [Pochva], where all the paintings were done using the unusual technique of hand stamps, when paint is applied to the canvas through a stencil with one's palms and fingertips. Since then, the artist has employed this technique exclusively—literally imprinting himself onto the canvas. In a recent interview with the *New Museum* [Novyi Muzei], he emphasized that, in addition to being "an attempt at self-enslavement" and the unification of the maximally individual with the maximally collective, this technique has another goal: to leave behind in the paint enough fragments of skin to facilitate his future resurrection (Novyi Muzei 2010). What the viewer sees here is not just the embodied metaphor of the artist who "lives in his art," but also a reflection of one of the axioms of Cosmism, the sphragistic principle, from the Greek work *sphragis*, meaning 'seal,' which has affinities with Russian Orthodox beliefs. According to the teachings of Saint Gregory of Nyssa (c. 335–after 394), the body, even after death, carries within it the stamp of the soul, which explains the Christian veneration of remains, icons, bones, and the grave sites of saints.

Cosmism and alchemy

As researchers have noted, in many of their objectives Cosmism and *sots-occultism* are close to the hermetic tradition (Glatzer-Rozenthal 1997). The proximity of

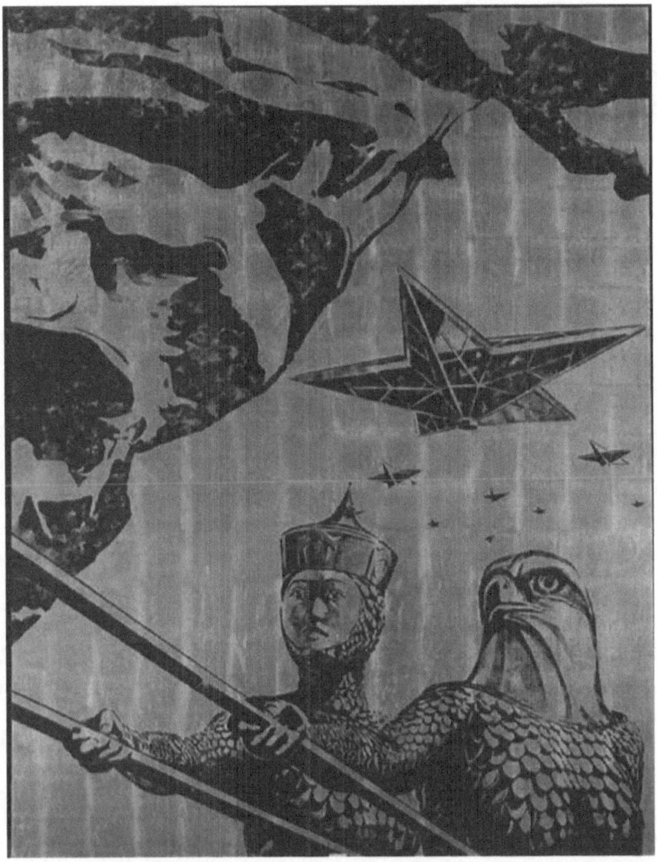

Figure 5.6 Cosmoparade, or the Victory Parade of 2937. A. Belyaev-Gintovt (2010). Gold leaf, red printing ink, handprint on canvas

alchemy and 'magical communism' consists in a connection between spirit and matter, an aspiration to mastery over matter, the transmutation of substances, immortality, and the resuscitation of the dead. However, unlike *esoteric* world traditions, Fedorov's Cosmism and Belyaev-Gintovt's Neocosmism are extremely *exoteric* insofar as theirs is a common cause from which the profane masses cannot be excluded simply because they do not belong to some chosen people or group of believers. The hermetic references in Belyaev-Gintovt's canvases have their source in the work of Evgenii Golovin (1938–2010), one of the most important figures in the Moscow metaphysical underground. From the 1960s to his death in October 2010, Golovin worked to acquaint the alternative public in both Moscow and Petersburg with European traditionalism and alchemy. Golovin wrote quite a bit about alchemy and the work of a number of poets and artists who dedicated themselves to the hermetic tradition, and all his writings have become required reading for New Eurasianists (Golovin 2003).

The exhibition *Victory Parade*, as well as some of Belyaev-Gintovt's earlier projects, demonstrates a genuine interest among contemporary artists in hermetically sealed modern utopias. The revolution associated with alchemy involves a world that is rotting, wallowing in evil; the revolutionary answer is imagined as a site of the fiery purification of matter. Belyaev-Gintovt's works are dominated by colors that reflect the various stages of the alchemical process: black (the nigredo stage) and white (the albedo stage), but also red and gold, which dominate the canvases at the *Victory Parade* exhibition.[46] Red and gold not only are the colors of empire, but also represent the rubedo stage. Rubedo, signifying 'red,' is the third and final stage in the alchemical process, the stage at which the adept achieves enlightenment and creates the philosopher's stone. Lead is a nigredo metal and silver is an albedo metal, while rubedo is represented by gold, which in alchemical terms signifies the seed of the sun, the true meaning of things. The world of rubedo is the ultimate goal of the alchemist and the revolutionary; it represents the transmutation of chaos in the nigredo stage into a golden order. Rubedo is also the ultimate tension, the boiling point, where all antagonistic substances come together to form a new alloy, a new quality that unites what cannot be united. The defining feature of rubedo is heat, and so, in the vessel where the process of transformation takes place, a great deal of pressure is generated. This stage is dangerous in that the adept may be unable to prevent overloading, in which case he may cause a new precipitation of nigredo, or a world of chaos and despair. For Belyaev-Gintovt, contemporary Russia is in a state of "complete nigredo," which arose after the explosion of the Soviet rubedo; it is a black world where the soul is dejected and miserable—a world that the artist must transform.

Another crucial stage of rubedo is its association with blood and with victory over death. In line with hermetic doctrine, a red tincture restores youth and vitality. In this context one is reminded first and foremost of the experiments of A. Bogdanov-Malinovskii (1873–1928), the author of *Red Star* and the director of the Institute of Blood Transfusion (Krementsov 2011). In his experiments related to the transfusion of blood from young to old revolutionaries, Bogdanov followed the traditions of European alchemy. For example, Marcello Ficino (1433–99), a humanist, philosopher, and astrologer of the early Renaissance, and the founder of the neoplatonic Florentine Academy, advised his friend to use the blood of the young to preserve the life of the old:

> Why shouldn't the old, who are bereft of any support, suck the blood of the young? I'm saying that those who are strong, healthy, gay, and prudent have the finest blood, and perhaps, a surplus. Let the old suck like a leech or a bloodsucker from an open vein on the left arm . . . during the waxing of the moon.
> (qtd. in Finkati 2011)[47]

In the context of hermeticism, it is no coincidence that Belyaev-Gintovt's enormous canvases are filled with images of flying saucers and rows of soldiers with the heads of birds, as well as levitating athletes. Carl Jung, who made use of the language of alchemy in his version of psychoanalysis, noted in *Psychology and*

Alchemy that images of dreaming accompanying the rubedo stage include birds soaring high in the sky, rockets, and spaceships, trampolines, fountains, fire and volcanoes, soldiers, horses, and giants and athletes (1997: 251–2). It is interesting in this regard that Belyaev-Gintovt has often mentioned in interviews regarding the exhibition *Victory Parade of 2937* that he portrayed nothing other than his dreams of Eurasia: "I'm only portraying my dreams. Dreams of an agro-cosmic civilization, which has achieved a state of harmony" (Karpekina 2010).[48]

Russian Cosmists share the alchemists' dream of the interpenetration of man and nature, nature and technology, and man and technology. The transfiguration of man's physical nature has become one of the main tasks of Cosmism. To that end, they have been developing, in addition to the theory of resurrecting fathers (*anastasy*), theories of autotrophy (Vernadskii 1993), longevity, transcendence of gender differences, transformation of mental thought into light energy, eugenics, and so on. According to Fedorov, technologization, which includes aviation, represents only a temporary stage in the development of science. Man must turn away from technological inventions to the transformation of his own organs, of his own body. Man must invest all his energy, not in artificial attachments to his organs, but in the organs themselves, to their improvement, development, and radical transformation to support life in a variety of environments. He writes:

> Man will have access to all the heavenly realms, to all the heavenly worlds only when he has transformed himself down to the most elemental materials—atoms and molecules—because only then will he be capable of living in any environment and assuming any form.
>
> (Fedorov 1982: 501)[49]

Fedorov referred to man's ability in the future to create for himself any kind of organ, which will in turn have the capacity to change in relation to his living conditions, as "full organicity."

Belyaev-Gintovt quite literally follows the ideas of Russian and Soviet Cosmists by creating images of "full organicity," of a new physicality, but he does so in a much more radical way than those predecessors who theorized and represented images of the "new man." The dream of uniting nature and technology is given its most obvious incarnation in the Eurasian army of sun-drenched super soldiers, Stalin's Falcons raised from the dead in fish-scales and feathers. This agro-cosmic miracle army can operate in any environment, armed with weapons in the shape of wheat stalks. The artist developed the theme of technological and organic weaponry in even greater detail in his drawings, which were not included in the final version of the *Victory Parade* exhibit. They feature deer and elephant rocket launchers with sickles instead of tusks, camels with satellites and antennae growing out of their backs, and missiles in the form of towers, which illustrate the potential of neoclassical architecture and beauty to serve as the greatest weapon of the Eurasian empire.

It should be noted that the utopian imagination and the interest in the past-future distinguishes the culture of the second modernism, but not only the projects of conservatives. For example, a utopian element is evident in Russian Prime

Minister Dmitrii Medvedev's modernization project. The complex of buildings on the campus of the Skolkovo Moscow School of Management is designed to demonstrate this elite institute's orientation toward the future. It is built from the perspective of the "upward gaze," which characterizes all cult-related buildings in traditional societies. In the case of Skolkovo, the perspective comes from Space; it is the perspective of Google Earth and a reproduction of Malevich's painting *Suprematism* [Suprematizm 1921/1927?] (Skolkovo website). In the works of Belyaev-Gintovt, just as in the works of the conservative revolutionary, there is reference to the "eye of God," but the New Moscow is planned as a pentagram. And if, in the opinion of Medvedev's modernizers, Russia's greatest contribution to world culture is Malevich, then for Belyaev-Gintovt it is the Soviet legacy, specifically Soviet Cosmism, early Soviet cinematography, monumental style, and science fiction. Thus, the *Victory Parade of 2937* can be seen as a response from the spokesman of the countercultural movement to the official projects—in this case, Medvedev's modernization—that assert the importance of the Soviet legacy in the fields of science and new technology.

Conclusion

The *Victory Parade* project is clearly distinguished by the artist's aggressive use of the inverted metaphor as an artistic device, popular among Russian avant-garde artists and postmodernists alike: the fairy tale becomes reality, Russia truly becomes the "homeland of elephants," Lenin is truly "more alive than the living," Stalin's Falcons return to formation, and beauty, to save the world, enters into union with atomic weapons.[50] It should be remembered here that the struggle against "tropical thinking" (*sensus tropicus*)—that is, the metaphorical approach to sacred notions that characterized the Old Believers—was opposed in its time to the influence of the Western Ukrainian tradition of the unconventional treatment of sacred texts, which was popularized in Muscovy before Patriarch Nikon's reform of liturgical books (Uspenskii 1994).[51]

The distance assumed by the artist toward his own great project is typical of all of Belyaev-Gintovt's artistic works and pronouncements. Nostalgia for the grand narrative is combined with the archaic literal reading of all ideological pronouncements, which instills confusion and suspicion in the viewer. The artist makes deft use of the 'double code,' remaining inside the genre while simultaneously distancing himself from it, allowing it to serve as a field of primitive consumption while meeting the demands of the aestheticizing "soldiers of taste."[52] The device of literalizing metaphors to produce a grotesque effect is something the artist applies widely not only in his paintings but also in his artistic manifestos and his appearances in the press. For example, consider the citation below from a recent interview regarding his last exhibit:

> I see communism with a non-human face. This is why people, animals, birds, and angels march in the parade in a single formation, symbolizing the generality,

> the universality of the grand Eurasian Victory Parade. The parade is a victory over chaos, a victory of centripetal over centrifugal forces.
>
> (Belyaev-Gintovt 2010)[53]

The Moscow and Leningrad undergrounds stood out initially by virtue of their interest in carnival and the mystery play, typical of cultures at the end of an era. The performances of Sergei Kurekhin (1954–1996), Vladislav Mamyshev-Monroe (1969–2013), and Petr Mamonov (b. 1951), and the malevolent vulgarity of the National Bolsheviks, headed by Eduard Limonov (b. 1943), Dugin, and Egor Letov (1964–2008), deride and destroy not only the world of Soviet official culture but also the world of our humanitarian traditions and the oppositional pathos of the dissident movement, as well as the glamorous excess and state capitalism that replaced them (Goscilo and Strukov 2010). Belyaev-Gintovt is inseparable from these groups, and so an alarmist approach to his work as pro-Putin or dangerously imperialistic is incorrect. The device of over-identification with totalitarian culture is often confused with direct identification, which fails to take into consideration the specific nature of this type of ideological critique. A more adequate and productive approach to the discussion of late-Soviet and post-Soviet countercultures, in my opinion, is that of Alexei Yurchak, who isolates and describes the method of mimetic critique and the device of over-identification, popular in the late 1980s in unofficial circles in the Soviet Union and in Eastern Europe. Yurchak insists that the countercultural method of indirect deconstruction is much more profound than the direct critique that was popular in dissident circles:

> By forcing one's public to doubt, to assume different opinions, and so on, they focus their public's attention on the fact that ideology works much more profoundly than it first appears, and that a direct contradiction of this or that ideological position may, upon further investigation, turn out to be just another aspect of that very same ideological system.
>
> (Iurchak 2009)[54]

The aesthetics and metaphysics of Soviet aviation, its heroism and physical discipline, its supramoralism and alchemical parades are, in their archeo-futurist form, as far removed as possible from liberal, free-market democracy, corporate state ideology, ethno-nationalism, and the narrow-minded mainstream that dominates post-Soviet Russia. The value and significance of Belyaev-Gintovt's new modernism, in my opinion, consists precisely in the endless search for an alternative to the rapid post-capitalist absorption of any countercultural movement, and in the attempt to create a project that inspires maximal discomfort and defies classification. Belyaev-Gintovt conducts his aesthetic and ideological search on the territory of established or only partly established utopias that stand in stark opposition to the 'spirit of the age,' with no chance of being mass produced or adopted as an official ideology.

Translated from Russian by Brian James Baer

Notes

1 Victor Pelevin's novel *Omon Ra* (1991) is a good example of this phenomenon. For an analysis of post-Soviet cinematic treatments of it, see Chapters 6 and 10.
2 A pivotal event in the transformation of the official image of Space was the legendary 'Gagarin-party' of 1991, which was the first rave in Russia to be held at the Space pavilion at the VDNKh park. Attendees numbered 250,000 people, among them the cosmonauts Valentina Tereshkova (b. 1937) and Georgii Grechko (b. 1931).
3 See the site of the Russian Federal Space Agency, Roskosmos.
4 [Pridet vremia—i pervaia delegatsiia inoplanetian, otkryt' kotorykh pomog dalekii polet Gagarina v 1961, sklonia prichudlivye zelenye golovy, poneset venki k mavzoleiu Il'icha.]
5 In this connection, see Chapter 8.
6 The word *imperium* means 'fullness of power' and is formed from the Latin verb *imperare*, 'to command.' Empire is the territory upon which the acts of imperium take place. Initially the concept of imperium signified the absolute executive (military, juridical, and administrate) and magisterial power (of counsuls, praetorians, and, in certain extraordinary circumstances, dictators) during the period of the Roman republic. After the fall of the republic, the emperor upon ascension to the throne was granted the highest degree of imperial power (*summum imperium*), which included military and other plenary powers and extended throughout the territory of the state.
7 Boris Mikhailovich Iofan (1891–1976) was one of the greatest Soviet architects. The founder of Soviet monumental classicism, during the Stalin period he designed the USSR pavilion at the World's Fair in Paris (1937) and New York (1939), the 'House on the Embankment' in Moscow, the Palace of the Soviets project, as well as a host of other buildings.
8 From the mid-1960s to the mid-1970s, representatives of Moscow's bohemian subculture would gather at the home of the writer Iurii Vital'evich Mamleev (b. 1931). This group included lovers of the esoteric, people who were interested in metaphysics and black magic, eroticism, alchemy and European mannerism, the 'other' side of reality, and the dark and secret aspects of the human psyche. They formed what would become known as the Moscow Metaphysical School. The leading figures of this school were, in addition to Mamleev himself, the poet and polyglot Evgenii Golovin (1938–2010), who was an expert in eastern philosophy, European alchemy, and French poetry, and Geidar Dzhemal' (b. 1947), a proponent of Islamic mysticism, 'fiery Islam,' who today runs the *Islamic Committee.* Following Mamleev's emigration, Golovin and Dzhemal' remained in the organization, and new people joined, such as the writer Venedikt Erofeev (1938–90), the Orthodox theologian Tat'iana Goricheva (b. 1947), the artist Anatolii Zverev (1931–86), and Vladimir Piatnitski (1938–78). Although Dugin did not take part in the legendary gatherings on Iuzhinskii perulok, he joined later when it was an already fully formed circle of like-minded metaphysicians at the very beginning of the 1980s; that is, after the routing of the circle in 1973 and Mamleev's emigration.
9 On the scandal around Belyaev-Gintovt's receipt of the prize, see Jonson 2015: 68–71.
10 It should be noted here that the theme of aviation is only marginally represented in new Russian art. Soviet aviation art that was once represented in the works of A. Deineka (1899–1969) and A. Samokhvalov (1894–1971) today is represented by Belyaev-Gintovt, G. Gur'ianov (b. 1961), and by the designer Ol'ga Soldatova (b. 1965).
11 Beuys, whose plane crashed in the Crimea, survived thanks to the local Tatars, who saved him by wrapping his body in felt (dried wool) and massaging it with melted butter. When he became an artist, Beuys used felt, which in Russia is used to make leg wrappings, along with butter and gold in his installations, linking the primordial quality of the materials with an avant-garde conception of performance.
12 On Deineka and aviation, see Chapter 2.
13 The mosaics of thirty-five cupolas in the Moscow metro station Maiakovskaia of 1938 on the theme 'Twenty-four Hours of Soviet Sky' and the seven mosaic panneaux in the

station Novokuznetskaia of 1943, created in the mosaic studio of the Leningrad Academy of Arts under the direction of V.A. Florov (1874–1942) according to sketches made by A. Deineka, represent the sky of the Soviet utopia. In these mosaic plafonds Deineka uses images of airplane mechanics, parachutists, skiers flying across the backdrop of the sky, and swimmers and gymnasts to develop the themes, so closely associated in official Soviet discourse, of the conquest of the air and sports. See Chapter 2.

14 The heroes of Soviet aviation were referred to as Stalin's falcons. The term was first used in a notice on Valerii Chkalov in the newspaper *Pravda* on 24 July 1936. The aviation parades of the 1930s, which took place on the Day of the Soviet Air Force, inaugurated by Stalin on 18 August 1933, were known as the Festival of Stalin's Falcons. See Gunther 1991, Kozlova 2010, McCannon 1998, Palmer 2005, Palmer 2006.

15 Sviatogor (the name of a proto-hero of ancient Russian epics) was the pseudonym of Aleksandr Agienko, the founder of the movement of poet-biocosmists. For more on biocosmism, see below.

16 [Esli religioznoe isskustvo, proslavliaiushchee gosudarstvo, bylo vsegda samym velichestvennym, samym sovershennym proiavleniem tvorcheskoi deiatel'nosti cheloveka, eto ob"iasniaetsia tem, chto takoe iskusstvo nikogda ne greshilo teoretichnost'iu. Khudozhnik tverdo znal, chto on izobrazhaet i zachem on izobrazhaet: blagodaria etomu mysl' ego byla iasna i opredelenna; dlia nee ostavalos' tol'ko podyskat' stol' zhe iasnuiu i opredelennuiu formu.]

17 The New Academy of Fine Arts was founded in 1989 by Timur Novikov (1958–2002). This movement, which touched upon every field of art, became the center of what was called the Petersburg Renaissance of the 1980s and 90s. In addition to the artists Georgii Gur'ianov (b. 1961), Bella Matveeva (b. 1961), Ol'ga Tobreluts (b. 1970), Andrei Medvedev (1959–2010), Vladislav Mamyshev (1969–2013), Egor Ostrov (b. 1970), and Stanislav Makarov (b. 1972), all the best-known figures of post-Soviet Petersburg joined the New Academy, including the rock groups Kino and Popular Mechanics, the Necrorealists, and representatives of alternative cinema and fashion.

18 FSB happens to be the acronym of the Secret Service [Federal'naia sluzhba bezopasnosti], established in 1995 to replace the KGB and its subsequent, short-lived reconfigurations.

19 [Rakety, zapuskaemye iz 'kolchana' grecheskogo boga, imeiut sugubo mirnoe znachenie—eto gradoboinye snariady i geodezicheskie shary, s pomoshch'iu kotorykh zhiteli smogut regulirovat' meteorologicheskie usloviia v svoem megapolise. Voobshche, sposobnost' vliiat' na klimat i okruzhaiushchuiu sredu vtselom iavliaetsia osnovoi protsvetaniia agrarno-kosmicheskoi derzhavy, stolitsei kotoroi i dolzhen stat' Novonovosibirsk.]

20 It is important to recall that the ballpoint pen is directly linked to the Soviet and American space programs. The zero-gravity pen, or the space pen, was created by the American inventor Paul C. Fisher (1913–2006) for use under extreme conditions, under water or in zero gravity. The space pen began to be used in the American space program in 1967 and two years later in the USSR. (Thanks to Vlad Strukov for pointing out this fact.)

21 [V kholodnom stile ischezaet total'naia osveshchennost', logicheskaia chitaemost' iskusstva. Paradoksal'nyim obrazom za iasnoi, chetkoi, prostoi formoi, ne predpolagaiushchei interpretatsionnykh ukhischchrennii, obnaruzhivaetsia nechto nepronitsaemoe, apelliruiushchee k inomu izmereniiu v iskusstve. Zritel', ostavlennyi odin na odin s etoi zhestkoi i kholodnoi formoi, vynuzhden dyshat' razriazhennym vozdukhom. On mozhet ispytyvat' radrazhaiushchee chuvstvo opustoshennosti i diskomforta. V obrazakh novoi klassiki net trepeta zhizni. Skoree, ona apelliruet k obrazam smerti, a tochnee—granitsy, no mozhet byt', imenno cherez etot kholod, stol' chuzhdyi vsem privychkam 'dukha sovremennosti', smozhet vernut'sia nechto zabytoe ili prostupit' eshche neizvestnoe v sovremennoi kul'ture.]

22 The category of the noumenal (from the Latin word *noumen*, meaning 'holy,' 'sacred,' 'divine') was introduced into scientific discourse by Rudolph Otto (1869–1937). See Otto 1958.

23 On the various forms of dandyism, see Vainshtein 2006 and Moers 1960.

24 At the basis of Gur'ianov's work is the question, so relevant in Russia today, of the movement from 'hard' to 'soft' power, achieved through seduction. His works treating the image of Aleksandr Samokhvalov (1884–1971) are imbued with the theme of 'Soviet eroticism.' It was in Samokhvalov's work that the Soviet underground first saw neoclassical motifs and homoerotic themes; it is due largely to his paintings that 'Soviet' came to be seen as an aesthetic resource.

25 [Geroi, po opredeleniiu, dolzhen byt' 'kholodnym.' Esli on ne otdelit sebia ot okruzhaiushchikh, esli on ne zamorozit v sebe teplye energii povsednevnoi chelovechnosti, on ne budet na urovne sovversheniia Nevozmozhnogo, t.e. na urovne togo, chto delaet geroia geroem. Geroi dolzhen uiti ot liudei. No za predelom sotsial'nogo uiuta bushuiut pronizyvaiushchie vetry ob"ektivnoi real'nosti, zhestokoi i vnegumannoi. Zemli i kamni vosstaiut na animal'nyi i vegetal'nyi miry. Agressivnaia rastitel'nost' raz"edaet mineral, a dikie zveri bezzhalostno topchut upriamye travy. Stikhii vne obshchestva ne znaiut sniskhozhdeniia. Mir sam po sebe—triumfal'noe pirshestvo veshchestva, nizhniaia ploskost' kotorogo slita s glybami donnogo kosmicheskogo l'da. Geroi kholoden, potomu chto on ob"ektiven, potomy chto on prinimaet ot mira estafetu spontannoi sily, beshennoi i nedobroi.]

26 The late Soviet counterculture was in demand in the sudden changes of epochs; with the end of perestroika, mediators of the transition became unnecessary, making the early death of its most important cultural heroes—Sergei Kurekhin (1954–96), Viktor Tsoi (1962–90), and Timur Novikov (1958–2002)—symbolic.

27 [Supramoralizm trebuet raia, Tsarstva Bozhiia, ne potustoronnego, a posiustoronnego, trebuet preobrazheniia posiustoronnei, zemnoi deistvitel'nosti, preobrazheniia, rasprostraniaiushchegosia na vse nebesnye miry i sblizhaiushchego nas s nevedomym nam potustoronnim mirom: rai, ili Tsartvo Bozhie, ne vnutri lish' nas, ne myslennoe tol'ko, ne dukhovnoe lish', no i vidimoe, osiazaemoe, vseoshchushchaemoe organami, proizvedennymi psikhofiziologicheskoiu reguliatseiu (t.e. upravleniem dushevnotelesnymi iavleniami), organami, kotorym dostupny ne trav lish' proziabanie, no i molekuly atomov vsei vselennoi dvizhenie, chto i sdelaet vozmozhnym voskreshenie i vsei vselennoi preobrazhenie.]

28 In patriotic circles, lines from the 1936 song 'Song of the Motherland,' with words by V.I. Lebedev-Kumach (1898–1949) and music by I.O. Dunaevskii (1900–55), remain popular today: "We love our country like a new bride; we will protect her like a loving mother."

29 For a survey of works casting Russia as bride, see Rutten 2010.

30 [Smert', mozhno skazat', est' anesteziia, pri koei proiskhodit samoe polnoe truporaz"iatie, razlozhenie i rasseianie veshchestva. Sobiranie rasseiannykh chastits est' vopros kosmotelluricheskoi nauki i isskustva, sledovatel'no, muzhskoe delo, a slozhenie uzhe sobrannykh chastits est' vopros fiziologicheskii, gistologicheskii, vopros sshivaniia, tak skazat', tkanei chelovecheskogo tela, tela svoikh ottsov i materei, est' zhenskoe delo.]

31 [Samoderzhavie v pervonachal'nom smysle est' diktatura, vyzvannaia opasnost'iu ne ot drugikh sebe podobnykh liudei, a ot sily slepoi, vsem bez iskliucheniia groziashchei smert'iu. Samoderzhavie . . . iavilos' totchas posle smerti pervogo otsa, soediniv vsekh v edinoi vole, v edinom zhelanii, vyzvannom utratoiu, smert'iu.]

32 [Nasha tsel'—Evraziiskaia imperiia . . . My predadim geroev drevnosti, velikikh imperostroitelei Evrazii, esli ne sozdadim eshche nechto bolee velikoe, chem oni. I lish' kogda my rasplastaem granitsy ot okeana do okeana, glaza nashikh mertvetsov okrasiatsia tikhim prozrachnym tsvetom: oni—nashi potomki—sdelali eto! Inache mertvye nas v pokoe ne ostaviat.]

33 [God nazad on uvidel na staroi kinoplenke parad fizkul'turnikov 1937 goda na Krasnoi ploshchadi. V roli postanovshchika etogo deistviia debiutiroval togda velikii

khoreograf Igor' Moiseev. 'Eto neveroiatnyi reportazh!'—vosklitsaet khudozhnik. 'Eto vygliadit velichavo, torzhestvenno. Ia vsego-navsego protransliroval v budushchee etu osvobozhennuiu energiiu ob"edinennogo naroda, velikii passionarnyi poryv k schast'iu. Konechno, to bylo tragichestkoe vremia. Takogo napriazheniia, no i takogo poryva k radosti, poryva v kosmos ia ne videl ne do togo, ne posle,' uveriaet Aleksei Beliaev-Gintovt.]

34 Patriotic publications stress the sacred dimension of Soviet parades (Diunov 2008).

35 In the ideology of Eurasianism space dominates time; historical time loses its power. Neoeurasianism has inherited the historiosophic idea of the Christian Empire of the Romans as a *katechon*, 'that which withholds,' preserving the *oikumen*, the universal community, in the face of the general apostasy that will follow the victory of the Antichrist. For Byzantines, the empire is an earthly icon of the heavenly order, which holds back the forces of chaos. It does not possess singular or fixed spatial-temporal characteristics; it can change its geographic position and reincarnate itself on the territory of other states (*translatio imperii romani*). In the Orthodox Christian tradition, the Russian government inherited the messianic katechonic significance of the Byzantine Empire.

36 [S kazhdym godom nashi parady fiskul'tury iavliaiutsia vse bolee veselymi, iarkimi i bogatymi. Vse bolee uverenno tverd shag molodezhi, i iarche gorit v glazakh ee radost' zhit' v strane, gde tak bystro i krasivo vospityvaetsia telo i tak ognenno, pobedonosno tsvetet v nem boevoi, geroicheskii dukh, ezhednevno vyiavliaia sebia v rabote na obogashchenie naroda, na oboronu rodiny, protiv vragov, pochti ezhednevno sverkaia smelymi podvigami na blago svoei strany. Vidia eti desiatki tysiach iunoshei i devushek, stroinymi riadami idushchikh k velikomu budushchemu, chuvstvuesh' volnenie, ot kotorogo serdtse gotovo razobrat'sia. Chuvstvuesh' i pechal'—ottogo, chto u tebia net mesta v riadakh etoi moguchei armii, chto ty uzhe ne v silakh idti v nogu s nei i, poravniavshis' s mavzoleem, kriknut' iskrennee 'ura!']

37 Gor'kii's article 'On the Parade of Athletes' was written about the parade that took place in Moscow on 1 June 1935. The article, however, was not included in the authorized collections of the author's work. It was first published simultaneously in the newspaper *Pravda* on 2 June 1935 (issue 180), under the headline 'M. Gor'kii on the Parade of Athletes,' and in the newspaper *Izvestiia* on 2 July 1935 (issue 153), under the title 'Joy and Pride.'

38 [Eto gigantskaia chasha, v tsentre kotoroi moskovskii Kreml'. Po mere udaleniia ot nego rastet etazhnost' zdanii. Esli posmotret' na plan etogo goroda budushchego, to my uvidim tri kol'tsa, peresekaiushchikhsia radiusami. V tochkakh peresecheniia vozvyshaiutsia vysotki. V pervom kol'tse 40 vysotok, vo vtorom—60 i okolo 90 dolzhno bylo byt' na peresechenii tret'ego kol'tsa. S liuboi tochki, s liubogo udaleniia viden tsentr etoi velichestvennoi kompozitsii—moskovskii Kreml'.]

39 [Prikrashchaetsia Paskha tol'ko u bludnykh synov, otriasaiushchikh prakh ot nog, t.e. pri prevrashchenii sela v gorod, chto nagliadno vyrazhaetsia udaleniem prakha umershikh za chertu poseleniia; v sele zhe, v nastoiashchem sele, ne zarazhennom gorode, etot prakh imeet vsegda tsentral'noe polozhenie.]

40 [Ischeznuv na Zapade . . . Paskha sokhranilas' v Rossii, i osobenno v Kremle pri grobakh sobiratelei, ili ob'edinitelei synov dlia dela voskresheniia otsov, v Kremle, t.e., v kreposti, oboroniavshei prakh ottsov i dolzhenstvuiushchei orudiia oborony obratit' v orudiia ozhivleniia prakha, kogda nebratstvo ischeznet. Koronatsiia, t.e. postavlenie stoiaschchego v ottsov-mesto na obshchee delo voskresheniia ottsov, ne imela by smysla, esli by sovershalas' ne v Kremle, a gde-libo v inom meste.]

41 The film was included in the festival of censored cinema, *The Shelf. Liberated Cinema*, which took place in the Kinoklub at Vinzavod from 29 August–20 September 2011.

42 [Odinakovykh liudei net i ne mozhet byt'. Eto burzhuaznoe ravenstvo. Samo poniatie sorevnovaniia snimaet poniatie ravenstva. Ravenstvo est' nepodvizhnost', sorevnovanie

est' dvizhenie. Ravniasia na luchshikh, pomogai otstaiushchim i dobeisia obshchego pod"ema. Luchshie–eto nashi vozhdi . . . Luchshie—eto te, kto tvorit mysli, nauku, tekhniku, muzyku . . . Eto vysokie umy . . . Te, kto boretsia s prirodoi, pobediteli smerti.]

43 Dugin has been very critical of Neocosmism as an occult theosophical movement (Dugin 2005: 32).

44 ['Krasnyi smysl' . . . sostoit v odnom-edinstvennom—v preodolenii smerti. Drevnie egiptiane, ispovedovavshie voskresenie Ozirisa, byli 'krasnymi.' Panteisty Indii, veriashchie v pereselenie dush, v neistrebimost' zhizni, byli 'krasnymi.' Iiusus, 'smertiiu smert' poprav i sushchim vo grobe zhivot darovav' byl 'krasnyim.' Nikolai Fedorov, propovedovavshii voskresenie iz mertvykh predkov, rasselit' ikh po planetam galaktiki s pomoshch'iu raket Tsiolkovskogo, byl 'krasnym.']

45 These are the soldiers and zoomorphs of the Eurasian empire, covered in scales, marching in unison. This bionic suit made of scales references both Soviet science fiction and archaic forms of protection using scales—the most ancient form of military armor. Even the image of Lenin, awkward, diminutive, and old, leading columns of young warriors, may be linked to the folkloric and mythological realm of Pushkin's Uncle Chernomor and the brave young warriors who defend the magic island (or the sacred space of the New Moscow in the works of Belyaev-Gintovt): "In scales, gleaming like gold/Thirty three warriors/All daring and bold/Young titans/All well-matched/and with them Uncle Chernomor/No defense is more sure/more courageous or diligent" (Pushkin 1960: 334–5).

46 It is worth noting that black, white, and red are the colors of the National Bolshevik Party (NBP).

47 [Pochemu nashi startsy, pochti lishennye vsiakoi podderzhki, ne dolzhy sosat' krov' iunstov? Ia govoriu o tom, kto silen, vesel, blagorazumen, imeet nailuchshuiu krov', i, vozmozhno, ee izbytok. Pust' zhe oni prisasyvaiutsia podobno piiavke ili krovopiitse k otkrytoi vene levoi ruki . . . vo vremia pribyvaiushchei luny.]

48 [Ia predstavil vsevo lish' svoi sny. Sny o nekoi agrarno-kosmicheskoi tsivilizatsiii, dostigshei garmonii.]

49 [Cheloveku budut dostupny vse nebesnye prostranstva, vse nebesnye miry tol'ko togda, kogda on budet vossozdavat' sebia iz samykh pervonachal'nykh veshchestv, atomov, molekul, potomu chto togda tol'ko on budet sposoben zhit' vo vsekh sredakh, prinimat' vsiakie formy.]

50 The phrase 'Russia is the birthplace of elephants' was a parody of Soviet patriotism, a critique of the state symbols and rhetoric concerning the Soviet Union/Russia's preeminence in the realm of scientific technology. The images of elephants in *Victory Parade of 2937* references imperial symbolism: the elephant in Roman tradition was a symbol of military triumph, and bronze sculptures of elephants adorned the Golden Gate of Constantinople.

51 The culture of the Old Believers was seen as a counterculture and is actively promoted by Dugin and his followers.

52 Belyaev-Gintovt's work has been highly esteemed by Russia's leading art critics (see Bobrinskaia 2004 and Borovskii 2008), even as some critics denounce what they view as its fascist values.

53 [Mne viditsia kommunizm s nechelovechskim litsom. Vot pochemu liudi, zveri, ptitsy i angely stoiat na parade v odnom stroiu, simvoliziuruia vseobshchnost', vselenskost' bol'shogo Evraziiskogo Parada Pobedy. Parada pobedy nad khaosom, pobedy tsentrostremitel'nykh protsessov nad tsentrobezhnymi.]

54 [Zastavliaia svoiu auditoriiu somnevat'sia, razdelat'sia vo mneniia i t.d. oni obrashchaiut ee vnimanie na to, chto ideologiia deistvuet gorazdo glubzhe, chem nam kazhetsia, i chto priamoe protivopostavleniie tomu ili inomu ideologicheskomu vyskazyvaniiu mozhet na poverku okazat'sia vsego lish' chast'iu vse toi zhe ideologicheskoi sistemy.]

Bibliography

Averintsev, S. (1973) "Zoloto v sisteme simvolov rannevizantiiskoi literatury." *Vizantiia. Iuzhnye slaviane I drevniaia Rus'. Zapadnaia Evropa*. Moscow: Nauka. 43–52.

Beliaev-Gintovt, A. (2009) *Chto takoe novaia ser'eznost'?* Available at: http://russia.ru/video/expert_6696/ (accessed 4 June 2012).

—. (2010) "Mne viditsia kommunizm s nechelovecheskim litsom." Available at: http://www.liberty.ru/foto/Aleksej-Belyaev-Gintovt-Mne-viditsya-kommunizm-s-nechelovecheskim-licom (accessed 4 June 2012).

Bobrikov, A. (2004) "Moloko i med." *Khudozhestvennyi zhurnal* 54. Available at: http://www.guelman.ru/xz/xx54/xx5415.htm (accessed 4 June 2012).

Boprinskaia, E. (2004) "Somnitel'naia suschchnost' iskusstva." *Khudozhestvennyi zhurnal* 54. Available at: http://xz.gif.ru/numbers/54/somnitelnaya/ (accessed 4 June 2012).

Borovskii, A. (2008) "Etaticheskii estetizm." Introduction to Beliaev-Gintovt, A., *Rodina-Doch'*. Moscow: Triumph Gallery.

Clark, K. (2011) *Moscow, the Fourth Rome: Stalinism, Cosmopolitanism, and the Evolution of Soviet Culture, 1931–1941*. Cambridge, MA: Harvard University Press.

Degot', E. (2008) "Mog li ul'trapravyi natsionalist ne poluchit'premiiu Kandinskogo," OpenSpace.ru (11 Dec 2008). Available at: http://www.openspace.ru/art/projects/89/details/6467/ (accessed 4 June 2012).

Dugin, A. (2005) *Konspirologiia*. Moscow: Arktogeia.

—. (2007) *Geopolitika postmoderna*. St. Petersburg: Amfora.

Duinov, M. (2008) "Filosofiia parade." *Novye khroniki* (7 May 2008). Available at: http://novchronic.ru/1303.htm (accessed 4 June 2012).

Engström, M. (2012) "Forbidden Dandyism: Imperial Aesthetics in Contemporary Russia." *Nordic Fashion Studies*. Peter McNeil and Louise Wallenberg (Eds.). Stockholm: Axl Books. 179–99.

"Evraziiskii soiuz molodezhi." (n.d.) "Manifest ESM." Available at: http://www.rossia3.ru/programma.html (accessed 4 June 2012).

Fedorov, N. (1982) *Sochineniia*. Moscow: Mysl'.

—. (1995) *Sobranie sochinenii v chetyrekh tomakh*. Vol. 2. Moscow: Progress. Available at: http://rudocs.exdat.com/docs/index-178807.html (accessed 4 June 2012).

Filippov, A. (2008) "Chuvstvennost' i mobilizatsiia. K probleme politicheskoi estezi-ologii." *De Futuro, ili istoriia budushchego*. Ed. D.A. Andreev and V.B. Prozorov. Moscow: Politicheskii klass. Airo XXI.

Finkati, V. (2011) "Laboratornaia alkhimiia i ispol'zovanie krovi." Sait Russkogo germet-ichskoko obshchestva. Available at: http://www.hermeticsociety.ru/researches/Fincati.htm (accessed 4 June 2012).

Giunter, Kh. (1991) "Stalinskie sokoly: Analiz mifa tridtsatykh godov." *Voprosy literatury* 11/12: 122–41.

Glatzer Rosenthal, B. (ed.) (1997) *The Occult in Russian and Soviet Culture*. Ithaca, NY: Cornell University Press.

Golovin, E. (2003) *Priblizhenie k Snezhnoi koroleve*. Moscow: Arktogeia-Tsentr. Available at http://evrazia.org/FTP/golovin1.pdf (accessed 4 June 2012).

Goralik, L. (2007) " . . . The Rosgorexport Sweet Cream Cheese: Symbolism and Symbols of the Soviet Era in Russian Branding Today," *Fashion Theory: Dress, Body and Culture* 3 (Spring): 13–32.

Gor'kii, M. (1949) *Sobranie sochinenii*. Vol. 27. Moscow: Khudozhestvennaia literatura.

Goscilo, H. and S.M. Norris. (eds.) (2008) *Preserving Petersburg: History, Memory, Nostalgia*. Bloomington, IN: Indiana University Press.

Goscilo, H. and V. Strukov. (eds.) (2010) *Celebrity and Glamour in Contemporary Russia: Shocking Chic*. London and New York: Routledge.

Grois, B. (1993) *Utopiia i obmen. Stil' Stalin. O novom. Stat'i*. Moscow: Znak.

—. (2003) "Kommentarii k isskustvu." Moscow: Khudozhestvennyi zhurnal.

—. (2006) "Bessmertnye tela." *Khudozhestvennyi zhurnal* 63. Available at: http://xz.gif.ru/numbers/63/tela/ (accessed 4 June 2012).

—. (2007) "Neligitimnyi kosmonavt." *Khudozhestvennyi zhurnal* 65/66. Available at: http://xz.gif.ru/numbers/65-66/kosmonavt/ (accessed 4 June 2012).

Hagemeister, M. (1997) "Russian Cosmism in the 1920s and Today." In: Glatzer Rosental, B. (ed.) *The Occult in Russian and Soviet Culture*. Ithaca, NY: Cornell University Press. 185–202.

Iung, K.G. (1997) *Psikhologiia i alkhimiia*. Moscow: *REFL-book*.

Iurchak, A. (2005) "Nochnye tantsy s angelom istorii. Kriticheskie kul'tural'nye issledovaniia post-sotsializma." *Kul'turnye issledovaniia*. A. Etkind. (Ed.) St. Petersburg: European University. Available at: http://berkeley.academia.edu/AlexeiYurchak/Papers/367195/ (accessed 4 June 2012).

—. (2009) "Mimeticheskaia kritika ideologii grupy Laibach i AVIA." *Chto delat'*. Available at: http://www.chtodelat.org/index.php?option=com_content&view=article&id=462:mimetic-critique-of-ideology-laibach-and-avia&catid=188:19-experience-of-perestroika&Itemid=446&lang=ru (accessed 4 June 2012).

Jonson, L. (2015) *Art and Protest in Putin's Russia*. London and New York: Routledge.

Karpekina, T. (2010) "Boevye slony Beliaeva-Gintovta." Sait Golos Rossii (1 June). Available at: http://rus.ruvr.ru/2010/06/01/8879227.html (accessed 4 June 2012).

Keys, B. (2003) "Soviet Sport and Transnational Mass Culture in the 1930s." *Journal of Contemporary History* 38.3: 413–34.

Klark, K. (2009) "Imperskoe vozvyshennoe v sovetskoi kul'ture vtoroi poloviny 1930-kh godov." *Novoe literaturnoe obozrenie* 95: 58–80. Available at: http://magazines.russ.ru/nlo/2009/95/kk8.html (accessed 4 June 2012).

Kommunisty Peterburga i Leningradskoi oblasti. (2007) "S dnem Sovetskoi kosmonavtiki, s dnem rozhdeniia KP." Available at: http://kplo.ru/content/view/108/5/ (accessed 4 June 2012).

Kozlova, I. (2010) "'Stalinskie sokoly': totalitarnaia frazeologiia i 'sovetskii fol'klor.'" *Antropologicheskii forum* 12. Available at: http://anthropologie.kunstkamera.ru/files/pdf/012online/12_online_kozlova.pdf (accessed 4 June 2012).

Krementsov, N. (2011) *A Martian Stranded on Earth: Alexander Bogdanov, Blood Transfusions, and Proletarian Science*. Chicago: University of Chicago Press.

Lapteva, E. (2011) "Gagarin-parti spustia 20 let." Sait Fontanka.ru. Available at: http://www.fontanka.ru/2011/04/17/020/ (accessed 4 June 2012).

Lemann, G. (2010) "Iskusstvo refleksivnoi modernosti." *Logos* 4: 87–108.

Livshits, B. (1989) *Polutoraglazyi strelets. Stikhotvoreniia, perevody, vospominaniia*. Leningrad: Sovetskii pisatel'.

Lotman, Iu. M. (1992) "Kanonicheskoe iskusstvo kak informatsionnyi paradox." *Izbrannye stat'i*. Vol. 1. Tallinn: Alexandra. 243–7.

McCannon, J. (1998) *Red Arctic: Polar Exploration and the Myth of the North in Soviet Russia, 1932–1939*. New York: Oxford University Press.

Miziano, V. (Ed.) (2007) "Progressivnaia nostal'giia. Sovremennoe iskusstvo stran byvshego SSSR." Special issue of *World Art Muzei* 33/34.

Moers, E. (1960) *The Dandy: Brummell to Beerbohm*. Lincoln: University of Nebraska Press.

Novyi muzei. (2010) "Interv'iu s khudozhnikom. A Beliaev-Gintovt." Available at: http://novymuseum.livejournal.com/2099.html (accessed 4 June 2012).

Oblacheva, M. (2001) "Novonovosibirsk," Sait Gosudarstvennogo muzeia arkhitektury. Available at: http://old.muar.ru/exibitions/exibit68.htm (accessed 4 June 2012).

Otto, R. (1958) *The Idea of the Holy*. Oxford: Oxford University Press.

Palmer, S.W. (2005) "Icarus, East: The Symbolic Context of Russian Flight," *Slavic and East European Journal* 49.1 (Spring): 19–47.

—. (2006) *Dictatorship of the Air. Aviation Culture and the Fate of Modern Russia*. New York: Cambridge University Press.

Panchenko, I. (2008) "Nesvoevremennyi utopist Iurii Olesha. Ideia 'novogo cheloveka' i sud'ba kinop'esy *Strogii iunosha*." *Novyi zhurnal* 250. Available at: http://magazines.russ.ru/nj/2008/250/pa13.html (accessed 4 June 2012).

Popov, S. (2004) "Imperskii avangard." *Khudozhestvennyi zhurnal* 54. Available at: http://xz.gif.ru/numbers/54/impersky-avangard/ (accessed 4 June 2012).

Prokhanov, A. (2002) *Gospodin Geksogen*. Moscow: Ad Marginem.

Pushkin, A.S. (1960) "Skazka o tsare Saltane." *Sobraniie sochinenii A.S. Pushkina v deviati tomakh,* Vol. 3. Moscow: Khudozhestvennaia Literatura. 312–37.

Roscosmos. Official site of the Russian Space Agency. Available at: http://www.federal-space.ru/main.php?id=26 (accessed 4 June 2012).

Rutten, E. (2010) *Unattainable Bride Russia: Gendering Nation, State, and Intelligentsia in Russian Intellectual Culture*. Evanston, IL: Northwestern UP.

Sal'nikov, V. (2003) "Diskobol. Uchebnyi fil'm peterburgskikh khudozhnikov." *Khudozhestvennyi zhurnal* 50. Available at: http://xz.gif.ru/numbers/50/diskobol/ (accessed 4 June 2012).

Siddiqi, A. (2008) "Imagining the Cosmos: Utopians, Mystics, and the Popular Culture of Spaceflight in Revolutionary Russia," *Osiris* 23: 260–88.

Skolkovo. Official site of the Skolkov Moscow School of Managers. Available at: http://www.skolkovo.ru/public/ru/campus-general/campus-malevich (accessed 4 June 2012).

Sviatogor, A. (1922) "Nashi utverzhdeniia." *Biocosmism* 1: 7–10. Available at: http://www.scribd.com/doc/104853/-14 (accessed 4 June 2012).

Toporov, V. (2003) *Peterburgskii tekst russkoi kul'tury*. St. Petersburg: Iskusstvo.

Uspenskii, B. (1994) "Raskol i kul'turnyi konflikt XVII veka." *Izbrannye trudy,* Vol. 1. Moscow: Gnozis. 333–67.

Vainshtein, O. (2006) *Dendi. Moda, literature, stil' zhizni*. Moscow: Novoe literaturnoe obozrenie.

Vernadskii, V. (1938) "Avtotrofnost' chelovechestva." *Russkii kosmizm: Antologiia filosofskoi mysli.* S.G. Semenova and A.G. Gacheva (Eds.). Moscow: Pedagogika Press. 288–303.

Part II

Film, animation, and computer games

6 Special/spatial effects in Soviet cinema

Birgit Beumers

Introduction

First patented in 1888, the film camera is, on the one hand, an optical device to capture movement—rather than static, posed images created by the photo-camera; on the other hand, it is a contraption that allows the user to look at the world through a lens. Such a lens can sharpen or blur the field of vision, or it can simulate depth by using two lenses and replicating the eyes' stereoscopic vision. Thus the film camera plays with our vision, tricking the viewer to surrender to optical illusions, while also recording movement in space.[1] In the earliest experiments with zoetropes and thaumatropes it was indeed the human eye, not the camera that tracked the movement of the separate images, filling the gaps, as it were. This combination of optical trick and recording movement still lies at the heart of the cinematic process today—suffice it to look at the use of motion capture in computer-generated images (CGI) and other special effects that make us believe we see a human body (or alien creature) when we actually see a digitally mastered creation.

In this chapter I analyze the techniques that have allowed filmmakers to create images of movement into other spaces, of (space) flight, and of an imagined life on other planets, and thus develop an aestheticized image of the cosmos. Instead of dwelling on ideological issues of the space race, I focus on the technical achievements of the Soviet film industry—sometimes even superior to Hollywood—when creating films about space and other planets. Soviet studios were not only capable of producing high-quality color films as early as the 1930s, using the three-color system developed by Pavel Mershin at Mosfilm (Maiorov 2011), but in the 1950s they also paralleled the American film industry's development of surround cinema with the system *Kinopanorama*—the Soviet version of *Cinerama* (Krukones 2010). Moreover, 3D technology was pioneered by the National Scientific Institute for Photography and Film (NIKFI) in Moscow, which received an Oscar in 1991 for continued work on the creation of stereoscopic films.[2]

Research on technological aspects and developments in Soviet cinema is a relatively novel area that has been emerging in recent English-language publications in the field, notably exploring sound systems (Bohlinger 2013; Kaganovsky and Salazkina 2014), camera use (Cavendish 2013a, 2013b), 3D and color technologies (Mayorov 2012a, 2012b), and *Kinopanorama* (Krukones 2010). This chapter attempts to redress the balance and set the aesthetic and thematic aspects of space stories in the context of technical achievements, therefore necessarily drawing occasional parallels to the US film industry without, however, aiming at a full-blown comparison of both industries.

In what follows I first look at early trick films, which created the illusion of space, making full use of cinema's role as a 'dream factory.' Through the case of Pavel Klushantsev I investigate Soviet images of space in connection with the projection of a utopian socialist society and the use of special effects to simulate the conquest of other planets, by means of which the Soviet Union sought to assert its superiority over both American space and American film industries. Once the space race was lost and the Americans had landed on the moon in 1969, the Soviet film industry lost interest in creating convincing images of life on other planets: the dream factory had been stripped of its purpose and the country no longer required uplifting visions of utopian planets and societies. Instead, special effects shifted into visual arts and art-house cinema in the 1970s, a period that also saw the rise of the sci-fi adventure film. In the late 1970s and 1980s, however, special effects almost disappeared from Soviet cinema because they could not be produced at the same level as in Hollywood owing to a lack of funding. I conclude the chapter by questioning the status of the filmic document as a piece of manipulated reality, exposed as 'fake' in a post-Soviet mockumentary, *First on the Moon*.

Film tricks: early (Russian) cinema and the flight of imagination

The first stereoscopic cards had been around since the 1860s, and the stop-frame camera arrived in the 1890s. It captivated audiences with tricks and visual manipulation, but also through the novelty of the moving image, as exemplified in the audience reaction to the Lumière Brothers' *Train Arriving at La Ciotat* (1895): audiences jumped aside as the train seemed to come off the screen.[3] These effects could be compared to those of panorama cinema in the 1950s or modern-day 3D images: they depend on the audience's habits and expectations. As Norman Klein has argued, an interest in effects that serve to fool or shock the viewer has long been part of the tradition of visual culture:

> Special effects add layers in space (on the screen and inside the theater). They also add layers to time itself, through stop-motion, or morphing. Most of all, they mix animation with live action. For example, animation is the art of compositing frame by frame. Essentially all special effects are animation inserted into a live-action space, or time.
>
> (Klein 2004: 223)

Special effects are no more than images manipulated through editing and animation, and my chapter analyzes the special effects that allowed filmmakers to show life on other planets: showing movement away from the earth and into other spaces, symbolizing a drive for liberation and freedom, while at the same time visualizing and populating that utopia.

The cosmos as a space to be colonized or where human life could be extended has interested philosophers and scientists for centuries, beginning with the second-century Syrian rhetorician Lucian's gazing through a telescope on the moon—in order to see earth. The optical device has always functioned as an extension into new territories and simultaneously as a reflection of the existing world. Travel to other planets has enriched human experience and wisdom, as is the case in Ludovico Ariosto's epic poem *Orlando Furioso* (1516), where it is on the moon that the English knight Astolfo finds Orlando's wit.[4] This dichotomy of "technical utopianism and the mystical tradition of Cosmism" has informed the Soviet space fad, as Asif Siddiqi has shown quite convincingly (2008: 262). With the arrival of film, it was possible to engage in flights of imagination, visualizing those unrealized dreams about journeys into other lands, onto other planets, and beyond. If—especially during the 1920s and 1930s—the concern with (aerial) flight was connected with technical progress and the invention of flying machines, then filmic images of space travel had challenged the imagination from the moment the still image began to move.

Indeed, the earliest experiments in moving images were concerned with motion, raising the human body from the ground and into space, using the flight of birds as a model. At the turn of the nineteenth century, this interest in movement—and in the human body briefly detaching itself from the ground and defying gravity—preoccupied scientists, artists, and photographers alike. In the 1880s, Étienne-Jules Marey (1830–1904) used a photo-camera to capture the movement of a human body and reduce it to patterns that complied with the laws of physics. Through chrono-photography, a technique of taking images of a body in motion at split-second intervals with a photo-camera, he tried to simulate motion through image sequences to understand the mechanics of the body: "Marey thought of the body as an animate machine whose motion was subject to the laws of theoretical mechanics" (Braun 1992: xvii). He re-synthesized the chrono-photographic record, which allowed him to slow down rapid movement. Thus, in *The Flight of Birds* [Le Vol des oiseaux] he created a collage of still photographs (1887) as well as a series of 3D models (1890) of the phases of a pigeon's and a seagull's flight that were later arranged in a giant zoetrope. This enabled him to derive solutions for problems in the field of aerodynamics. Similarly, the earliest—albeit amateur—experiments in drawn animation in Russia refer both to the movement of dancers and to the flight of birds. Aleksandr Shiriaev's (1867–1941) paper film with depictions of birds reduces the flight movement to abstract lines that seem like birds' wings subjected to multiple exposure and, when animated, generate chaotic movement in space (Beumers, Bocharov, Robinson 2009).[5] Thus even before film could roll, photographers sought ways to represent flight; they were fascinated not only by the physical feat of a jump, a spin, or proper flight, but also by the camera's capacity to record movement and by the use of optical devices to simulate it. Thus the fascination with flight is simultaneously driven by a scientific interest and the desire to imitate flight to trick the viewer into believing in its possibility: as Siddiqi puts it, it is "a combination of technology, fantasy, and liberation" (2008: 264).

Alongside early cinema's attempt to capture movement with a largely statically positioned camera, cinematographers also worked imaginatively with the medium. Filmmakers could play tricks, stopping the camera to make changes on the set; or, by reversing motion, they could adopt a slower or faster pace, in addition to being able to use double exposure. The master of such trick films was without a doubt Georges Méliès (1861–1938), who deployed such tricks to create the illusion of space and space flight. One of Méliès' earliest films, *The Astronomer's Dream* [La lune à un mètre 1898], depicts the moon in a fantastic, romantic fashion, showing it both as a circle with a face and as a sickle. *A Trip to the Moon* [Le Voyage dans la lune 1902] presents a narrative: a group of gentlemen-astronomers inspect the moon through a viewing device before boarding a rocket that takes them to the moon. While they travel, the moon acquires human features and they land smack in the moon's face. On the planet they descend into the moon's inner world, where they encounter strange creatures: a potential enemy has to be fought off before the gentlemen can make a successful return to earth, and their lunar rocket dips into the sea before the passengers reach the mainland. Such details—the depiction of the impact on the moon, the landing in the sea—show a great deal of prescience about technology: it is precisely in the sea that rockets would land half a century later. At the same time Méliès clings to fairy-tale ideas of anthropomorphized planets ("the man in the moon") when he shows the rocket hitting the moon's face, the stars as cabaret dancers, or Jupiter as a grumpy old man. These images reflect the fantastic side of his depiction. In the colored version of *A Trip to the Moon* (1903),[6] Méliès uses rich colors for lunar life, which he juxtaposes with the black-and-white dullness of the earth, where the astronomers are dressed in pale-colored outfits. Life on the other planet is exotic, rich, and colorful, but it also represents a potential threat that makes earth a safer place, after all. Tricks (stop-frame) and special effects (drawn animation combined with live-action) are deployed to create another space: an alien world, where even the underwater world into which the rocket plummets looks familiar and comforting.

Film, then, can be used to trick, to make believe, to simulate, and to delectate the audience with an experience of what has not happened yet or what may never happen. Through its possibility of visual tricks, the medium of film invites the 'illustration' of the fantastic and imaginary, and the combination with the fantastic in turn creates scope for imagined views of a utopian and scientific world: science fiction. It is no coincidence that the first films based on fairy tales used special effects and animation for the magic sequences, whether by Méliès or Władysław Starewicz in his *Christmas Eve* [Noch' pered rozhdestvom 1912], which shows Ivan Mozzhukhin's devil on the roof with the witch Soroka as they play with the moon's sickle. Later in the film, through the use of double exposure and animation, the devil is shrunk to pocket size so that he can help steal the tsarina's slippers. Both effects were created by applying a stop-frame technique within a live-action film. The use of tricks to project dreams onto the silver screen was widespread in the early years of cinema, when animation and stop-frame live action served to manipulate the image to render the supernatural elements of the narrative or to visualize dreams of technical progress: the flight to another place or transportation by magic, as, for example, in *Christmas Eve*, where the devil takes the blacksmith to St. Petersburg on the witch's broom. The technical means for this manipulation of images were inherent in the mechanism of the film camera, allowing the filmmaker to edit or to stop the film from rolling.

In the 1920s, Lev Kuleshov experimented with the juxtaposition of images to create the impression of a non-real world. He researched the effect of montage to manipulate, for example, city geographies by inserting close-ups of houses into streets of a different location; or by combining the handshake of two actors with the face of another actor; or by showing a girl in a dressing room and juxtaposing images of her with shots of lips, eyes, and nose belonging to a different actress. He thus determined that the content of a frame changes its meaning depending on the context: "In the combination of the filmed fragments the important thing is the dependence (the relationship) of the first filmed fragments on the second and this dependence is the principal organizational element in the construction of a film" (Kuleshov 1922). Kuleshov argued that the content of one frame is not as important as two frames joined together: he juxtaposed a shot of Ivan Mozzhukhin from an old film, where he expressed the pains of love, to a frame showing a bowl of soup in a close-up—and the emotion in Mozzhukhin's face turned into hunger; when he was placed next to a coffin, his facial expression became one of grief. Kuleshov realized the possibility of creating a different (imagined, artificial) space by filming people in Moscow and transposing the shots onto images of the White House. Various technological opportunities of film—drawing, double exposure, editing—allowed filmmakers to create illusions of another reality, and soon they would use these skills to show other planets. These technical possibilities began to fascinate filmmakers, who, rather than artists, were engineers of the image: the 1920s saw not only experiments in montage and sound, but also the development of drawn animation, which could be combined with live-action shots—a technique of combination shots for which Aleksandr Ptushko would gain fame in the 1930s.

Soviet filmic images of the future

In the context of the rapid industrialization in the 1920s, but also in the broader context of mechanization and the rationalization of labor (Taylorism), engineers and scientists appear more and more frequently on screen—from Prite, the pioneer of electricity, in Lev Kuleshov's *Engineer Prite's Project* [Proekt inzhenera Praita 1918] to Podobed in his *Death Ray* [Luch smerti 1925], a film concerned with the invention of powerful rays while exploring the theme of industrial sabotage. Iakov Protazanov's *Aelita* (1924) centers on the imagined journey of engineer Los to planet Mars and is preoccupied with the idea of taking the revolution to outer space. Aleksandra Ekster's Constructivist set designs for life on Mars reflect the aesthetic interests of the time, combining technology with artistic vision. A similar idea—taking the revolution to other lands and planets—lies at the heart of *Interplanetary Revolution* [Mezhplanetnaia revoliutsiia 1924] (Fig. 6.1), an animated film by Nikolai Khodataev, Zenon Komissarenko, and Iurii Merkulov initially conceived as an insert for *Aelita*, where the Bolsheviks export the revolution to Mars, whence the world's capitalists have escaped, but the planet is already in the hands of the revolutionaries. While offering a visual parody of *Aelita*, the film remains thematically within the tradition of a global confrontation, pitting capitalism vs. communism in a manner typical of political cartoons of the late 1920s. *Interplanetary Revolution* used to great effect the innovative technique of transposition [perekladka] to move objects from frame to frame to create a sense of movement.

In the 1930s, technology continued to play a major role in cinema, but more in the form of tractors and machines, telephones and new inventions, workers and engineers. Yet flight was represented on screen as well, thanks to the major impulse that came from Konstantin Tsiolkovskii (1857–1935), who was the first to conceive the multi-stage rocket. A flight to the moon is shown again in Grigorii Aleksandrov's *Circus* [Tsirk 1936], where the circus number of a flight to the moon—a sickle with a face—accomplished by the American Marion Dixon is offset by its Soviet counterpart, by the engineer Martynov (Sergei Stoliarov): the "flight into the stratosphere," repeating the title of an essay by Sergei Korolev published in 1934.[7] Stoliarov also played the rocket launcher in Vasilii Zhuravlev's *Space Voyage* [Kosmicheskii reis 1935], which shows a space flight that would take place on 9 August 1946. The film drew on Tsiolkovskii's construction designs for rockets and told the story of the boy Andriusha—an inventor of a telescopic slingshot—who sneaks onto the spaceship traveling to the moon. The launch pad

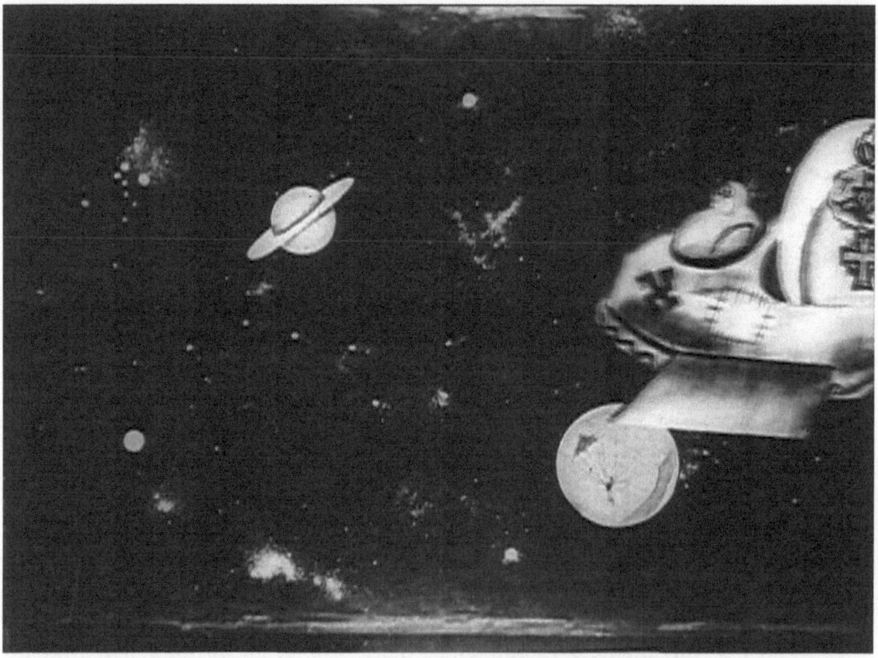

Figure 6.1 Still from *Interplanetary Revolution* (1924), an animated film by Nikolai Khodataev, Zenon Komissarenko, and Iurii Merkulov. DVD

of the rocket is reminiscent of the Constructivist curved slopes leading to elevated tribunes in *Aelita*, while the huge bridge that serves to launch the rocket and the hangar clearly reflect Stalin's Grand Plan for the reconstruction of Moscow and especially designs for the Palace of Soviets. While the engineering design is quite ambitious, the images of the moon are rudimentary. The greatest problem for the film's crew was the depiction of weightlessness, which looks clumsy. Instead of floating in space, the actors perform some acrobatic jumps. Despite the use of special suits and demonstration of technical detail such as liquid fuel properllers, multi-stage rockets, and atmosphere re-entry, all derived from Tsiolkovskii's concepts, the actual body weight betrays the illusion of zero gravity. It is precisely in the context of the representation of space and weightlessness that the work of Klushantsev would become most important—but not until after the war.

Soviet cinema lands on the moon: sci-fi and popular science film

During the Cold War, the Soviet Union accelerated its space program: the launch pad at Baikonur opened in February 1955; the first Sputnik satellite circled earth in 1957, followed by a capsule carrying the dog Laika;[8] the moon probe Luna 1 was launched in 1959; and in April 1961 Yuri Gagarin became the first man in space. He was followed by the first woman in space, Valentina Tereshkova, and the first space walk, performed by Aleksei Leonov in 1965.

Before, during, and after these milestones, the film pioneer Klushantsev experimented in various endeavors to achieve an accurate representation of cosmic space on screen. Obsessed with outer space, he demonstrated an incredible inventiveness and ingenuity for creating special effects, which he used in his popular science documentaries and his one (science) fiction film. Klushantsev was the first filmmaker to show a meteorite shower, northern lights, orbiting planets and stars, and the surface of other planets in a convincing manner. He achieved this through technical feats, such as combined animation and nature shots; reverse shots (e.g. for an attack by plants, he reversed motion so that the capture of the tentacles looks real); underwater footage; and shots from above or below along a vertical axis. Through these technical inventions he also managed to create the impression of weightlessness—the zero gravity that people could only experience after 1959 and for 25 seconds at a time on a parabolic flight (the so-called 'vomit comet'). Considering that NASA aircraft were used to create similar effects for Hollywood films in the 1960s, what Klushantsev accomplished with his cameraman Anatolii Lavrentiev and designer Iurii Shvets from the late 1950s onwards through meticulous frame construction is a genuine feat, and it is not surprising that his effects would be pilfered in a number of Hollywood B-movies in the early 1960s, such as *Battle Beyond the Sun* (1959), Curtis Harrington's *Voyage to the Prehistoric Planet* (1965), and Peter Bogdanovich's *Voyage to the Planet of Prehistoric Women* (1968).

In his *Northern Lights* [Severnoe siianie 1946] he created the effect of the aurora borealis in the sky; in *Meteors* [Meteority 1948] (Fig. 6.2) he showed the movement of planets using a method he called "luminostaging":

> Lavrentiev and I used the luminostaged method of shooting which was later patented in his name. We had to show the solar system with the movement of the planets and orbital paths of meteors. Animation wouldn't have worked here, since all of the action had to be of a floating nature and animation only worked for sharp, quick action. . . . We coloured the celestial bodies with paint and illuminated them with ultraviolet light. . . . On film it appears that we filmed only those celestial bodies, and all of the unlit mechanisms remain invisible.
> (Barker and Skotak 1994a: 82)

In his subsequent film, *Road to the Stars* [Doroga k zvezdam 1957], he showed a rocket launch in such detail that even the Soviet inventor of the rocket, Sergei Korolev, not only approved it, but remained stunned.[9] Most importantly, Klushantsev found ways of creating a sense of weightlessness on screen. One method was a vertical shot of the actor suspended above the cameraman, with the ropes hidden by the actor's body, thereby achieving a sense of floatation. Another, more complex method functioned as follows (Fig. 6.3):

> We manufactured a latticework barrel four metres in diameter and six meters long out of wooden beams. We cleared out the workshop so that the barrel—which represented the cabin of the spacecraft—could roll back and forth across

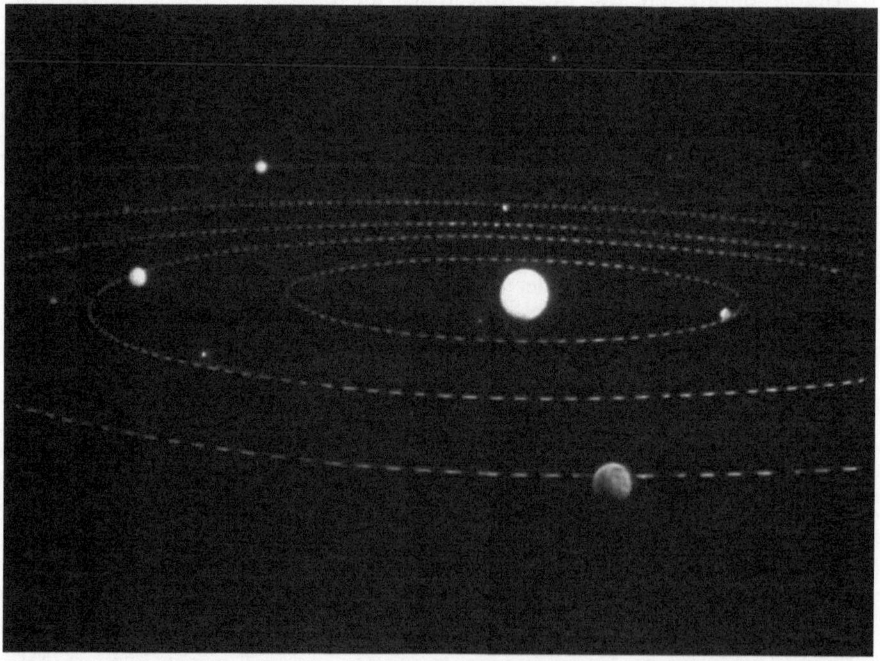

Figure 6.2 Still from Pavel Klushantsev's *Meteors* (1948), showing the use of luminostaging. DVD

the floor. At the end of its path we made ramps so it didn't smash into the walls but would roll up the ramp, and roll back. We mounted set decorations—all of the lighting equipment, the cameraman and his camera—inside the barrel. The cameraman was in a special long box, fastened with straps, so that during the rolling of the barrel, he would be first right-side up, then upside-down. Two actors worked in the scene. One was firmly fastened to the roof in a chair solidly attached to the set. The other was able to freely walk or jump on the floor, walls or ceiling of the 'cabin'. The camera angle was always in the same position and the rotations weren't observed. Our cosmonaut freely crossed from the floor to the wall, then to the ceiling and walked about. For greater effect the shot was overcranked and the action seemed more slow and floating. In any case, it made the onlooker and the film workers feel perfectly nauseated.

(Barker and Skotak 1994b: 77–8)

Klushantsev's *Road to the Stars* coincided with Soviet and American preparations for space flight, but preceded the successful launch of a Sputnik in October 1959, at a time when the Americans failed to get a rocket into orbit. The competition between the United States and the Soviet Union took place in terms not only of achievements in space travel, but also of cinematic advances and visual language.[10] Indeed, Klushantsev's depiction of weightlessness inspired Stanley Kubrick's

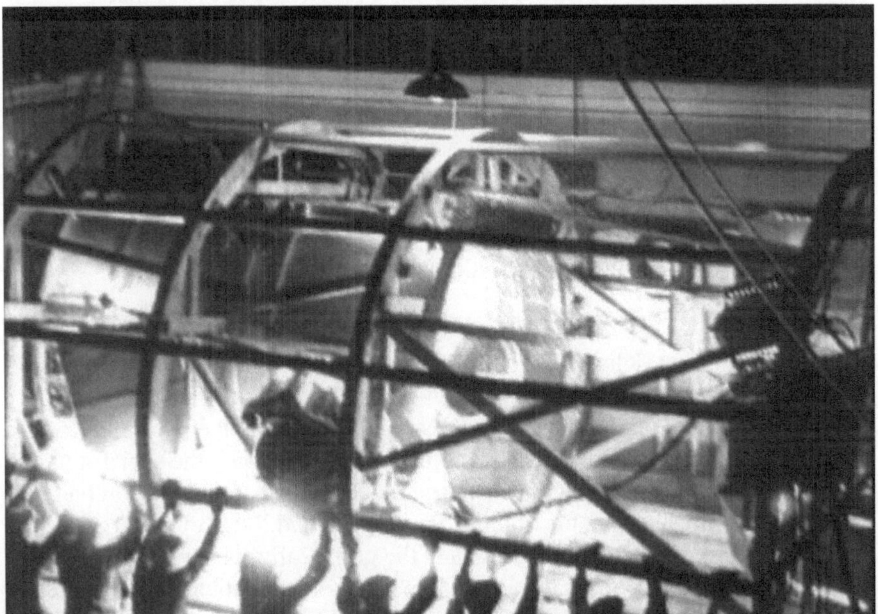

Figure 6.3 Klushantsev's latticework barrel used to create the effect of weightlessness. Courtesy of Viktor Bocharov, author's personal archive

Space Odyssey 2001 (1968), but, even more importantly, many of Klushantsev's tricks were lifted and assimilated by Hollywood filmmakers. Klushantsev's only fiction film, *Planet of Storms* [Planeta bur' 1961], showed numerous facets of an imaginary planet visited by a space crew. Klushantsev used special effects to depict a lunar landscape, hovering vehicles, a robot, volcanic eruptions, and underwater scenes. Purchased by producer Roger Corman and re-edited, the film featured some beach babes (Mamie van Doren, Mary Marr, Paige Lee) in a cave on the planet to make it more palatable to US audiences, where it was released as *The Voyage to the Planet of Prehistoric Women* (1968, dir. by Peter Bogdanovich). Klushantsev's special effects were simple yet impressive: he moved the lunar vehicle from behind by an invisible crane that was hidden by stones and rocks. The Robot 'John' was built by Aleksandr Nadezhin and manned by a weightlifter capable of lifting two astronauts. Different perspectives were used for combined shots: the space ship Sirius was filmed at 1/5 scale and from a 50-meter distance, while the ship's lower part was filmed in full size at a distance of 75 meters (Barker and Skotak 1994b: 80). Puppeteers from the Leningrad Puppet Theater designed and operated the pterodactyl bird. Underwater shots were created using thick fog that looked like silt, dispersing when people moved through it. Moreover, fish tanks were placed on the set through which the camera would film to create the impression of sea life.

Stunning effects were also used in *Moon* [Luna 1965], where Klushantsev showed a lunar vehicle [lunakhod] that looked exactly like the spider-legged lunar vehicle used on the Apollo missions (Fig. 6.4). As Klushantsev recalled,

> Shvets and I decided to film with a camera lying on its side. That way, the camera's motion wouldn't have to be on a vertical plane. It could ride horizontally on rails in a cart. We built the moonscape set on its side. It was 16 meters high. We had to carry out the shot in slow motion. We placed a specially prepared reduction gear in the camera, shooting roughly two frames per second. The vertical position of the moon's surface caused us to wait half a day before filming the crater Copernicus, to get the beams of light which occur during sunrise, when it is horizontal.
>
> (Barker and Skotak 1994b: 81)

In the film *Moon*, Klushantsev fully reveals his obsession with a perfect and convincing image of 'the other planet' as well as his idealism—his belief not only in finding oil, water, and minerals on the moon, but also in the possibility of life there, and of a better society. What he projects onto the moon is the socialist utopia that had not worked out in his own country. Indeed, Klushantsev himself had narrowly avoided arrest during the Stalinist Purges, and had lived for a year under the Leningrad Siege, so he had seen first-hand the grim reality of the communist project. In this context, his belief in a better life, where space stations on other planets are decorated with grandmothers' pots, plants, and flowery curtains, explains the utopian quality of this film, which combines with its scientific detail.

The case of Klushantsev shows the Soviet film industry's advanced position in creating special effects, which were exploited by the American dream factory.

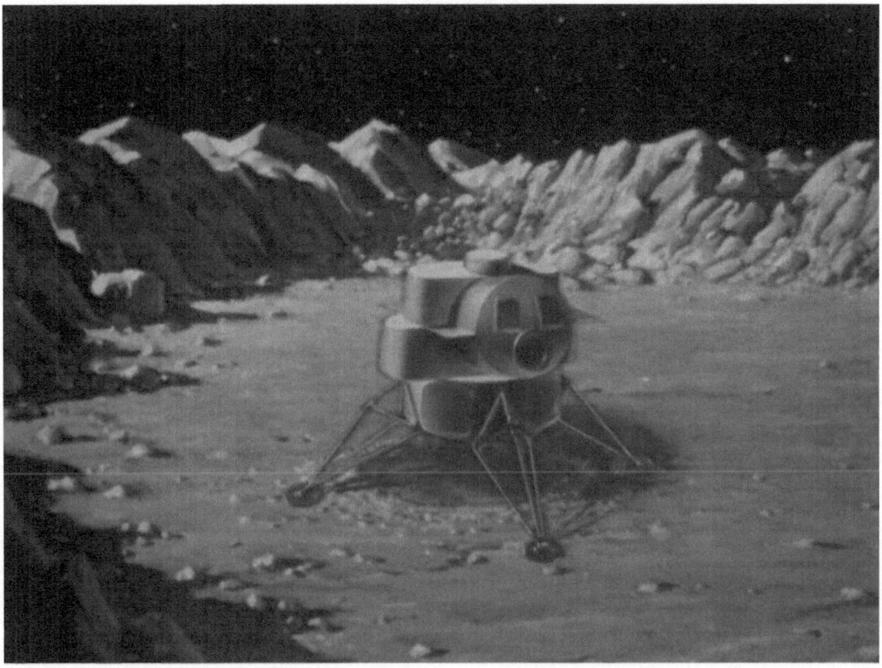

Figure 6.4 Still from Pavel Klushantsev's *Moon* (1965), showing the 'lunakhod.' DVD

They were undervalued by the Soviet system, which dismissed Klushantsev from his job at Lennauchfilm in 1974 because his visual creations showed, after all, the one achievement in which the Soviet Union had not succeeded—the landing on the moon. In the 1970s, this failure was too painful a reminder for the subsequently defeated superpower. While capable of creating an illusion of life in space, cinema could also remind viewers of what the USSR had not attained in reality.

The silver screen has always invited the visualization of man's unachieved dreams through special effects and tricks. Though Soviet cinema had created a utopian and varnished version of reality during the 1930s, it largely neglected the achievements of its pioneers: in color, stereoscopy, panorama cinema, and the work of Klushantsev—especially the methods he developed to create the impression of weightlessness. The regime, fully aware of the myth-making function of cinema, nonetheless refrained from using it to this end in a field where it strove for true superiority: technology.

Following a series of failed missions and the death of Sergei Korolev in 1966, the Soviet space program stalled: the Politburo shut down the moon program after the American landing on the moon. Popular science films became a thing of the past, while sci-fi was relegated to entertainment for children, as evident in the works of the fantasy writer Kir Bulychev and the sci-fi films of Richard Viktorov.

Life on other planets: the Soviet take

Soviet cinema of the 1950s produced a significant science-fiction film on space travel, Aleksandr Kosyr' and Mikhail Kariumov's *The Sky Is Calling* [Nebo zovet, Dovzhenko Studio 1959], which contains some stunning special effects while remaining quite conventional on the narrative level. The plot, essentially adhering to Soviet revolutionary romanticism, revolves around the competition between an American and a Soviet spaceship (called *Typhoon* and *Rodina* [motherland], respectively), both bound for Mars. The Soviet crew has to rescue the American astronauts who risk the mission and their lives when they want to reach the planet first. Roger Corman and Francis Ford Coppola re-edited the film and released it in the United States under the title *Battle beyond the Sun* in 1963, changing the plot to a conflict on earth as it stands divided between North and South. Noteworthy here is the work of production designer Iurii Shvets, who previously had worked on Aleksandr Ptushko's *New Gulliver* [Novyi Gulliver 1935], where combined shots were used to juxtapose the Lilliputians to the 'giant' humans, and also on *Space Voyage*. *Battle beyond the Sun* contains a scene of weightlessness when the cosmonauts walk on the planet Icarus, which is filmed against black cosmic space as a backdrop. However, the special effect of people moving in supposed weightlessness remained unconvincing: the construction of a transparent platform to allow movement in space was all too obvious. The space station itself is captured against a background that is in motion, while the actual ship remains static in the foreground. The Soviet space station replicates earth: there are flowers, climbing plants, fruit, and vegetables in abundance. Indeed, the characters walk normally, standing with their feet on a platform (and on the spaceship), dressed in overalls and wearing helmets. This image ties in with the nostalgia for earth that prevails throughout the space journey: one cosmonaut even takes a little bag of soil from earth with him on the mission. This attachment to earth through soil, greenery, and domestic trinkets would be repeated in Tarkovsky's *Solaris*.

Undisputedly, Klushantsev made groundbreaking discoveries in using special effects to represent outer space on screen, to visualize other planets, and to engage with the phenomenon of weightlessness. His work was extraordinary to such an extent that many of his effects—and two Soviet films, *The Sky is Calling* and *Planet of Storms*—were re-edited for American audiences because US studios could not yet achieve such convincing effects. However, it was Stanley Kubrick's *2001: A Space Odyssey* (1968) which revised and refined standards for the image of outer space on screen. No doubt Kubrick was influenced by Klushantsev and/or by the Americanized, re-edited versions of Soviet space films. His original contribution consisted of adding another dimension of cosmic narrative in the innovative design of the spaceship, which made cosmos not a space where we import bits of home to make it feel comfortable, but a space of a minimalist design, for which Kubrick used largely primary colors in a setting informed by and influencing modern interior design of the time. Moreover, Kubrick perfected some effects, in particular the technical accuracy of weightlessness. In a manner reminiscent of Klushantsev, he created the effect of loss of gravity:

> To compensate for the weightlessness of outer space, the ship's crew compartment was envisioned as a centrifuge that would simulate gravity through the centripetal force generated by its rotation. A 30-ton rotating 'ferris wheel' set was built by Vickers-Armstrong Engineering Group, a British aircraft company . . . . The illusion of astronauts floating in space was created by hanging stunt performers upside down with wires from the ceiling of the studio, often for hours at a time.
>
> (De Met 1999)

While Soviet special effects made their way into Hollywood films in the 1960s, after the Soviet failure in the mission to the moon, the production of science fiction and popular science films in the USSR was neglected. Those few sci-fi films that did appear mostly targeted the young generation. At that time Soviet audiences had no access to American sci-fi movies, which in the 1970s began to deploy CGI graphics to perfect effects of weightlessness and other cosmic phenomena. The first computer-generated images were used in *Westworld* (1972) for the infrared point of view of a robot. Subsequently, the formation of Industrial Light and Magic (ILM) by George Lucas developed a motion control camera for *Star Wars* (1977). These American space adventure movies were often set on space stations, which were packed with objects from earth and often replicated home. The link to earth was further strengthened through messages that functioned as visual and acoustic reminders of home, of the past, and of the power of memory: alongside curiosity about other planets, there is a longing for home, and journeys are frequently undertaken not for the sake of gaining new knowledge, but to bring peace to other planets and civilizations.

This tendency of maintaining a link to earth is reflected in Richard Viktorov's *Moscow-Cassiopea* (1973) and *Teens in the Universe* [Otroki vo vselennoi 1974], which portray space missions of teenage Pioneers, who reach Cassiopeia to save the planet from the threat of destruction by an alien force that threatens its population with the removal of feelings to create the perfect man.[11] The starship [zvezdolet] has a modernist design typical of the 1960s, in steel and glass, with long corridors and windows, doors shutting and opening at the push of a button. Magnetism cancels the weightlessness, allowing the teenagers to move normally, except for one scene in which they deliberately try out weightlessness. This sequence is shot in a small cabin, where one of the boys is always in contact with a wall, a ceiling or a chair, making clear that this scene was shot in a rotating container or with a tilted camera. The film also includes a scene of a space walk when one of the boys accidentally ejects into space. Yet what is important for Viktorov is not the technical side of the film, but the story and the ideological message: that even a lazy boy can be integrated into the collective. The mission to restore feeling to the inhabitants of other planets is accomplished, placing emotion over technology. Quite an original addition to the spaceship is the leisure chamber, which takes the teenagers back home. Nostalgia and a sense of domesticity, a longing for childhood, seem almost out of sync with the Soviet ambition for progress and continuous emphasis on technology and rationality. Moreover, the 'other' planet resembles earth—unlike the Planet of Storms, which looks uninhabitable and exotic. *To the Stars by Hard Ways* [Cherez ternii k zvezdam 1980] continues the theme of space travel from the earlier films, but now space journeys are by public transport from the newly built terminal 'Sheremetyevo 2' in Moscow, the destinations encompassing a range of planets.

Metaphysics in space

As the Soviet space program stalled, the interest in science within visual arts shifted toward the experience of cosmic space as a mystical experience, returning to the ideas of Cosmism that had inspired the earliest space experiments in science and in its artistic representation (Siddiqi 2008: 280–83). This trend can be seen in Andrei Tarkovsky's *Solaris* [Soliaris 1972], based on the novel by the Polish science-fiction writer Stanisław Lem. The painter Mikhail Romadin (1940–2012) created the design for the space station, which echoes the style of his canvases and book illustrations: he is preoccupied with different perspectives and angles that destabilize vision, creating a mere illusion of scientific design. Men travel into space not to engage with and participate in the mechanical world that has dominated Soviet visions of the cosmos, but to take with them the frills and ornaments, the clutter and décor of their earthly life: books, Brueghel's *Hunters in the Snow*, Bach's music, and a plant. Weightlessness in Tarkovsky is visually not so different from the levitation scenes in *Mirror* [Zerkalo 1975] and *Nostalgia* (1983), where floatation in space stands for a different, heightened awareness of the self: "the characters' weightless state and fluid movements seem to remind them of their precarious state between the cosmic void and the alien flows of Solaris" (Bird 2008: 65).

The characters are neither at home nor in space, but in the flow between two spaces, where Tarkovsky is interested in the state of the in-between. Not unlike Lucian gazing through the telescope to look at the earth, Tarkovsky is interested in the other—cosmic—space as a mirror for the self: the ocean displays human consciousness, offering a mirror of a different kind from the shiny and reflective surfaces that are typical of modernist space stations. The frequent use of shining surfaces, mirrors, and water in his films speaks to this point. Scientific journeys into space thus primarily serve to visualize the experience of the self, delving into the mysterious depths of the human psyche. The conspicuous absence of special effects as a purely technical concern emphasizes Tarkovsky's preoccupation with metaphysical issues and his interpretation of special effects as a destabilization of social and moral laws.

If sci-fi films tended to engage with the thirst for discovery and the desire to bring good human values to other civilizations (i.e., to rescue a mission and/or to save a planet), then this 'missionary' role of the Soviet space expedition is rejected in *Solaris*: human life is but a minute particle in the cosmic system. Vlad Strukov has suggested that *Solaris* undercuts the aviator myth in Soviet culture: "Kris and Burton [sic] fail as potential aviation heroes: they do not fulfil their task. Moreover, they engage in a series of personal crises at the expense of their duty towards the state" (Strukov 2008: 63). Tarkovsky refrains from glorifying technology, showing instead its dehumanizing effect and exposing the absurdity of man's ambition to understand, not to, say, dominate, space. Here he agrees with Lem, who saw humankind as a "speck of dust" in the universe. Tarkovsky therefore sets two-thirds of the film on earth, and his spaceship is a messy heap of metal; only the ocean is visually attractive. Taking Strukov's comment a stage further, we can see Tarkovsky as going against the trend of his time, and the aestheticization of scientific and geometric forms, as, for example, in the kinetic art of Francisco Infante-Arana (b. 1943) and Lev Nussberg (b. 1937) or the geometric works of the Hungarian-French father of Op-art, Victor Vasarely (1906–97).

The post-Soviet return of special effects

Special effects created the illusion of space travel: as a trick image they projected an alien space, a space for a utopia, or the space of a journey into the self. In all these instances, the filmmaker used space to project, to invent, and to imagine life through optical tricks (the technical feat) and illusions as a mirror-image of life on earth (the spiritual investigation)—either as a reflection of the Self or a different Other altogether. During the 1970s and 1980s, special effects did not advance in the Soviet film industry, leading to a significant hiatus in technological developments in comparison with the rapid development of technical novelties in the USA. It took Russian cinema until Timur Bekmambetov's 2004 film *Night Watch* [Nochnoi dozor] to generate special effects at a level sufficiently sophisticated for films to compete once more with American cinema—at least on the domestic market. Similarly, Fedor Bondarchuk's *Inhabited Island* [Obitaemyi ostrov 2009] displays the full range of special effects, using them purposefully and making them integral to the story. Having technically reached a level to create special effects in a satisfactory way, Russian cinema still has another issue to address: the ethical issue of manipulating images to create utopian worlds and fake histories, without returning to Socialist Realist methods of 'varnishing' reality.

The invention of the medium of cinema has always been closely linked with tricks and effects that can be achieved by manipulating images. Cinema has the power to make visible what the eye cannot (yet) see—using close-ups; it can transport the viewer into spaces that are not (yet) accessible—through combination shots; and it can therefore take the viewer into space long before the first cosmic journeys. If Vertov stated that the camera is the perfect eye and can perfect reality through editing and manipulating images, then in modern cinema the camera is used to make visible the artificiality and fakeness of images, and not necessarily perfect them. The camera can map not only territories and countries,[12] but also the cosmos, extending the concept of human conquest through visual assimilation to hitherto unknown spaces. The lens of the camera is therefore a powerful tool, and, aided by the editor's scissors, it selects and adapts images and presents them to the viewer as documents of empirically verifiable events or transports the audience into utopian lands. At the same time, the filmmaker can manipulate the viewer into experiencing what is simulated, into believing half-truths, into accepting for real that which is faked. As the industry masters the effects, alternative cinema questions the role of the moving image.

In Aleksei Fedorchenko's mockumentary, *First on the Moon* [Pervye na lune 2005] (Fig. 6.5), what appears to be a documentary cinematic image is faked; at the same time, the film establishes that the same cinematic image is the only source for what actually happened. Thus, all visual (cinematic) evidence is at the same time the only valid document, yet also invented, doctored. In other words, we have no other means than cinema to visualize history, but it is always a falsified document, so we might as well use it for the purpose of invention, optical illusions, and tricks. *First on the Moon* begins in Chile in 1938, where a meteorite impact is captured on film, and the celluloid 'record' is preserved in the UN film archive. Fedorchenko then presents a history of rockets from the Middle Ages to recent times before cutting to film chronicles from 1936, showing the obsession of pioneers and scientists with rocket-designer Tsiolkovskii's findings. The newsreel *Soiuzkinokhronika 54*, dating from November 1936, shows a group of pioneers grinning at the viewer following a short performance in honor of Tsiolkovskii. It features, among others, today's literary advisor of the Theater of the Young Spectator in Fedorchenko's native Ekaterinburg, Oleg Loevskii, in the role of the lecturer who tells scientists about the need for man to become physically hardened. The viewer realizes that these chronicles are neither filmed with the objects 'caught unawares,' nor were they made in 1936.

Indeed, Fedorchenko references the practice of 'creating' evidence—historical and biographical—through film, a method related to the practices of secret police investigations, as suggested by Cristina Vatulescu: "Filmmakers took on more than the look of the secret police: they were instrumental in its major projects, such as identifying the new profile of the socialist criminal/enemy and molding the public response to it" (Vatulescu 2010: 77). Exploring the influence of the police file on literary and cinematic practices of the 1920s and 1930s, Vatulescu argues that "the filmmaker should not only portray but actually emulate the secret police agent's work and approach to reality" (Vatulescu 2010: 78). Unsurprisingly, then, Fedorchenko's film continues by intercutting archival chronicles (in black and white) and comments from survivors (in color) with more black-and-white footage that the film crew allegedly found in the secret police archive. The KGB archivist makes a pertinent comment: "Everything that happened was filmed; therefore it had happened," thus underlining the power of the documentary filmmaker. Fedorchenko fools the viewer with his alleged filmic evidence: almost 90% of the film consists of non-archival (i.e., fictional) footage. As Vatulescu argues, the secret police file creates evidence that is passed off as real and creates a semi-fictional document, or documentary fiction. The line between document and fiction is removed in the film, inasmuch as the police both collect and fake evidence about the objects of investigation. Fedorchenko's film raises ethical issues about the status of the (moving) image as document, and proposes, through mockery and exaggeration, that all images are manipulated. In so doing it suggests, by extension, that we can take either the American landing on the moon as cinematic fiction, or his fictionalized images as the truth.

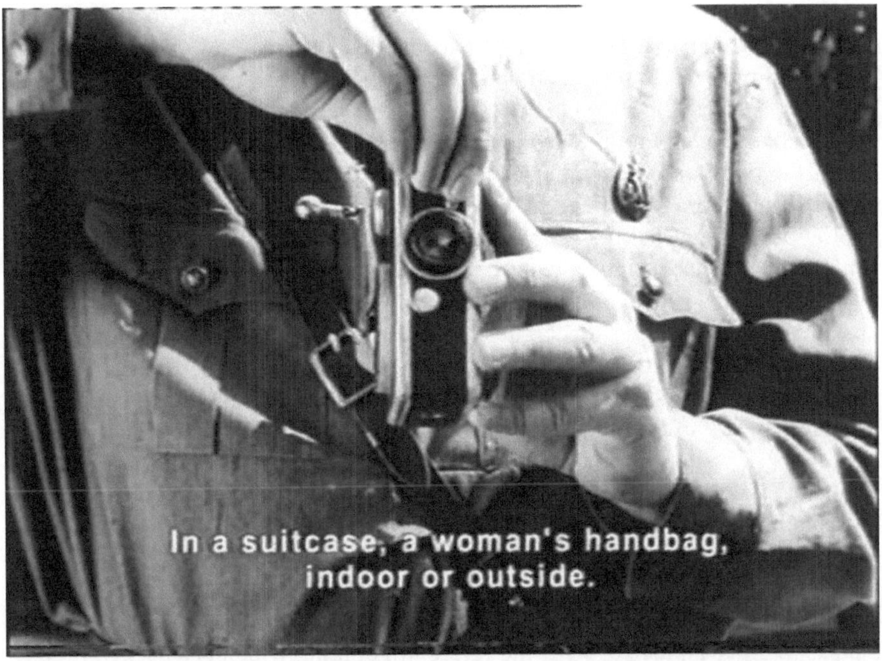

Figure 6.5 Still from Aleksei Fedorchenko's *First on the Moon* (2005): the use of surveillance cameras. DVD

First on the Moon 'investigates' Ivan Kharlamov, a target of secret police (NKVD) observation since 1935, who is filmed with a hidden SK29 camera. Comments from contemporaries are collated, and their short biographies are provided in the style of 'information for your consideration' of the famous spy thriller *Seventeen Moments of Spring* [Semnadtsat' mgnovenii vesny 1972, dir. Tat'iana Lioznova]. There are numerous references to other Soviet films, in particular to the 1936 *Space Voyage*, which is shown at a local cinema. The viewer is constantly reminded that if there is film, then the events occurred—even if a present-day engineer tells the film crew that such rocket designs could not have existed in the 1930s and this is all a hoax, and even if a group of young men reconstruct the rocket according to the original drafts and launch it (but it burns out). As suggested in the opening sequence, such a space flight was impossible in the 1930s, but the filmic evidence exists: the candidates' training exercises and experiments with animals have been filmed, so it happened (as the KGB archivist argues). According to this logic, then, the film *Space Voyage* would further confirm the viability of space flight in the 1930s. Or would it? All the traces of such a journey have disappeared. Or have they? The films were destroyed by the NKVD after the mission's failure; so what are we (the survivors, the audience) watching here? The film asserts that the launch of the rocket took place on 16 March 1938, and the fireball hit Chile on 24 March. Kharlamov made his way from Chile across the Pacific Ocean to Mongolia, then to Chita, where he was admitted to a psychiatric clinic from which he subsequently fled, having lost the power of speech. So even if the 1938 Soviet mission to the moon had been successful, the regime—governed by camera surveillance and control—would have destroyed the evidence. So does that mean that without filmic evidence, the mission did not take place?

On the one hand, Fedorchenko comments on the ultimately self-destructive nature of totalitarian regimes that wipe out files, and therefore events and people.[13] On the other hand, he mocks the utopian vision that this regime inspired while simultaneously undermining it. As one critic noted:

> It might appear that Fedorchenko tried to create a grandiose grotesque, a Russian analog to Paul Verhoeven's under-appreciated masterpiece *Starship Troopers* (1997), in which the 'cosmic' environment is used to analyze the mechanisms of total propaganda and the birth in society of the germ of fascism. (Kovalov 2006)

Ultimately, the archivist is right, and in more ways than one. First, cinema both documents and invents facts, and the camera is a tool for exposure and myth-making. Second, we believe (only) what we see—whether what we see are engineered feats, special effects, or trick films. If we are shown life on another planet, we have an idea of what the future might hold, which is why we are reassured by the analytical and scientific approach of popular science films.

Fedorchenko draws our attention to early experiments with the medium, linking them specifically to Soviet history of the 1930s, when images were manipulated for political purposes—a process that had started in the 1920s with Vertov's and Sergei Eisenstein's ideas of montage and that ultimately made the filmmaker a selector of reality, choosing 'evidence' from a range of narratives. However, experiments that involved the manipulation of images had started much earlier, and long before the advent of the camera.

Special effects, then, allowed filmmakers to project visions of life—maybe to be achieved on earth during the socialist utopia—onto other planets. The journey into space was not only illusionary, but also ambitious in terms of the technological standards not yet reached on earth (while ignoring, ironically, the achievements in cinematic art). Following the collapse of socialist ideologies and the infrastructure of Soviet film production, cinema—in self-reflection—critically assessed the status of the image in general as a source for manipulation, dismantling its myth-making capacity and leaving special effects where they truly belong: to action and science-fiction films and to the Russian 'majors' that have been producing CGI animation and 3D films since 2004.

Notes

1 For an animated explication of some early optical devices, see the app "Walking Pictures" developed by Beumers and Julia Bracegirdle with Aerian Studio, 2013.

2 This system set two images side by side on 70 mm, and was limited in its application, for it required a special projector. It has been used largely by the Estonian animation studio NukuFilm.
3 For example, as recorded by Georges Sadoul (*Histoire générale du cinema*, vol. 1, 1946) and a number of other classical sources. This audience reaction has been debated in recent scholarship.
4 One of the poem's plotlines is linked to the war of the Saracens against the Christian emperor Charlemagne. Charlemagne's paladin Orlando neglects his duty and elopes with his love interest, and upon losing her, goes mad. Astolfo journeys to foreign lands and planets in search of a cure for Orlando's madness.
5 Aleksandr Shiriaev was a dancer and choreographer at the Mariinskii Theater in St Petersburg, assistant to Marius Petipa. Around 1900 he acquired an amateur film camera and recorded dances. Later (1906–9) he also made stop-frame puppet animation. While experimenting with the camera he made several paper films, including a recording of the "Buffoon's Dance" from *The Nutcracker* and "Cakewalk." Without indication of frame lines, he made a short paper film of the flight of birds, which was digitized by Aardman Animation and first shown in Pordenone at the Giornate del Cinema Muto in 2008 (see Beumers, Bocharov, Robinson 2009).
6 The color version was restored in 2011 by Lobster Film's Serge Bromberg and has been used extensively in Martin Scorsese's *Hugo* (2011).
7 See Chapter 8 for a discussion of this film.
8 For a discussion of the use of animals in the Soviet space programe see Chapter 8.
9 As stated in *The Star Dreamer* (Zvezdnyi mechtatel'), a documentary film produced by Sonia Vesterholt and Mags Bostrup, Vesterholt Film (Denmark) and MIRIS (Russia), 2002.
10 See the discussion below of Roger Corman's acquisition of Soviet films: the US film industry seemed to have no qualms in relying on Soviet effects for their film distribution sector.
11 A concept that features in Vladimir Maiakovskii's *The Bedbug* (Klop 1929) and in Evgenii Zamiatin's dystopian novel, *We* (My 1921). It also appeared in American films in the 1950s, e.g. *The Invasion of the Body Snatchers* by Don Siegel.
12 For a discussion of the assimilation of (Soviet) territories, see Widdis 2003; for the role of filmmaker, see Vertov 1922.
13 For an example of how people were effaced from photographic records, see King 1997.

Bibliography

Barker, L. and R. Skotak. (1994a) "Klushantsev: Russia's Wizard of *fantastika*." *American Cinematographer*. June: 78–83.

—. (1994b) "Klushantsev: Russia's Wizard of *fantastika*." *American Cinematographer*. July: 77–82.

Beumers, B., V. Bocharov, and D. Robinson. (Eds.) (2009) *Alexander Shiryaev: Master of Movement*. Gemona: Le Giornate del Cinema Muto.

Bird, R. (2008) *Andrei Tarkovsky. Elements of Cinema*. London: Reaktion.

Blackford, J. (2011) "Red Skies: Soviet Science Fiction." *Sight & Sound 9* (July): 44–8.

Bohlinger, V. (2013) "The development of sound technology in the Soviet film industry during the first Five-Year Plan." *Studies in Russian & Soviet Cinema* 7, 2: 189–205.

Braun, M. (1992) *Picturing Time: The Work of Etienne-Jules Marey*. Chicago and London: University of Chicago Press.

Cavendish, P. (2013a) "The delirious vision: the vogue for the hand-held camera in Soviet cinema of the 1920s." *Studies in Russian & Soviet Cinema* 7: 1: 5–24.

—. (2013b) *The Men with the Movie Camera: the Poetics of Visual Style in Soviet Avant-garde Cinema of the 1920s*. Oxford: Berghahn Books.

DeMet, G. D. (1999) "The Special Effects of *2001: A Space Odyssey*," *DFX* (July). Available at http://www.palantir.net/2001/meanings/dfx.html (accessed 22 October 2011).

Kaganovsky, L. and M. Salazkina. (Eds.) (2014) *Sound, Speech, Music in Soviet and Post-Soviet Cinema.* Bloomington: Indiana University Press.

King, D. (1997) *The Commissar Vanishes: The Falsification of Photographs and Art in Stalin's Russia.* New York: Metropolitan Books.

Klein, N. M. (2004) *The Vatican to Vegas: A History of Special Effects.* London, New York: The New Press.

Kovalov, O. (2006) "First on the Moon." *KinoKultura* 11. Available at http://www.kinokultura.com/2006/11r-firstmoon1.shtml (accessed 20 October 2010).

Krukones, J. (2010). "Peacefully coexisting on a wide screen: Kinopanorama vs. Cinerama, 1952–66." *Studies in Russian and Soviet Cinema* 4.3: 283–305.

Kuleshov, L. (1922) "Amerikanshchina." *Kinofot,* 25–31 August. Trans. Richard Taylor and Ian Christie, 1988: 73.

Maiorov, N. (2011) "Stanovlenie i razvitie otechestvennogo stereokino." *Mir tekhniki kino* 19: 33–51.

Mayorov, N. (2012a) "A first in cinema . . . stereoscopic films in Russia and the Soviet Union." *Studies in Russian & Soviet Cinema* 6: 2: 217–39.

—. (2012b) "Soviet colours." *Studies in Russian & Soviet Cinema* 6: 2: 241–55.

Siddiqi, A. A. (2008) "Imagining the Cosmos: Utopians, Mystics, and the Popular Culture of Spaceflight in Revolutionary Russia." *Osiris* 23: 260–88.

Strukov, V. (2008) "Virtualisation of Self and Space in Tarkovsky's *Solaris.*" In *Tarkovsky*, N. Dunne (Ed.). London: Black Dog Publishing. 58–79.

Taylor, R. and I. Christie (Eds.) (1988). *The Film Factory. Russian and Soviet Cinema in Documents 1896–1939.* London and New York: Routledge.

Vatulescu, C. (2010) *Police Aesthetics. Literature, Film and the Secret Police in Soviet Times.* Stanford, CA: Stanford University Press.

Vertov, Dz. (1922) "We: Variant of a Manifesto." In *Kino-Eye: The Writings of Dziga Vertov,* A. Michelson (Ed.) Berkeley: University of California Press, 1984. 5–9.

Widdis, E. (2003) *Visions of a New Land: Soviet Film from the Revolution to the Second World War.* New Haven: Yale University Press.

7 Leaving the house of dreams

The myth of flight in Russian films of the 2000s

Julian Graffy

The prevalence and symbolism of flight imagery in Russian and Soviet cinema

Aerial heroism, aerial records, aerial enthusiasm[1]

Film and manned flight are almost exact contemporaries, and both are frequently represented as key manifestations of early twentieth-century modernity.[2] It is therefore unsurprising to see aeroplanes in early films, in Russia as in other countries. Accordingly, I begin by surveying early Russian and Soviet film to identify those major concerns that have surfaced in recent film, analyzed in the main section of this chapter. As early as in Eduard Puchalski's 1916 comedy, *Antosha Ruined by a Corset* [Antoshu korset pogubil], for example, the sudden appearance of a plane overhead causes a rapt crowd to gather in admiration of the pilot's prowess. Aviation, after all, was deemed a masculine endeavor in both the late-tsarist and the Soviet era, its riskiness and challenge to nature and traditions of envisioned human possibility accounting for its dramatic appeal and, subsequently, for its political significance. For the unheroic Antosha (Anton Fertner), on the other hand, the crowd's upwardly distracted gaze provides an opportunity to rid himself, at least temporarily, of the eponymous article of incriminating female clothing.

With the Revolution, that appeal and the image of the flier as hero acquired official endorsement, as propaganda immediately incorporated the flier into the rhetoric of revolutionary optimism. In the "Aviators' March," written in 1920 by Pavel German and Iulii Khait and later adopted as an anthem by the Soviet Air Force, Soviet airmen can "conquer distance and space," since their "reason" has given them "arms of steel" in the wings of their planes, and "the peace of our borders" breathes in every plane's propeller.[3] Flight continued to be associated with excitement and liberation throughout the 1920s: in *Bed and Sofa* [Tret′ia Meshchanskaia 1927, dir. by Abram Room) the new lodger, Volodia (Vladimir Fogel′), seduces the heroine, Liuda (Liudmila Semenova), by taking her out of her confining basement flat on a breathtaking plane flight high over Moscow on an Agit-Plane of the Society of the Friends of Aviation and Chemical Construction of the Russian Soviet Federal Socialist Republic.[4] In the symbolic system of vertical representation, he elevates her to the status of an airborne citizen—a status inaccessible at the time to the vast majority of Soviet citizens, and especially to women, but one urged by the state, however theoretically, as a measure of technological progress in an international context.

By 1930, a speech at the end of Aleksandr Dovzhenko's *Earth* [Zemlia 1930] rhetorically represented the sky itself as the domain, not of saints and angels, but of the "Bolshevik aeroplane," and the following decade saw a number of records set by long-distance Soviet fliers and 'stratonauts,' who flew their balloons high into the stratosphere.[5] This in turn led to official recognition, most notably to Stalin's famous toast at the Kremlin on 2 May 1935: "To our brave fliers! They are the proud falcons of our army."[6] That predominantly male 'avian' army[7] was intended to showcase Soviet supremacy in the skies as an index of the USSR's power and technological as well as ideological pre-eminence. Throughout the 1930s cinematic fliers performed extraordinary feats of patriotic heroism in the context of the period's darkening international political mood. To take but one example, the Air Force officer heroes of Abram Room's *Squadron No. 5 (War Begins)* [Eskadril′ia No. 5 (Voina nachinaetsia) 1939], set on an air-base near the Western border, respond to a German attack using their flawless German to masquerade successfully as Nazi officers, penetrate a secret German underground airbase, perform an act of sabotage, radio back to the motherland, and escape to fight another day—a scenario reprised in less melodramatic mode in Tat′iana Lioznova's 1973 cult series *17 Moments of Spring* [17 mgnovenii vesny], which, however, ignored aviation, to focus on a more intellectual mode of sabotage, in tune with the 1970s.[8]

The Great Patriotic War soon became and has remained a staple Russian film genre, and the films made on the theme, whether during the war itself or in later decades, provided dozens of aviation heroes who performed scores of heroic acts. While most of these were conventional, there was still room for narrative and formal innovation. One of the most dramatically and visually compelling cinematic achievements of the late-Soviet period, Semen Aranovich's *Torpedo Carriers* [Torpedonostsy 1983], adapted by Svetlana Karmalita from the wartime stories of Iurii German, sets the heroism of an air force regiment based in a small garrison on the Northern Front against the small dramas of the daily lives of the fliers and their families. Though habitually the war films of these later years lauded the exploits of male fliers,[9] thus providing a continuity with the films about the record-breaking airmen of the 1930s, two notable films made by female directors challenged these gender stereotypes by reminding viewers of the achievements and sacrifices of Soviet women aces. Larisa Shepit'ko's *Wings* [Kryl′ia 1966] lamented the state's inability to provide a continued role in aviation in the post-war years for a female WWII fighter pilot, now reduced to running a school. And in 1981 the former fighter pilot and Heroine of the Soviet Union Evgeniia Zhigulenko, who thirty years after the end of WWII had graduated from the All Union Cinema Institute (VGIK), directed '*The Night Witches' are in the Sky* [V nebe 'Nochnye ved′my'], a story about the wartime exploits of a squadron of female pilots responsible for the bombing of enemy aerodromes.[10] So, though many of the most successful Soviet flier films were made in the 1930s, the genre proved open to reconceptualization and experiment to the end.

A parallel if minor line was devoted to space exploration. As early as 1912, the brilliant animator Władysław Starewicz made *Flight to the Moon* [Polet na lunu],[11] while the new Soviet industry produced a flight to Mars in Iakov Protazanov's *Aelita* (1924). The moon was once again the destination in 1935, in Vasilii Zhuravlev's *Cosmic Voyage* [Kosmicheskii reis]. It shows how an ancient professor, accompanied on the spaceship *USSR 1. The Iosif Stalin* by a young female assistant and a boy-inventor stowaway (the solidarity between the generations signaling ideological continuity and referencing future explorations), beams back the signal "USSR" to waiting crowds in his Moscow Institute. Then, in the post-war years, the real achievements of Soviet space technology led to a cornucopia of films about space travel and the lives of great space scientists. Given the competitiveness of the superpowers during the Cold War, this line remained vital to the end of the Soviet period. It reached its apogee in the works of Pavel Klushantsev, notably *The Way to the Stars* [Doroga k zvezdam 1957] and *Planet of Storms* [Planeta bur′ 1961] which was released on 14 April 1962, the day of Gagarin's Moscow encounter with Khrushchev. (These films are discussed in detail in other chapters in this volume.[12])

Aerial doubt, aerial indiscipline

Yet, from the very beginning, this optimistic, self-glorifying impulse in Russian and Soviet culture was challenged by a contradictory desire to accentuate the darker elements of aerial flight. At its simplest, this trend manifested itself in an insistent association of aviatory achievements with danger and tragedy. One of the first flier heroes in Russian culture was the protagonist of Aleksandr Blok's poem "The Aviator," written in 1910–January 1912. Initially described as "released into freedom," he disappears from earthly sight and seems to aim his plane at the sun itself, anticipating the Futurist opera *Victory over the Sun* [Pobeda nad solntsem 1913], which hailed the aviator as the consummate hero of a new world striving for fearless, cosmic self-identity. When Blok's aviator reappears, however, he has merely broken a "miserable" world record, and, owing to problems with his plane, he plunges to his death. The poet bemoans his demise, explaining it in turn by the airman's desire for fickle public acclaim, a crazed death-wish, or the poisonous vision of future wars. Blok's manuscript dedicated the poem to one of the first Russian fliers, Vladimir Smit, who had fallen to his death before the poet's eyes on 14 May (old style) 1911, on the first day of the St. Petersburg Aviation Week at the Kolomiazhskii Aerodrome (Blok 1960: 33–34, 506–7). The motif of the wounded aviator appeared in Iakov Protazanov's 1918 film *Jenny the Maid* [Gornichnaia Dzhenni] and the trope of the flier falling damaged or dead from the sky after a crash or a mechanical failure recurred with remarkable frequency in the films of the Soviet period, particularly those about the Great Patriotic War. Perhaps the most famous of these is Aleksandr Stolper's *The Tale of a Real Man* [Povest′ o nastoiashchem cheloveke 1948], in which the pilot Aleksei Meres′ev (Pavel Kadochnikov) loses both legs after his plane is shot down behind enemy lines but eventually returns bravely to the skies.[13]

More destructive still of official optimism was a rejection of the very premises that underlay the Soviet cult of flight. A corrosive sense of irony toward such utopian dreams pervades *Aelita*, in which, for all the extravagance of the representations of life on Mars, those images are eventually revealed as merely the listless imaginings of the film's scientist hero. Restored to sober earth, he declares, in the film's final words, "Enough of dreams. A different, real work awaits us all."

On other occasions the fault lay within the flier himself as someone attracted to daredevil stunts and refusing to conform to the Soviet principle of discipline. Beliaev (Ivan Koval′-Samborskii), the handsome young hero of Iulii Raizman's 1935 film *Fliers* [Letchiki], is just such a man. Tempted to show off, he crashes his new plane in a test flight. He is banned from flying and when he proves unable to reform he is elided from the film. But this rebellious impulse is a danger that Soviet films about flight can never defeat completely. For instance, a compulsion toward risk and transgression bordering on a death-wish is also central to the representation of the flying ace at the center of Mikhail Kalatozov's *Valerii Chkalov* (1941). In a key scene, wise Stalin (Mikhail Gelovani) finds it necessary to remind the untameable hero, in words that echo Maiakovskii's response to the suicide of the poet Esenin, that "dying is hard, but not so difficult, comrade Chkalov. . . . But we support those people who want to live, live as long as possible."[14] The film ends in a triumphant transcontinental flight and a glorious return to Moscow, but the real-life Chkalov died in a plane crash, in the year after the final events depicted in the film.[15] Controlling one's instincts (as well as nature) to serve the state according to Party precepts constituted one of the chief lessons promoted by Soviet and especially Stalinist ideology and shaped countless Socialist Realist narratives. The ability to learn that lesson proved more widespread in film than in real life.

Late-Soviet stirrings

In the final years of the Soviet Union, when all its values came under radical scrutiny, Soviet cinema's gaze was torn between examining the real problems of the present and exposing the 'glorious' Soviet past through narratives steeped in mockery, bitterness, and accusations. The young director Maksim Pezhemskii attempted to deal once and for all with the cult of the Soviet flier in his sardonic 1990 short film, *Comrade Chkalov Crosses the North Pole* [Perekhod tovarishcha Chkalova cherez Severnyi polius]. But the impulse toward rejection is not the only motive for looking up into the skies in the films of this period. Images of a more abstract kind of flight were used by Pezhemskii's contemporaries to express the sense of unease and foreboding as well as the desperate desire for change that pervaded the late-Soviet consciousness, and to discover a potential source of hope and meaning. Aleksandr Sokurov's *Days of Eclipse* [Dni zatmeniia 1988] opens with a cinematic descent to fallen earth, a plunging 'cosmic' shot accompanied by eerily otherworldly music that places the Turkmen town of Krasnovodsk in which the film is set and the lives of its protagonists in the eye of some otherworldly intelligence. Throughout the film the doctor protagonist, Malianov (Aleksei Ananishnov), who is dedicating his life to medical research as if he were a hero out of Chekhov, is visited by a succession of signs and portents. One of the most striking of these is an angelic but sick boy, who, despite Malianov's protestations, does not believe that the doctor can cure him, because he is "too busy" with his writings. The promised connection between heaven and earth is severed. The boy flies away as mysteriously as he came, accompanied by the words "come, my angel, spread your wings," uttered by his unseen father. Though the heavens are a potential source of knowledge and comfort in *Days of Eclipse*—a film also engaged with questions of religious faith—the hero is too preoccupied to benefit from them.[16]

The sense of possibilities that are never realized characterizes another film of the period, also set in a border area of the Soviet Union, this time in the Far North: in Nikolai Dostal′'s *Cloud Paradise* [Oblako rai 1991], the dreamer hero, Kolia (Andrei Zhigalov), vaguely aware that there must be more to life than is offered by a dreary Petrozavodsk Sunday, seeks desperately for companionship and meaning, as everyone around him treats him with indifference, condescension, or dull aggression. He expresses his accumulated frustration in a clumsy, wistful song:

> There's a cloud-paradise in the sky, go and guess where it is,
> It's impossible to find, recognize, and understand it.
> Please tell me what color it is and where.
> You're probably in a place where it's very hard to find you.
> But I'm trembling from the thought that I'll never find you,
> I want to forget, to run away, to twist and turn.
> Where are you, where are you, cloud-paradise—
> Don't drop my leaves of hope.[17]

But this dissatisfaction and his very openness to the desire for change serve as a mark of Cain for Kolia. By the end of the film he is cast out of the smug Soviet collective and his song has gained a somber new verse:

> I cure myself with anguish and grief, in my soul I strive for heaven.
> I seek my happiness in the thick fog.
> Perhaps someone, some time, will suddenly open my eyes,
> And I'll see it through the bad weather.
> But it's not there, and again, I don't tire of repeating,
> Living in hope of a meeting with you . . .
> Where are you, where are you, cloud-paradise
> Don't drop my leaves of hope.[18]

Whereas for the heroes of the early Soviet period the heavens were devoid of metaphysical significance, in the waning years of the Soviet Union, both Sokurov's Malianov and Dostal′'s Kolia—men at opposite ends of the intellectual and social scale—betray a desperate hope that the spiritual link to the heavens can be restored (see Introduction). It is nevertheless significant that neither man finds an answer and that both of them are left in earthbound solitude at the end of the film.

Flight imagery in post-Soviet cinema: re-interpreting the past

Flight imagery in films of the Soviet years was mainly associated with the achievements of the present and hopes for the future. The most famous cinematic hero of the 1990s, Danila Bagrov, in the opening sequence of Aleksei Balabanov's film *Brother* [Brat 1997], was also set upon his path out of provincial tedium and into a dynamic future by hearing the words of the song *Wings* [Kryl′ia], with its refrain "Where are your wings, which I liked?"[19] Though a preoccupation with flying will remain, cinema of the subsequent decade, by contrast, associated flight imagery predominantly with the Soviet past.[20] The rest of this chapter examines some important films made between 2003 and 2011 in which aerial or cosmic flight plays a focal role in the plot. They range in genre across the spectrum of recent Russian cinema, from commercial films aimed at a young audience to art-house productions. They are connected not only by a preoccupation with flying but also by examining that interest through the relationship of fathers and sons (both real and symbolic), which links them not only to an important strand of Soviet film history, but also to one of the Russian cinematic master-texts of the of the early twenty-first century.[21]

In two recent commercial films oriented toward youthful audiences, *Black Lightning* [Chernaia molniia 2010, dir. by Aleksandr Voitinskii and Dmitrii Kiselev] and *We Are from the Future* [My iz budushchego 2008, dir. by Andrei Maliukov], a fantasy flight (achieved in the latter case without the aid of technology) takes thoughtless modern young men back into the Soviet period, where they find values that have been lost to the contemporary Russian world. Space travel in these films becomes time travel, real flying gives way to imaginary flying, and illusions about the Soviet past remain intact. Clearly, the goal of both films is entertainment, with an admixture of didacticism and an approving nod to sacrosanct Soviet traditions.

A contrasting group of four recent art films looks back in more thoughtful mode to assess the moral and cultural legacy of the Soviet space program. Two of them, Aleksei Uchitel′'s *Dreaming of Space* [Kosmos kak predchuvstvie 2005] and Aleksei German Jr.'s *Paper Soldier* [Bumazhnyi soldat 2008], are shot through with melancholy and irony; the achievements of the cosmonauts are moved to the margins of a broader engagement with the Soviet past. Iurii Kara's more conventional biopic, *Korolev* (2007), examines the tragic effect on the space program of Stalin-period paranoia. By contrast, Aleksei Fedorchenko's mockumentary, *First on the Moon* [Pervye na lune 2005], uses and abuses Soviet space triumphs by celebrating space flights that never took place. Standing apart from all of these films, Boris Khlebnikov and Aleksei Popogrebskii's *Koktebel′* (2003) uses pervasive imagery of flight in the service of a nuanced depiction of exhausted Russian masculinity and an interpretation of the relationship of the first two post-Soviet generations.

Youthful time travel

Much of recent mainstream Russian film production, like that of cinema industries the world over, is aimed at the young people who account for most of the box-office receipts. Consequently, most of the young protagonists of these films are shown as lively and energetic, though there is often something infantile about their vigor—evidence that the uncertainty and confusion that reigned in the country twenty years ago have marked more than one generation. Perhaps predictably, then, in the improbable cases of time travel in recent films about young people, they seek reassurance and certainty in the Soviet past.

Dima, the working-class Moscow University student hero of *Black Lightning*, is given an old black Volga GAZ-21 by his tram-driver father. At a lecture on business strategy and ethics, Viktor Kuptsov, a super-wealthy businessman, whose very name suggests his commitment to making money,[22] challenges him to put aside all altruistic impulses and to concentrate on getting rich. When he follows this advice Dima inadvertently causes the death of his own father, which leads to a profound sense of guilt and the desire to atone. Appropriately, it is the GAZ-21 that provides the means. The dastardly Kuptsov, it transpires, is trying to get hold of a mysterious "nanocatalyzer," which will enable him to drill under the foundations of Moscow to steal the diamonds that apparently lie below (callously sacrificing the lives of millions in the process). Because of some skulduggery caused by thwarted love back in the Soviet past, this scientific wonder is hidden in Dima's car, enabling the vehicle to fly.[23] What follows is a preposterous and sentimental tale, articulated through the trite wonders of CGI. Following in the tradition of Superman, Dima, the 'Black Lightning' of the film's title, flies through the Moscow skies to pluck a child from a burning building or to prevent a baby from being crushed to death by robbers who have hijacked a security van. In the film's closing sequence his Volga engages in aerial battle with Kuptsov's white Mercedes, now similarly enhanced.[24]

In the flier films of the 1930s, such as *A Town Under Attack* [Gorod pod udarom 1933, dir. by Iurii Genika], *The Motherland Calls* [Rodina zovet 1936, dir. by Aleksandr Macheret], *The Deep Raid* [Glubokii reid 1937, dir. by Petr Malakhov], and *Squadron No. 5* (1939, mentioned above), aerial combat between Soviet and enemy forces made for stirring patriotic drama. It was treated with visual reticence and high moral seriousness, for the fate of the nation depended on its outcome. Here, Dima's unconvincing yet predictable victory is a personal adventure, which completes his moral restoration, wins him the girl, and is filmed with a succession of startling special effects. But more interesting are the cultural codes and the cinematic context in which the struggle is enacted. Having been temporarily suborned by a false father, Dima must now return to the quieter values of his real father, which he had earlier spurned. That these have their source in the years of The Thaw is apparent both from the reminiscences of Dima's parents of their youthful courtship and from the car chosen for his aerial exploits.

The most famous flying car in Soviet film history took the champion weaver Tania Morozova to Moscow and the glorious future in Grigorii Aleksandrov's 1940 musical, *The Radiant Path* [Svetlyi put′]. The Volga Gaz-21, on the other hand, is a potent symbol of the Soviet past. Mass production of the car began in 1957, and it remains perennially associated with the values of the Thaw period.[25] Yuri Gagarin was given a customized model on his return from space.

In Soviet film history it is famously the car stolen by Innokentii Smoktunovski, as Detochkin, in El′dar Riazanov's *Mind That Car!* [Beregis′ avtomobilia 1966], and the vehicle in which Tat′iana Doronina sings the plangent super-hit "The Earth Is Empty without You" [Opustela bez tebia zemlia] in Tat′iana Lioznova's *Three Poplars on Pliushchikha* [Tri topolia na Pliushchikhe 1967]. According to Muireann Maguire, the model appeared in over two hundred and fifty Russian and Soviet films (Maguire 2010). And within the narrative of *Black Lightning* it is explicitly associated with Dima's father, who has bought it as a present for his son. It is while they are sitting in the car that his father explains to him his philosophy of moral intervention. Later, when the hooded crusader starts appearing in the Moscow skies, Dima's young sister tells him that the Black Lightning is their dead father. And it is the beautiful old Soviet car that enables Dima to atone for his moral lapse and to make his filial amends. It is equally appropriate that Kuptsov drives a Mercedes, a car much favored by wealthy 'new Russians' ever since it became possible to buy and to flaunt them after the fall of the Soviet Union. Indeed, this is not the first cinematic struggle between a Mercedes and a less glamorous Russian vehicle. There is an earthbound version of this narrative, in which a simple Russian car trumps a white Mercedes and old Russian values are seen to be more rooted and enduring than modern materialism, in Dmitrii Astrakhan's kitsch masterpiece of triumphant Russian feeling, *Everything Will Be OK* [Vse budet khorosho 1994].[26] In the same way, in *Black Lightning*, the meretricious values of post-Soviet Russia are defeated by those of the Soviet past, which can no longer be defended by their initial proponents and thus have to be *re-discovered* by the generation that follows the go-getters of the 1990s. In this context, the fact that the special effects of the twenty-first century are applied to flight not in a super-cool plane or a spaceship but in a fifty-year-old car encapsulates the moral position the film attempts to articulate. In short, flight in *Black Lightning* connects modern technology (and cinematic special effects) to older values (both ethical and aesthetic), restores today's Russians to a simpler, morally purer Soviet past, and thus forges moral and aesthetic links between the Soviet era and a contemporary generation that has learnt to be appreciative of its imperial legacy and its uncomplicated worldview.[27]

The heightened interest in the Soviet past that marked the cinematic production of the perestroika years was but a part of Russian society's attempt to understand how it had arrived at its parlous state. Disillusionment with the entire Soviet experiment led directors to concentrate their attention on the 1930s and the final years of the Stalin regime, periods in which, they felt, the seeds of the disaster had been sown. Some of the most interesting historical films of the late 1980s resorted to the trope of time travel to examine the nature of connections with the past and the latter's role in the present. The heroes both of Vladimir Khotinenko's *A Mirror for the Hero* [Zerkalo dlia geroia 1987] and of Valerii Ogorodnikov's *Prishvin's Paper Eyes* [Bumazhnye glaza Prishvina 1989] are transported from film sets that are part of productions depicting 1949 into events of precisely that era, and their participation in those events helps them to understand the true nature of the Stalinist past.[28]

Twenty years later, the violence and bloodshed engendered by the revolutionary experiment are extensively on display in Andrei Smirnov's epic film *Once There Lived a Simple Woman* [Zhila-byla odna baba 2011], but cinematic scrutiny of the darker side of the Soviet legacy has become considerably rarer. With a change in Russian society's attitudes to the Soviet period, and a new interest in emphasizing the positive, the sites of historical retrospection have changed. Contemporary 'nostalgia' has concentrated on two periods that can still yield a positive reading: the youthful optimism of The Thaw (as in *Black Lightning*) and, above all, the heroism of the Soviet people during WWII. Dozens of films and television serials about the War have been made in the last decade.[29] One of them takes the conceit deployed in *Black Lightning* one step further, by allowing its young Russian heroes to fly unaided through the air into the Soviet past.

At the start of Andrei Maliukov's *We Are from the Future* [My iz budushchego 2008] four arrogant and unthinking young St. Petersburgers are digging for Nazi war trophies in the countryside outside the city to make money by selling them to collectors. They are accosted by an old woman, whose preternatural age (we later learn that her soldier son was born in 1917, making her perhaps 110) should have alerted them to her significance as an agent of intervention in their fates. She asks them to find the son, who was lost in these parts in 1942. Of course, their promise to do so is insincere, but when strange things begin to happen they follow her suggestion and decide to cool off in a nearby lake. This is where the 'flying' sequence takes place in the film. One after another the young men strip off and hurl themselves high into the air from a diving board before plunging into the inviting waters. A storm sets in and they are almost drowned. But (as their nakedness suggests) this is the water of their rebirth. They emerge, terrified, in August 1942, in the thick of the fighting outside Leningrad. The main body of the film concerns their moral reformation, effected by the tough demands of their sergeant major (Boris Galkin) and the example of bravery and stoicism offered by their new Soviet fellows. They fulfill their promise to the old woman by tracking down her son, who in turn (in the logic of folktales) helps them to return to the present. In a brief, silently eloquent coda, they now look askance at a group of Nazi skinheads. Once again, as in *Black Lightning*, the flight sequence frames the film's ethical debate. The young heroes are morally transformed by a flight that exposes them to the (now positively coded) values of the Soviet past. And once again it is an encounter with an earlier generation that changes their attitudes. It is instructive to contrast the narrative of *We Are from the Future* with a more conventional example of 'time travel' back to WWII in the famous sequence near the end of one of the key films of The Thaw, *Il′ich's Gate* [Zastava Il′icha 1964, dir. by Marlen Khutsiev], in which Sergei, one of the film's heroes, dreams of a meeting with his father, a soldier who had died in that war. When he asks his father how to live, the older man is unable to advise him.[30] In the infinitely simpler moral landscape of *We Are from the Future*, however, the 'fathers' that the young heroes meet in the wartime sequence (fathers in terms of their relative ages in the symbolic dream terrain of the film's central episodes) are only too ready to resolve the young men's moral dilemmas and to turn them into real

(Soviet) men.[31] Thus, in *We Are from the Future* as in *Black Lightning*, flight into a mythologized, simple past restores coherence to Russian history and wholeness to the Russian male character.

Re-imagining Soviet space flight: dreams of escape

Aleksei Uchitel′'s *Dreaming of Space* begins in the autumn of 1957, while the first sputnik is circling the earth, and Pavel Klushantsev's *The Way to the Stars* has captured the popular imagination. It ends with Yuri Gagarin's walking the red carpet in Red Square on 14 April 1961, about to be congratulated by Nikita Khrushchev on his successful flight through space two days earlier. Made from a script by Aleksandr Mindadze (b. 1949), it is set in a small town in the far north, a place significant less for being on the margin of the Soviet Union than for its location on the edge of the outside, Western world. This border strip is ubiquitous in *Dreaming of Space*, functioning as the setting of the hero's endless bike rides, of his al fresco love-making with his waitress girlfriend, and of his friend's training swims in the cold Arctic waters. It is the locus of encounters with foreign vessels and foreign sailors, a place marked, significantly, in red on a map in the swimmer's modest room.[32]

As was often the case in Mindadze's scripts for the several films he made with Vladimir Abdrashitov (b. 1945) in the late-Soviet years, the film's narrative examines the consequences of an encounter between two men with different sets of values and different degrees of awareness about the Soviet system—an encounter leading to a crisis in the life of the more conventional hero. The enigmatic swimmer, German, is tough, driven, strong, charismatic, his (gradually revealed) determination to escape abroad making him ready, if necessary, even to steal his friend's girl.[33] The hero, Viktor, known to everyone through the infantilizing diminutive Konek [little horsy], is weak, impressionable, naïve, and dangerously childish—in his enthusiasm for his new hero, he blabs German's story to a kindly informant—and all of those qualities render him a more typical Soviet citizen of the period.[34] True to the age-old traditions of Russian literature (Onegin and Lenskii, Bazarov and Arkadii, Oblomov and Shtol′ts), these antithetical individuals cherish emotional allegiances to opposing places and systems. The English phrase that German is doggedly trying to learn associates the West with freedom, symbolized by his contraband possessions—a transistor radio that plays jazz, a cigar, an English phrasebook—and by the distant lights of the *Lake Michigan*, the Western 'mother ship' anchored off shore, which, when he finally swims out toward it, sails treacherously off into the dim distance. The Russia in which Konek is destined to remain is, by contrast, a place of knife fights in seedy dining rooms [stolovye], squalid barracks [baraki] and constant observation by stool pigeons and volunteer patrols.

Yet, for all their differences, the two men are also, paradoxically, potential doubles. In thrall to German's mystery, Konek begins to model himself on his hero, growing the same mustache and wearing the same cap and leather jacket, clothes that link them visually with the flier heroes of Soviet cinema.[35] When German makes him a gift of a transistor radio, he takes it to the dining room, coolly playing "Rock

Around the Clock." But when he tries, with the same brazen words that German had used there, to order a cup of strong tea, the waitress ignores him. After he leaves he is beaten up by the volunteer patrol in whose eyes "a transistor is politics." Perhaps mercifully for him, Konek lacks the charisma to be German.

The film repeats the doubling motif throughout. Both young men are caught up in evolving relationships with two young women, sisters who work as waitresses in the restaurant where Konek is a cook. The Kazakh town of Kustanai, first represented, in the obfuscating tales that German tells Konek, as the place from which flights are to be launched into space, later becomes the site of a labor camp. Konek's encounter with eloquent, self-confident German is doubled with a 'real-life' encounter with the diffident young Yuri Gagarin, a man whose character is cast as nearer to his own. One role model is replaced by another, less dangerous one.

Most importantly, the film is about two kinds of journey: an escape into emigration (a one-way ticket) and a journey (with a return ticket) into space. Emigration and space travel in the first Soviet film about space exploration, Protazanov's *Aelita*, were linked by the simple expedient of having the same actor play the scientist émigré and the scientist traveler to Mars. Both journeys were negatively coded as dereliction of Soviet duty. Here, however, German's attempted emigration is seen as a doomed but noble act of desperation, perhaps the only sane response after his eyes are opened (as he describes it) in the Kustanai labor camp. And this tragic element in the film turns the journey into space into a pale, safe surrogate for aspiration to an uncertain life abroad.

The 1957 sputnik flight is twice glimpsed by Konek and Lara (though not by the film's viewers) as they are making love in the open air. Neither occasion is shown as auspicious. On the first, they have to run humiliatingly from a border patrol; on the second, Lara has already betrayed him with German and Konek is about to leave her for her more go-getting sister. German has explained his passion for swimming to Konek by implying that he is an astronaut training for a space flight—a story that provides a useful cover for his real intentions but is believed by Konek, since the ordinary citizen had seen physical fitness as an accouterment of the Soviet hero ever since the parades of *fizkul′turniki* in the Stalinist 1930s. In this respect, transcending their drab lives with an impossible dream, both German and Konek become, in their different ways, doubles of the eponymous hero of Ilya Kabakov's installation *The Man Who Flew into Space from his Apartment*.[36] But their most extended conversation about this takes place, bathetically, on a tram, as the lights of a space rocket's take-off are mimicked by the sparks flying from its overhead wires. The sense of anticlimax that pervades this encounter is echoed in the following scene, in which Konek and the kindly stoolpigeon, Kirych, pedal furiously on bicycles. Gagarin himself is encountered not at an airbase, but on the traditional Soviet means of transport, the train, and is shown as merely another shy, self-deprecating young man who makes even Konek seem worldly. Asked by Konek (who has inherited the idea from German) when he is going "on a rocket to the stars," Gagarin replies that he "cannot even dream" yet of space flight. Indeed, when the four central characters do invoke Soviet aerial aspirations by alluding to the Aviators' March with cries of "Vyshe!" (Higher!), they are "flying," belittlingly, on an attraction at the local amusement park, minutes after passing, with childish enthusiasm, through the deceptions of a hall of mirrors.

Dreams of space and myths of flight in Uchitel′'s film are a kind of consolation prize for those who cannot achieve a greater freedom by making the ultimate escape from Soviet reality. In this sense the oppositions between characters are less important than their shared experiences and desires. The film proceeds as a relay race of sacrifice and partial awakening, which leads, through a succession of rhyming scenes, from the (named but unseen) political prisoner Valerka, whom German had met in captivity, via German and Konek to Gagarin. Yet Konek (as German implicitly realizes) lacks the consciousness and the desperation to follow German's example, and so it is his fate to remain in the Soviet geographical and mental space, and to encourage the young Gagarin to find a safer form of flight. The extremists, Valerka and German, die violent deaths; the more conventional Konek and Gagarin survive (at least within the trajectory of the film).

In retrospect, of course, the journey abroad has turned out to be far more attractive to former Soviet citizens than the (imagined) journey into space, immensely more capable of fulfilling their long-frustrated desires. In the popular Russian imagination, enthusiasm for space flight reached its height in response to the pioneering feats of the 1950s and 1960s. For all the recent celebration of the fiftieth anniversaries of those achievements, it has now (at least temporarily) lost all enchantment. As the ambiguous English word suggests, *space* is empty, uninhabited, blank. So *Dreaming of Space*, the English title chosen for the film, is not just a mistranslation of the Russian, but also a fundamental misrepresentation of Mindadze's and the film's concerns. There is almost no dreaming of space in the film, but there is, to translate its title literally, some sense of *Space as Presentiment*, of space flight as a neutered expression of a hardly acknowledged desire to fly across the borders of the Soviet world. Natal′ia Sirivlia has called Mindadze's script a "fable about the attraction of a child-people, educated in the socialist barracks, to universality, beauty, and freedom" (Sirivlia 2005: 20). But the "child-people" of Uchitel′s film are not yet ready for the risks and challenges of that freedom:

> And the myth of space, crowned with bravura by the documentary sequences of Gagarin's passage across Red Square, becomes here not a presentiment of freedom but a means of affirming everlasting slavery: there is no reason to rock the boat, run off somewhere; it's warm here and they feed you, and if you want to assuage your masculine self-esteem, here's a rattle for you, a cause of national pride, a belief in a great act of heroism that you can achieve without falling out of the bear hug of the system.
>
> (Sirivlia 2005: 23)

Sacrifice and death

Mindadze and Uchitel′ used the first space flights as a background to their interrogation of the extent to which Soviet mentalities changed during the Thaw period.

They concluded that the pride that Soviet citizens felt in the achievements of Gagarin was poor recompense for remaining in a state of perpetual ignorance about the modern world. Both Kara's *Korolev* and Aleksei German Jr.'s *Paper Soldier* move attention to the (real and fictional) people actually involved in the space program, and both films show the high human cost of Soviet scientific achievement.

One of the enterprises commemorating the centenary of the Soviet space program's 'father,' Sergei Korolev (1906–66), the eponymous film was released on 12 April 2007, the forty-sixth anniversary of Gagarin's flight.[37] Though Korolev's identity had remained secret during his lifetime, a version of his story had been told through the character of Andrei Bashkirtsev in the hugely successful 1972 film *The Taming of Fire* [Ukroshchenie ognia, dir. by Daniil Khrabrovitskii].[38] Whereas this film had made no allusion to the darker side of Korolev's life, to his arrest and his time in the GULag, *Korolev* opens with the arrest, accompanied by violence and the theft of property, on 27 June 1938. What follows is a story of interrogations and beatings, with accusations of running an "underground organization," imprisonment, trial, and life in the camps, interspersed with flashbacks to show Korolev's genius and romantic enthusiasm for his studies. Eventually, in 1940, he is transferred to one of the famous *sharashki* of the kind described by Aleksandr Solzhenitsyn in his novel *The First Circle* [V kruge pervom 1968], prison research institutes in which brilliant political prisoners were employed to continue their fundamental research. A brief coda takes us to the successful firing of a rocket from the Tiura-Tam launch range in Kazakhstan in 1957.[39] Though Kara had insisted that "an artist has the right to have his own idea of the protagonist,"[40] the film came under sustained attack, both from those who pointed out its historical inaccuracies and from those who bemoaned what they saw as its betrayal of the glorious Soviet past. Admirable as it may be to devote much of the film to reminders of the brutality experienced by Soviet citizens in the Stalin years, the decision to tell Korolev's story in the form of a conventional biography and to concentrate on filling in the 'blank spots' [belye piatna] of Soviet history and individual lives give the film a curiously dated air. The aesthetics and the crusading desire to expose past ills evoke films of the perestroika years. Kara tells his viewers almost nothing about the dramas and achievements of Korolev's later life as designer in chief of the Sputnik program, while his debilitating lack of cinematic imagination leaves them bemused. In sum, as the film's reception has indicated, *Korolev* makes more allusions to the traditions of the Soviet cinematic biopic and the perestroika exposé film than to the genre of the Soviet space film, and in so doing it undermines both its hero and his project.

There is a similarly dark reading of Korolev's life in *'The Cedar' Pierces the Sky* ['Kedr' pronzaet nebo 2011], a serial made for Russian television by Aleksandr Barshak. The film's fictional space researcher hero, Sergei Lykov (Igor′ Petrenko), first encounters his idol Korolev (Igor′ Skliar) in prison after his arrest on absurd charges, and a later episode shows them at work with other scientists in the *sharashka*. What is most remarkable about *'The Cedar' Pierces the Sky*, which cinematographically is completely unmemorable, is that a film intended for a popular audience, made with star actors for a leading Russian TV channel, continues to represent the story of Soviet space research as one of disappointment, betrayal, and tragedy rather than of comradeship and heroism, a reading emphasized by setting seven of the film's eight episodes in the years before the death of Stalin. In addition to the arrest and interrogation of Korolev, and the ruination of the fictional hero's life, the film's narrative includes the banning of the book *Rockets, Their Structure and Application* [Rakety, ikh ustroistvo i primenenie] by Georgii Langemak and Valentin Glushko soon after its 1935 publication and their subsequent arrest; the visit of Soviet scientists to Peenemünde after the German defeat and the battle with the Americans over the transfer of Werner von Braun's equipment and technical documentation; and the constant interference and distrust of Soviet politicians.[41] Though *'The Cedar' Pierces the Sky* does take the story of the Soviet quest to put a man in space up to its successful conclusion, its triple ending places that success in an ambiguous context. At the conclusion of the last episode, Lykov, living in Kazakhstan after his release from a second term in the camps, renounces his interest in space research, telling his wife that she is his cosmos and his children that they are his sputniks. At the same time Gagarin makes his successful flight and meets Khrushchev at the Kremlin (a sequence also present in *Dreaming of Space*). But in a brief coda set twenty years later, at the end of the Soviet period, Lykov's now grown-up son prepares for training as a cosmonaut. What the father has not been able to achieve, the son will do for him, aspiring to heal the wounds of the past.[42]

While in German Jr.'s *Paper Soldier* the space program is again center stage, its (invented) narrative is also used for a broad examination of Thaw sensibilities, even suggesting their survival into the twenty-first century. Set later than *Dreaming of Space* and *Korolev*, in the spring of 1961, at the Baikonur cosmodrome in Kazakhstan and in a Moscow training center for astronauts, the film focuses on the period leading up to the first manned space flight, formally constructed in the manner of a countdown, with intertitles and a narrator's voice announcing "Week 6," "Week 5," "Week 4" . . . . But German, like Mindadze and Uchitel′ before him, shows little interest in the technology, the spectacle, or the personal triumph of the space flight, preferring, as he put it, to "turn my camera away from the huge metal mass" of the space rocket and to reconstruct the life of one of the people in the background of this great event (German 2008: 48).

As the film opens, Baikonur is revealed as a place of mist, emptiness, and stagnant water, where cows, sheep, and horses roam and a haughty camel stands and stares. The cosmodrome is still under construction, its few buildings marking a train terminus in the broad, open steppe, a veiled suggestion that it is the end of a line. In contrast to the heroic rhetoric of the film's narrator, the people employed on the base wander around arguing listlessly about their mundane concerns, haggling over a bath or complaining about the lack of soap. A crazed little boy spits at the hero, Dania, while an old man tries to sell him a portrait of Stalin. This grim visual environment recalls the psychologically devastating, nuclear-seared settings of the missile town of Kapustin Iar in Svetlana Vasilenko's fiction.

The ominous mood is heightened by a report of an accident: a rocket's parachute has failed to open, and its canine 'parachutist' sits bloodied and disconsolate on the ground. The would-be astronauts and the doctors monitoring them—both in Baikonur and in their snowbound Moscow training center—talk openly about their fears. The young men vomit, faint, and have nightmares; they know that they are called "laiki," after the dog who died in space. German's attention to the suffering of dogs involved in the Soviet space program is factually based, inasmuch as Laika, the first space dog, died on 3 November 1957, and a further seventeen dogs would share her fate. The extreme contrast between the bleak atmosphere of *Paper Solider* and the cheerfully knowing tone of another film about dogs in space, the animated *Belka and Strelka. Star Dogs* [Belka i Strelka. Zvezdnye sobaki 2010, dir. by Sviatoslav Ushakov and Inna Evlannikova],[43] which takes nothing at all seriously, only underscores the sober melancholy of German's vision. Not only dogs suffer; one of the recruits burns to death in a training chamber, leading both the hero's wife and (in a dream) his dead father to question the ethics of preparing young men for this probable fate.

One of the men we watch being trained and tested is destined to be the first man in space and to become, for a while, the most famous person on the planet. Another, his reserve on this occasion, will follow in his footsteps four months later. The film calls them simply Iura (short for Iurii) and German, and pays them scant attention. Until they put on their orange spacesuits and start practicing the high-flown speeches professing loyalty to Party and proletariat that they have been trained to deliver on landing, they are indistinguishable from their eager, fresh-faced rivals. When the epoch-making flight finally takes place, it is scarcely noticed in the back of the frame, relegated to the status of a backdrop for the death and mourning of the film's hero.

German's real concerns emerge through the experiences of the fictional participants, especially Daniil (Dania) Pokrovskii (Merab Ninidze), the central character in the film, a doctor engaged in the physical preparation of the would-be astronauts. He had once wanted to be an astronaut himself and he continues to profess his belief in the space program, which, he insists, will soon establish a base on the moon, to which enormous liners will fly by the end of the century. Dania is charming and popular but tormented by self-doubt; he has a complex about not living up to his dead father, whom he encounters in a dream sequence, once again reprising the model of posthumous cross-generation encounters set in *Il'ich's Gate.* He is racked by ill health and lacking in will.[44] At the end of the film he dies a shockingly sudden death that completes his identification with the protagonist of the 1959 poem by Bulat Okudzhava, which gives the film its title.[45]

His wife, Nina (Chulpan Khamatova), is also a doctor, but shares none of Dania's enthusiasm and illusions. Her father (like both of Dania's parents) had died in the camps and she refers to Lenin as a bloodsucker. She cannot stand the fuss around the space program, questioning why there have to be "Sputnik bicycles, Sputnik electric shavers."[46] She reminds Dania that the rockets that are being constructed are intended to deliver bombs. She resists the request to join him in Baikonur, and when she finally goes there she wanders into a scene of horror. The huts of a nearby camp for Wives of Enemies of the People are being torched as an "undesirable reminder of the errors of the past," while its remaining guard dogs are being shot.[47] The shadow of the past threatens to extinguish the promises of the glorious future. Death is pervasive, for the Soviet Union is a 'zone.' As Evgenii Slivkin notes,

> In another episode, the cosmonauts, training in coarse, knitted suits, look like zeks in a labor camp. Parallels and editing cuts of this type in the film unite the cosmodrome and the camp into a single zone, and in the semantic field of the film, space flight is read as the preparation to escape from this zone—escape that is possibly doomed to tragic failure.
>
> (Slivkin 2011: 66)

German's portrayal of the contradictory moods of the Thaw generation is broadened further through Dania's relationships with his Moscow friends, all of them as anxious and neurotic as he. At a dacha picnic in celebration of the defense of his dissertation, they argue with him about the very point of flights into the cosmos and about the nature of the contemporary intelligentsia, which one of his friends calls a "weak and vulnerable creature, capable only of lamenting and getting upset," comparing it to a "beautiful, meaningless vase." This judgment leads Dania to wonder why, unlike their forefathers, his generation does not know why it is living: "We are all somehow sympathetic but meaningless. . . . Why do we chat so much? Feel such a desire to reflect, discuss, to run away from important, necessary work?"[48]

Like *Dreaming of Space*, *Paper Soldier* has a coda that takes place several years after the main narrative. In the earlier film, this conclusion is set in April 1961 and shows Konek running out of cheering Moscow crowds to congratulate his hero, Gagarin, thereby allowing the film to end on at least an ostensible note of optimism and achievement. The conclusion of *Paper Soldier*, set ten years later, in 1971, during a second country picnic, offers no such solace. The Thaw is over and the Brezhnev stagnation has begun. Gagarin has burned to death in a training accident, invisibly becoming a double of Dania and another of the film's eponymous heroes.[49] The statue of him made by one of Dania's friends is oddly lifeless, emphasizing the finality of death and the fickleness of memory. Suicide, emigration, abandonment, loneliness, and failure have marked their lives. A generation's optimism has been destroyed. As one of them says, in words that can be applied to the whole Soviet experiment:

> Sometimes I think that we thought up that epoch ourselves. It seemed to us then that we could take a few steps and everything would work out, but nothing worked out.

The frustrations and thwarted dreams of the Thaw generation parallel those of the space engineers. Like Uchitel′ before him, German has used the myth of Soviet space flight to provide a gloomy assessment of the broader hopes and disappointments of the Khrushchev years. And when Liubov′ Arkus, in her conversation with him, made precisely this point, he suggested an even broader existential reading:

> Despite the fact that the film announces 1961, this story is nevertheless about us and our time. For me the past is only an excuse. And though—I shan't conceal it—it was very flattering to hear from you about the subconscious of an epoch, I nevertheless consider that the film is not about the past. For

> me the main thing was to transmit the sensation of endless solitude in which we all exist, one way or another. And for me this film is above all about the impossibility of taking a step.
>
> (German 2008: 52)

Space flight in *Paper Soldier* is a passion of inexperienced young men and a distraction for their elders. The most thoughtful and sensitive character in the film, Dania's wife, Nina, is soberly dismissive of its meretricious attraction. By removing it to the background of both the cinematic frame and the narrative, German (like Iakov Protazanov three-quarters of a century earlier) insists that there are more pressing existential concerns down here on earth. In so doing he calls into question both the intentions and the achievements of Soviet modernity.[50]

Mystification

A different but equally challenging and original recent cinematic engagement with the Soviet space program and its cult is provided by Aleksei Fedorchenko in *First on the Moon* [Pervye na lune 2005], a film that purports to tell the story of the first Soviet flight to the moon, which it dates, provocatively, to 1938, a year which saw, on the one hand, the last of the major Moscow show trials and the execution of Bukharin, Rykov, and Iagoda, and, on the other, the institution of the title *Hero of Socialist Labor*.[51] Both the sinister and the heroic aspects of the epoch are reflected in the film.

Seductive though probably exaggerated stories quickly grew up about the film's reception, as audiences and critics vied with one another to boast of the time it took them to realize that they were being tricked and thus to reveal the extent of their ignorance and gullibility.[52] For *First on the Moon* is cast in the tradition of the 'mockumentary,' a term and genre as yet unknown in the Russian cinematic tradition. There is no sign of the moon in *First on the Moon*, and no steps are taken there except by the professor and his crew in interpolated extracts from Vasilii Zhuravlev's *Cosmic Voyage*, but for audiences who come forewarned to the film or who know enough about the history of space exploration to doubt the premise of its title, the film offers other, more subtle pleasures. Superficially an exposure of the cant that surrounded the space cult, it is, in fact, like *Paper Soldier*, a thoughtful and ambitious examination of the role of space exploration in both official Soviet propaganda and popular consciousness, and, more broadly, of Soviet and post-Soviet mentalities.

The film opens with what seems like documentary footage of the surface of the moon, but is described by the voice of the narrator as a "United States film chronicle" of peasants digging in the mountains of Northern Chile, where a "meteorite" (or perhaps the remains of a spacecraft) allegedly fell to the ground in March 1938. A woman and two men describe what they saw. This section is followed by a brief history of rocket technology over the centuries, in China and across Europe, rendered through visual illustrations which are just as fanciful as *First on the Moon* itself, before the film settles into a recapitulation of the Soviet space project. This reconstruction of a world that did not exist—yet might have done—is effected with extraordinary sophistication.[53] Stylizations of Soviet film newsreels are interspersed with dramatic reconstructions and present-day interviews with survivors and contemporary researchers, who attempt, unsuccessfully, to replicate the great flight.

The memories of the one surviving 'space pilot' [kosmoletchik], Khanif Fattakhov (played by both Anatolii Otradnov and Aleksei Slavnin), are a key source, but he dies in the course of filming. Fattakhov describes the rigorous selection program, stressing that all those chosen are "sportsmen and women, members of the Komsomol, handsome" [sportsmeny, komsomol´tsy, krasavtsy]—a phrase typical of Fedorchenko's allusive, polysemic approach. On the one hand, it suggests the values and the discourse of the Stalin period—a 1929 article in the journal *Sovetskii ekran* about the Chuvash actress Tania Iun introduces her as "a sportswoman, a gymnast, a record breaker" [sportsmenka, gimnastka, rekordsmenka][54] – and the documentary footage of its parades, in which attractive young *fizkul'turniki* showed their paces to Party leaders (Fig. 7.1). On the other hand, Russian viewers recognize it as a play on a now famous line from one of the most popular film comedies of the 1960s, Leonid Gaidai's *The Woman Prisoner of the Caucasus. Or The New Adventures of Shurik* [Kavkażskaia plennitsa. Ili novye prikliucheniia Shurika 1966], in which the words "a sportswoman, Komsomol member, and simply a beauty" [sportsmenka, komsomolka i prosto krasavitsa] are applied to the prisoner-heroine, Nina. Fattakhov describes the training of the short-listed four—a young Captain, Ivan Kharlamov (Boris Vlasov); a champion runner, Nadezhda Svetlaia ('Hope Bright') (Viktoriia Il´inskaia), whose name, consistent with the ambiguity and the ludic principle that lie at the heart of *First on the Moon*, associates her both with the 1930s and with The Thaw;[55] a circus dwarf called Mishka Roshchin[56] (Viktor Kotov), selected when the scientists feared that they would be able to launch only a small capsule; and Fattakhov himself (Fig. 7.2). We learn, too, about the brilliant "general constructor," Fedor Suprun (Andrei Osipov), who first took his rocket drawings to the Kremlin in 1928 and who, by 1933, had one and a half million people working on his secret project. When the filmmakers report what they have learnt about the early lives of the would-be space pilots, they once again evoke the 1930s through dense cinematic allusions. Khanif had been a street orphan [besprizornik], as had the heroes of the first ever Soviet sound film, Nikolai Ekk's *A Ticket to Life* [Putevka v zhizn´ 1931], while Kharlamov was wounded fighting basmachi in Southern Turkestan in 1930, a fact that in both plot and setting recalls Mikhail Romm's famous Eastern, *The Thirteen* [Trinadtsat´ 1936], itself a version of John Ford's *The Lost Patrol* (1934). The film's pervasive reliance on allusions, both visual and thematic, to the feature films of the period reminds viewers of the important role that feature film material plays in 'authenticating' our sense of the look of past decades.[57]

The remarkably persuasive evocation of the thirties testifies not only to Fedorchenko's absolute knowledge of the look and feel of the period, but also to his extraordinary attention to detail. "Documentary footage" from Soiuzkinozhurnal is interspersed with a sequence from a training film about how to use a secret camera and with shaky hand-held footage purportedly taken by that camera. All

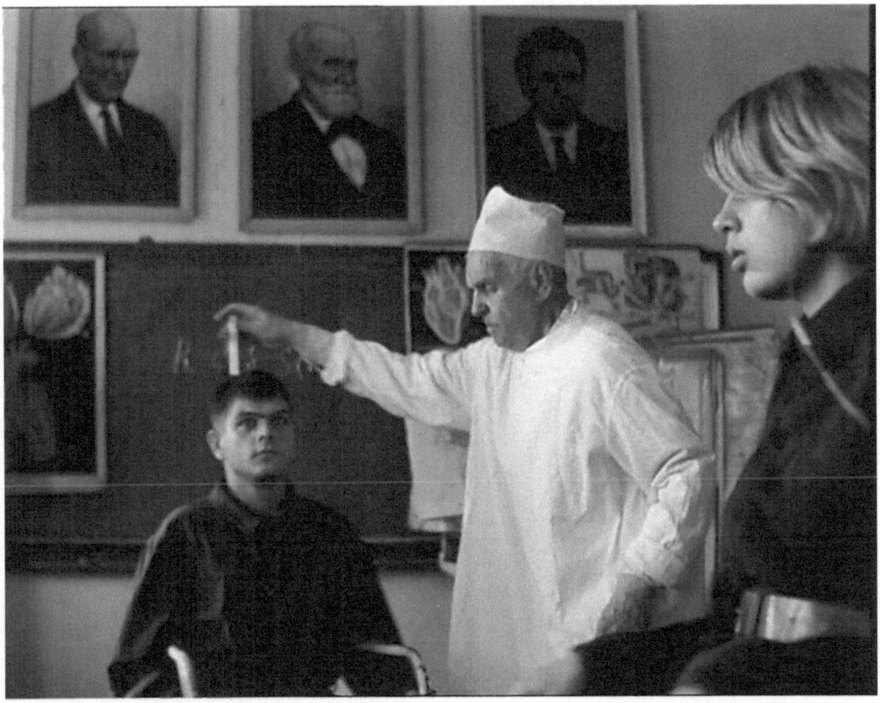

Figure 7.1 Training the space pilots. Film still from Aleksei Fedorchenko's *First on the Moon* (2005)

this material is in black and white, but it is of differing technical quality—a shortcoming admitted and apologized for in a preliminary intertitle, cleverly intended to stress authenticity. Some of the sequences—on Physical Education Day [Den′ fizkul′turnika] for example—even look familiar, but, as the director noted, "this is a feature film—there are almost no documentary shots in it" (Fedorchenko 2005). The persuasiveness of the impersonation is eloquently described in a brilliant engagement with the film by the critic Oleg Kovalov:

> In filming the pseudo-documentary, Fedorchenko does not imitate an arbitrary 'flow of life'; instead, he imitates the normative aesthetics of official film-journals—educational, instructional, and other types of applied films intended for use in 'official work.' He reproduces precisely this method of staging . . . it is distinctively 'an imitation of an imitation.'
>
> (Kovalov 2006)

Further 'authenticity' is supplied by the sounds of the period, with military marches and "Ever Higher" (described as "the emblem for people who were young in the 1930s," von Geldern and Stites 1995: 257) on the sound track, and by the

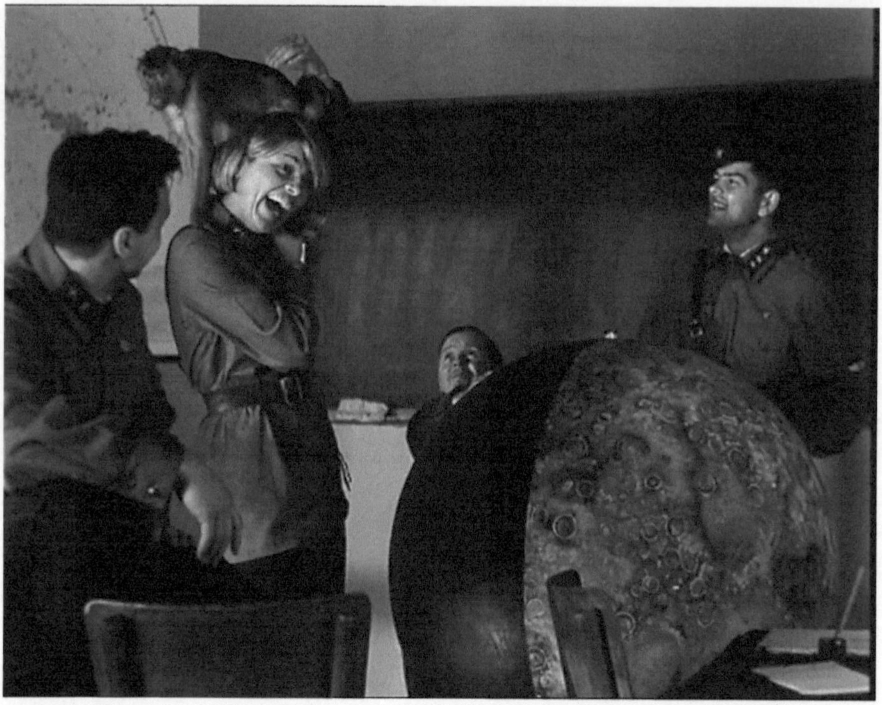

Figure 7.2 The space pilots. Film still from Aleksei Fedorchenko's *First on the Moon* (2005)

participants' faces, bodies, clothes, demeanour, and gestures. Kharlamov receives a medal alongside Valerii Chkalov and relaxes at a masquerade ball that looks just like the carnival night celebrations in Aleksandr Medvedkin's *New Moscow* [Novaia Moskva 1938]. The trainees are taken to see *Cosmic Voyage*, and Mishka Roshchin writes a poem of which he later remembers but a single line: "I greet you, great epoch" [Ia privetstvuiu tebia, velikaia epokha].

When contact is lost with the rocket and the mission is assumed to have failed, all signs of this disaster must be eradicated. As the voiceover reports: "The project was closed down, the people disappeared; it now seems that the time itself never existed." It is in the investigation / reconstruction of this side of the story that the terrifying dark side of the epoch is evoked. All the film footage that can be found is incinerated, and all the participants made non-persons: Nadezhda Svetlaia and others are poisoned in their beds by a sinister group of men wearing gas masks. Khanif lives to tell the tale only because he goes to ground in the Crimea, while Fedor Suprun, sensibly fearing for his life, disappears.

The rest of the story, which is intercut with the main 1930s narrative, concerns the filmmakers' attempts to track down the survivors and those who have encountered them so as to learn about their later lives and fill in the 'blank spaces'—a task beloved of real-life Russian documentary filmmakers from the 1980s onwards. The predominantly color footage in this part of the film recreates the lives of Soviet citizens in later decades. Once again, Fedorchenko carefully distinguishes the clothes, the behavior, the language, and the attitudes of diverse periods. Mishka Roshchin finds employment in a troupe of circus dwarfs in Ul′ianovsk, where he is known as "the human nucleus"; Fattakhov works among exotic, fantastical creatures in the St. Petersburg Zoological Museum; the NKVD cameraman is tracked down to his hospital bed; and Suprun, after a thirty-year search, is found working in a cemetery outside Capetown.

The most exacting of the searches is for Ivan Kharlamov. The filmmakers discover that in August 1939, the month of Marshal Zhukov's decisive victory against the Japanese in the battle on the River Khalkhyn-Gol, a Japanese prisoner reported on a wild-looking man they had encountered in the steppe. Found there eighteen months after his spaceflight, Kharlamov, we learn, was taken to a psychiatric hospital and interrogated by the NKVD. Once again, as in *Dreaming of Space* and *Korolev, 'The Cedar' Pierces the Sky* and *Paper Soldier*, Fedorchenko shows the worlds of space exploration and of imprisonment and punishment, of the cosmodrome and the 'zone,' to be inextricably linked. An old woman who had been Kharlamov's nurse and fallen in love with him describes how he escaped from his NKVD investigators. A member of the circus troupe of dwarfs subsequently recalls how Kharlamov had turned up at the Ul′ianovsk circus in 1946 in search of Mishka Roshchin, and (in one of its greatest comic coups and most striking cinematic allusions) the film shows footage of the patriotic post-war circus number invented for him, in which, as Aleksandr Nevskii, he defeated the Teutonic Knights, to the music of Prokof′ev.[58] At the end of 1955 Kharlamov left the circus and was never seen again.

So, through the detail and authenticity of these 'imaginary biographies,' *First on the Moon* evokes, decade by decade, the hopes and fears, the achievements and the constraints of the Russian twentieth century, and it does so, according to Fedorchenko, "without mockery – with irony":

> It's not a postmodernist film. We had another inner attitude to the material and to the epoch. We made the film with love for these people and with sympathy for their strivings, for their belief in the building of Communism and in the fact that we would be first on the moon. It was respect for these ideals that forced us to engage in open mystification. We told ourselves that, for all the problems of the socialist system, our fathers and grandfathers were honest people.
>
> (Fedorchenko 2005)

The "love and sympathy" for the aspirations of Soviet citizens and the distinction between their honesty and the demands of the state invoked by Fedorchenko can be compared to the post-post-modernist "new sincerity" that was first detected in Russian culture by Dmitrii Prigov and Mikhail Epshtein. A recent article by Alexei Yurchak stresses that this new sincerity does not lack irony (Yurchak 2008: 259) and shows that it can include a positive interpretation of some aspects of the Soviet past, including the achievements of Gagarin and other Russian astronauts. Yurchak describes these impulses as part of a broader "contemporary phenomenon of reassessing the history of Soviet modernity, exploring its meaningful aspects, and separating the original ethical aspirations of that modernity from the political regime which relied on them for its ends" (Yurchak 2008: 276).

If Fedorchenko's statement seems less than totally persuasive to some viewers (including this one), it is because the sequences displaying "belief in the building of Communism" in his film are outnumbered by those showing the state's ruthlessness in constraining and destroying the lives of its people. Nevertheless, the film should equally not be reduced to an(other) act of "historical exposure," for there is an additional dimension to Fedorchenko's film, a dimension that sheds light both on contemporary attitudes to Russia's past achievements in space technology and on the broader vogue for interrogating the Soviet past. *First on the Moon* is also a wonderfully intricate stylization of the kind of documentary (bearing series titles such as "Secrets of the Age" or "A Historical Detective Story") that has been a staple of Russian television channels for the last twenty years.[59] These investigations have long since established their rhetorical tropes and formulae, all of them painstakingly observed and simulated here. To heighten the illusion of authenticity, an unseen narrator describes, in the first-person plural, the minutiae of the filmmakers' search as they talk to wizened and toothless survivors and get permission to search in (part of) the FSB archive, where an archivist tells them, in defiance of everything we have learned about the Soviet manipulation of history, "Everything that happened had to be filmed. If it was filmed, that means it happened"—a notion operating on the meta-level, inasmuch as, following Kant's categorical imperative, it must also apply to Fedorchenko's film. The filmmakers peruse old newspapers, in Spanish and Italian, French and English, and take Fattakhov back to the now deserted cosmodrome. They make a painstaking reconstruction of Suprun's rocket at 1/38th the original size, but it explodes even more quickly than the original. Here and elsewhere they draw attention to their own successes and failures, suggesting that, like the makers of 'real' documentaries, they are as interested in the minutiae of their own reconstructive quest as in the story they purport to tell.

First on the Moon is much more than an interrogation of the Soviet space program. It is not only a historical detective story and a compellingly ambiguous reconstruction of an epoch, but also a faithful reconstruction of the fascination with that epoch in recent decades and Russia's painstaking, self-conscious, yet self-congratulatory efforts to tell its stories. One heroic enterprise rhymes with another, for Fedorchenko's investigation of mentalities is not confined to those of the Soviet period. Choosing flight as his central metaphor and the basis of his narrative line enables him to construct a dense web of allusion to Soviet myths (of heroism, comradeship, and aerial supremacy) and to several periods of Russian history: to the 1930s, when Soviet airmen broke long-distance flying records and rescued the crew of the Cheliuskin from the polar ice; to the Thaw years, when the real achievements of Soviet space technology revived interest in the aerial world and moved it up into the cosmos; to the late 1960s, when pervasive disillusionment and apathy followed the premature end of that Thaw; and to the post-Soviet period, when all these epochs have come under close, if sometimes self-serving, scrutiny. More perhaps than any of the other films discussed in this study, *First on the Moon* attempts a global survey of the mentalities of the Soviet period and of the shifting ways in which they have been interpreted and marketed in the post-Soviet years.[60] What links all these five recent films about the space program, despite their variations in intention and tone, and what distinguishes them from the cinematic *Bildungsromanen* about contemporary young men discussed earlier, is their readiness to question myths and to highlight the tragic elements in the Soviet past. What connects all seven films discussed above is the fact that the site of meaning and value they choose to interrogate is in the Soviet past (which can provide the intergenerational connections on which they so often rely). For Russian filmmakers, as for Russian citizens more broadly, the long Soviet past is a capacious country, which can yield whatever complex and absorbing narratives they seek to find there, from horror and brutality to optimism and moral certainty. Though some younger filmmakers at last are providing the same density in stories set in the present, the post-Soviet period has yet to produce its own compelling myths.

Aerial dreams of a Russian Icarus

In one of the most important Russian films of recent years, the dreams about flight are played out in the undramatic Russian present. The child protagonist with whom they are associated has no imaginative connection with the Soviet period and no recourse to its imagery. But his aerial obsession also has its source in the past, since it is based upon his pride in his father's earlier career as an aircraft engineer. Thus the boy's visions are significant both on the existential and on the historical plane.

Most discussions of Boris Khlebnikov and Aleksei Popogrebskii's *Roads to Koktebel* [Koktebel' 2003, hereafter *Koktebel'*] have centered, understandably, on the film's portrayal of the relationship between a father and his son, a concern shared by several important Russian films of the early twenty-first century.[61] What is also remarkable, however, is the saliency of height and flight to the articulation of that relationship, permeating both the film's visual imagery and its dialogue from beginning to end.

In the opening scene of the film, father and son seem to emerge from the ground, from darkness, from a tunnel in which they have slept, and this earthbound state is later revealed to be symbolic of how low they have fallen from their earlier life in Moscow. In the first words of the film, the boy asks whether they are going to walk to Crimea ["Tak v Krym poidem?"]. The father replies that they will go by train, and adds, "If you don't want to take a train, we'll go by plane." The contrast between (actual) walking and flying (only dreamt of) will pervade the film. The antithetical relationship between lowness and height, between earth and air, between stasis and movement is further developed in the next sequence, in which the two are seated on the floor of a goods carriage. Finding a worm in an apple, the boy, prompted by his transformative imagination and his inherited desire for a scientist's understanding of the way the world works, asks his father what kind of butterfly it will become. This is also the scene in which their destination, Koktebel′, a town in Crimea best known for its Russian cultural associations, is mentioned for the first time, so the link between flight (through the butterfly) and ideal goal (Koktebel') is established from the very start of the film.

Koktebel' is constructed as a series of encounters on the father and son's journey south. The first is with a railway inspector and his daughter, who engages the boy in conversation. In statements peppered with words semantically associated with height [vys′], he tells her that he does not attend school, since his father, who has a higher education [vysshee obrazovanie], teaches him everything. He himself "can see everything from high up" [A ia umeiu videt′ vse s vysoty]. As he utters these words, Sandor Berkeshi's normally static camera provides a high aerial shot of the boy's vision, in which trees, a road, a barn, and a field are reduced to the patterning on a knitted sweater, just as they were in the long overhead sequence which opens Aleksandr Zel′dovich's film *Moscow* [Moskva 2000].[62]

The low, empty landscapes that father and son had seen from the train are echoed in the wide field they cross after leaving the railwayman's hut, and throughout the film the Russia they traverse will be flat and open. So the boy's interest in flying and height reflects his aspiration to replace the mundane horizontal plane along which they are forced to travel with the imaginative potential of the vertical. His attention is drawn to a bird soaring high in the sky and he asks his father to identify it. When earlier asked to name the butterfly, the man had dissembled by inventing a plausible-sounding Latin name, but here he can rely on real knowledge. "It's one of the falcon family [odin iz sokolinykh]," he tells the boy, and proceeds to explain the "very good aerodynamics" that enable it to soar on an air current [tok vozdukha] and to glide without flapping its wings. But a falcon cannot glide for long, unlike an albatross, which "glides for many kilometers." Though knowing little about the mythical bird, the boy immediately transfers his allegiance to it, and asks his father whether there are albatrosses in Koktebel′. Despite the negative response, hereafter the bird and the town are joined in the boy's mind as symbolic expressions of his desires. His rejection of the falcon [sokol], with its Soviet and specifically Stalinist connotations, for the albatross, a bird that does not fly over Russia, is a measure of the ardency of his need to move forward into the present and out into the world (Fig. 7.3).

Imagery of height and flight also permeates their next encounter, with Mikhail Aleksandrovich, a distrustful old man who agrees to let them mend his roof. It is here, with the father's revelation that he used to build planes, that the boy's obsession with flying emerges as an expression of his filial loyalty, a determination not to let his father forget the successful life he had led before his existential fall. Drawings of the albatross and information about its capacities that he finds in an illustrated book in the old man's attic nurture the boy's fascination, and later, standing at a high window, he 'sees' himself tiny in the nocturnal world.

But in this sequence, as in the previous one, the boy's abstract, poetic identification with high places is undermined by a more earthly treatment of the image. During their stay with the railway inspector, his daughter had unknowingly parodied the boy's obsession with soaring flight by leaving her ghetto blaster incongruously trapped in the branches of a tree while she was using the outdoor

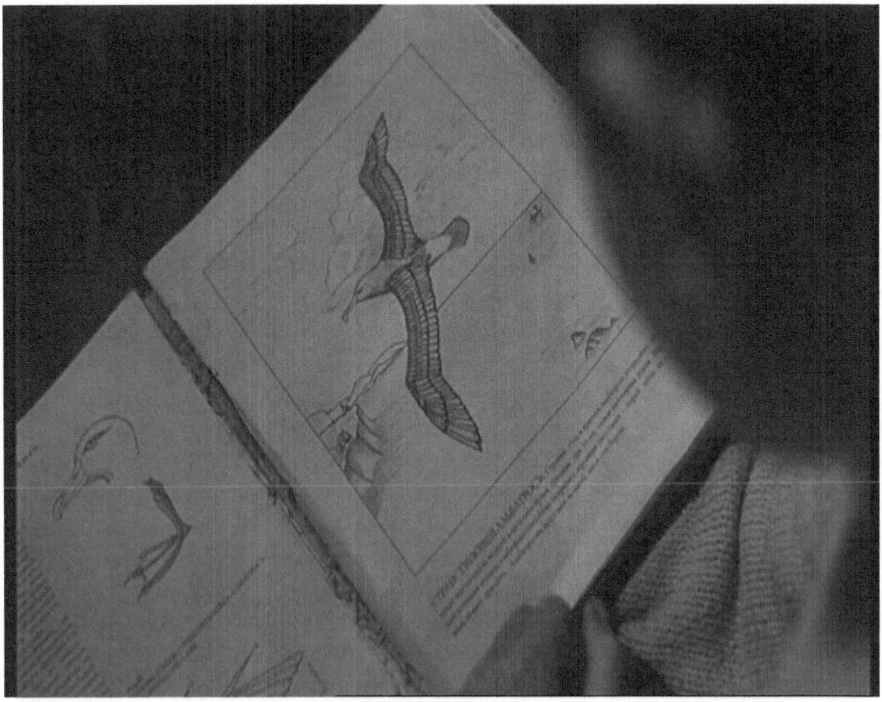

Figure 7.3 The boy reads about the albatross. Film still from Boris Khlebnikov and Aleksei Popogrebskii's *Koktebel'* (2003)

toilet. Now his father's earthbound simulation of the boy's lofty dream—and a mocking echo of his own aerial past—consists of standing on the precarious rooftop and shouting "I'm flying!" [letaiu] as he hurls down sections of the old iron roofing, which curl and buckle on the wind but ineluctably fall to the ground (prefiguring both the boy's own aerodynamic experiments and his eventual 'fall to earth'). Images of vertical struggle and of falling also pervade the poems by Pushkin and Lermontov of which the old man recites extracts to the boy.[63]

If the boy's imagination associates him with height, the mundane reality stresses his disabling shortness. When the foolish old man draws a gun on them, the boy draws himself up as tall as he can and, raising his arms further, attempts to protect his father's body. But the man fires over him, wounding the father. This scene is then rhymed, even more humiliatingly, during their third and last shared encounter, with the doctor, Kseniia Viktorovna. The boy watches the woman hang out the washing, which flaps in the air like a bird's wings. In embarrassment that she will pick up and hang up his underpants, he makes a grab to do it himself. But the line is too high, and the smiling woman has to help him, 'cutting him down to size' in the process.

The subjective image of Koktebel' mutates in the doctor's house, where the boy finds another symbolic book—an atlas that contains no mention of the

Koktebel′ of his reveries, either in the map of Crimea or in the index. His reaction to this betrayal is expressed in vertical terms. First he constructs a high pile of detritus and proceeds to knock it down. Then he takes a series of photographs in which the normal vertical axis is reversed—an upside-down house, an upside-down tree, seen as if from a great height, and, finally, his father and Kseniia Viktorovna upside down by the treacherous washing line. When he later accusingly tells his father that Koktebel′ does not exist, the man explains that in 1944 Stalin deported the Crimean Tatars,[64] and, since Koktebel′ is a Tatar word, the town was renamed Planerskoe in 1947, in honor of its big glider base. The name renders the town's association with flight tantalizingly explicit for the boy. Moreover, when the father explains why the terrain, with its constant wind and its rising air currents [voskhodiashchie toki vozdukha], is right for gliding, his choice of words connects the location with the falcon the boy had seen earlier. According to the father, if the boy stands on the summit of the range of hills outside the town, where there is even a glider's monument [pamiatnik planeru], and throws a piece of paper into the air, it will fly "upwards and far away" [vverkh i daleko] (Fig. 7.4). Transfixed, the boy asks whether the glider can fly a long way [daleko], to which his father replies that, depending on the skills of both constructor and pilot, it may fly several kilometers. "Like the albatross?" asks the body. At this key stage, Koktebel′, great height, gliding, and the mythical bird are all connected in the boy's mind, indivisible elements of what his imaginary conceives as paradise. Whereas the father had relied on the power of technology, the boy puts his trust in nature. For the moment, he forgets that, historically, Koktebel′ is also a place of treachery, and, as if rehearsing future efforts, he walks to the top of a nearby hillock with a paper kite and releases it into the air. It immediately falls to the ground, for the hill is insufficiently elevated and the winds are not strong enough. For his purposes, it seems, this is not the place.

This failure precedes a crisis that occurs when the father's growing affection for Kseniia Viktorovna (Agrippina Steklova) leads him to suggest a postponement of the rest of their journey till spring. Once again, Koktebel′ has been betrayed. Stalin's treachery had been expressed linguistically, in the elision of the town's name, and now, in words that shock both his father and the viewer, the devastated boy takes the Tatar word and damagingly re-semanticizes it: "Kokt*ebel′*, Kokt*ebel′*, Vy tut *ebetes′*, a chto mne zdes′ delat′?" [Koktebel′, Koktebel′! You're fucking here, but what am I to do here?]

As the boy sits, alienated and alone, on a hilltop that is the highest place for miles and looks despairingly at the flat landscapes around Kseniia Viktorovna's house, he realizes that he cannot rely on his father and decides to set off for Koktebel′ without him. In the middle of the night, he is picked up and lifted aloft by a boisterous truck driver—an action that intimates that this man will help him in his quest. The next time we see him, the boy is sitting high in the cab of the truck, a position that enables him to float above the flat plains that dominate even on this final leg of his journey. Prophetically, however, as they approach a road sign announcing that they are at last entering Crimea, it is partially obscured by a large passing seagull. Asked where he wants to be dropped off, he remembers the words he had heard

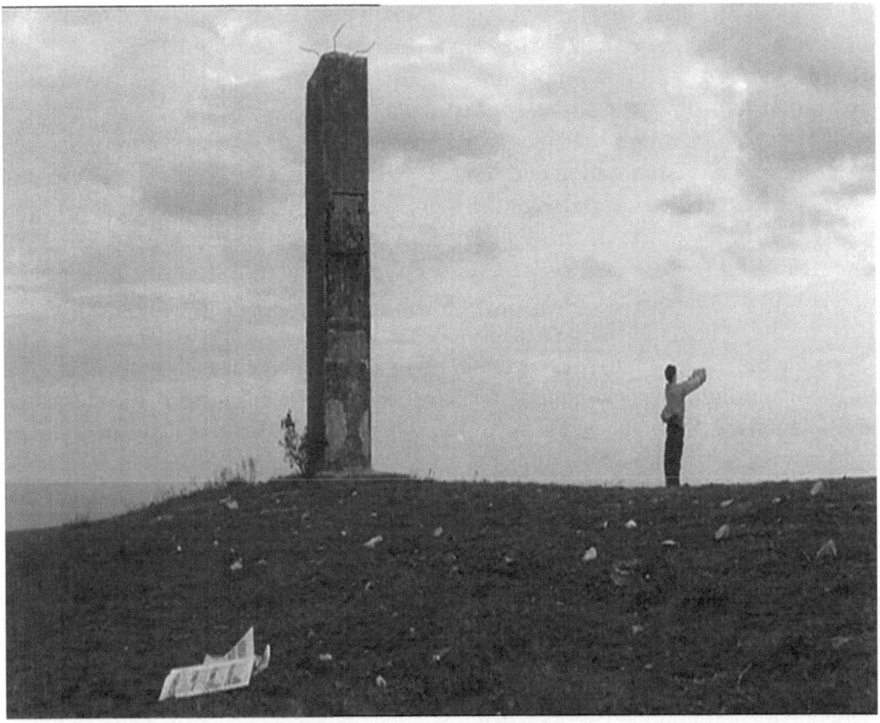

Figure 7.4 The boy at the Glider's Monument. Film still from Boris Khlebnikov and Aleksei Popogrebskii's *Koktebel'* (2003)

from his father, "In the hills, where there's a wind" [v kholmakh, gde veter], and demands to be left at the monument.

At last there is a real hill, and we watch his tiny figure determinedly climb to its summit, only to be disappointed, for the monument commemorating flight has been vandalized. All that remains is the metal strips that had supported the glider and the ugly concrete base. Nevertheless, since the ground is elevated and the wind is strong, the boy tears a page from an atlas and releases it into the wind. It falls promptly to the ground, perhaps as a punishment for his stealing the book from Kseniia Viktorovna, his rival for his father's affections. Only on the fourth attempt does the paper blow across and out of the frame, yet even then the shot is curtailed, and we do not witness its flight "up and far away" [vverkh i daleko].[65] Though the boy has achieved his aim, it seems far from dramatic and strangely unsatisfying, without any of the poetry that infused his earlier dreams of high places.

This bathos is compounded by his walk round Koktebel′, which has become a tawdry post-Soviet resort that deflates the boy's idealistic expectations. The camera, now significantly reduced to the boy's ordinary height, offers a succession of close-ups of overweight tourists' bottoms. A faded crooner in a bar grinds out a song about 'Roads' that, with its lines—"And they did not give me a take-off time/And in the end my path became a runway"—mocks his aspirations and his epic journey.[66] Most dispiritingly, there is no albatross. The denizen of the Pacific and southern oceans gives way to a malevolent imposter-seagull, the most banal of seaside birds, which attempts to steal his food and pecks him on the arm. In frustration at the thwarting of all his hopes, the boy grabs it by the throat, but eventually he releases it into the skies. This final action, redolent of compromise and loss of illusion, of maturation and of acceptance of his earthbound mundaneness, reveals the psychological terms of the journey made by the boy, which constitutes a *Bildungsroman* or rite of passage. The film ends with a final overhead shot from the great height of the boy's aspirations, but, crucially, not from the boy's point of view. It establishes the perspective of omniscient fate or inevitability: he has become the object, not the owner of the gaze.

As the boy looks out to sea, his father rejoins him and the visual rhymes between them—their light tops, haircuts, and silent frontal gaze—suggest reconciliation and tolerance of the other's failings. The modern-day Daedalus had not colluded wittingly in his Icarus-son's dreams of flight, which he himself had already abandoned. The boy soared only in his imagination, and the banality of human disappointment has saved him. In *Koktebel'*, as in many of the other films discussed in this chapter, the dream flight represents immaturity, lack of worldliness, inexperience—all qualities that must eventually be abandoned in the transition to adulthood.

Conclusion: leaving the house of dreams

Khlebnikov and Popogrebskii's *Koktebel'*, then, narrates the eternal story of maturation as the loss of illusion, and, for all the differences in setting and protagonists, this is also the story of the films about the Soviet space race discussed earlier in this chapter. As narratives of painful entry into adulthood, they can all be seen as part of a more general intellectual and societal phenomenon in Russia, in which the utopian belief in change that animated the late 1980s and early 1990s has been abandoned in the face of harsh economic and political reality. To continue to dream of flight, it would seem, you need to be one of the naïve, cocky young men of *Black Lightning* or *We Are from the Future*. Other cinematic myths still flourish, of course—about the Soviet experience of the Great Patriotic War, about The Thaw, and, in the present, about tough young men, be they soldiers, businessmen, or gangsters. But some of the most thoughtful and ambitious Russian filmmakers have turned away from their ideal visions of flight and abandoned the house of dreams.

Notes

1 I wish to express my very sincere thanks to Professor Helena Goscilo and Dr Vlad Strukov for their immensely helpful and productive comments on an earlier version of this chapter.
2 See also Chapter 6.
3 For the text of "Vse vyshe," see von Geldern and Stites 1995: 257–8.
4 Aviakhim, the Society of the Friends of Aviation and Chemical Construction, had been formed in May 1925, on the amalgamation of the Society of the Friends of

the Air Fleet [Obshchestvo druzei vozdushnogo flota], founded in March 1923, and the Society of the Friends of Chemistry [Dobrokhim], set up a year later. Formally 'volunteer' organizations, they were in fact officially run and had the aim of providing elementary programs of military training and chemical and aviation technology for the masses. Closely linked to the Red armed forces, they had a predominantly male membership. In January 1927 they would be combined with the Military Scientific Society [Voenno-nauchnoe obshchestvo] to form the Society of Friends of Defense, Aviation and Chemical Construction [Osoviakhim]. On the history of these societies, see Odom 1973, from which these details are taken.

5 On Soviet aerial achievements in these years see Palmer 2006, especially 219–46. The achievements of the stratonauts influenced the triumphant circus number performed at the end of Grigorii Aleksandrov's musical comedy *Circus* [Tsirk 1934] and can also be seen in the documentary footage in Andrei Tarkovsky's *Mirror* [Zerkalo 1974].

6 ["Za nashikh khrabretsov-letchikov! Eto—gordye sokoly nashei armii]," published in *Pravda*, 6 May 1935; quoted from Dushenko 2002: 558. The sobriquet was later applied to the Soviet airmen of WWII—see for example the 1941 poster by P. Vandyshev and L. Torich, which bears the legend "Glory to the Heroes of the Patriotic War! Glory to Stalin's Falcons!" [Slava geroiam otechestvennoi voiny! Slava stalinskim sokolam!] and can be viewed at http://www.a2m.ru/forums/viewtopic.php?f=2&t=160 (accessed 23 January 2012).

7 In the late 1920s and 1930s there was a considerable effort to train and showcase female pilots. For example, the 1938 *Motherland* [Rodina] flight involved a crew of three women who flew from Moscow to Komsomolsk, aiming to set an international women's record for distance flight.

8 For an overview of major Soviet films representing aviation, see Palmer 2006.

9 There were a few notable exceptions, including the following films featuring female leads and showing women in a heroic manner: *Airmail* [Vozdushnaia pochta 1939, dir. Dmitrii Poznanskii], and *Fliers* [Letchiki 1935, dir. Iulii Raizman].

10 On the 'Night Witches,' see Goscilo and Hashamova 2012: 8.

11 See Chapter 8 for a discussion of his work.

12 On the achievements of Soviet space technology, see Siddiqi 2010. On the reflection of Iurii Gagarin's space flight in Soviet cinema, see also Sibirtseva 2011. The passing of fifty years since the flight briefly rekindled Russian cinematic enthusiasm but it is somehow symptomatic that Petr Parkhomenko's *Gagarin. The First in Space* [Gagarin. Pervyi v kosmose 2013] was not released until a year after the anniversary.

13 On the heroes of the Soviet pilot films of the Stalin period, see Kaganovsky 2008, especially 89–115.

14 [Umirat′ tiazhko, no ne tak trudno, tovarishch Chkalov . . . A vot my za takikh liudei, kotorye khotiat zhit′, zhit′ kak mozhno dol′she . . .] (Valerii Chkalov 1941: 45).

15 On the fate of the real-life Chkalov, see Bergman 1998.

16 For more on the vertical imagery and the metaphysical concerns of *Days of Eclipse*, see Graffy 2011.

17 [Na nebe oklako-rai, podi ego ugadai, / Naiti, uznat′ i poniat′ nevozmozhno. / Kakogo tsveta i gde, skazhi pozhaluista mne, / Skorei ty tam, gde syskat′ ochen′ slozhno. / A ia ot mysli drozhu, chto nikogda ne naidu / Khochu zabyt′, ubezhat′, zakrutit′sia. / Gde ty, gde ty, oblako-rai / List′ia nadezhdy moei ne roniai.]

18 [Toskoi i grust′iu lechus′, dushoi na nebo ia rvus′. / V gustom tumane ishchu svoe schast′e. / Byt′ mozhet kto-to, kogda, mne vdrug otkroet glaza, / I ia uvizhu ego skvoz′ nenast′e. / No net ego, i opiat′, ne ustaiu povtoriat′, / Zhivia nadezhdoi o vstreche s toboiu . . . / Gde ty, gde ty, oblako-rai / List′ia nadezhdy moi ne roniai.]

19 [Gde tvoi kryl′ia, kotorye nravilis′ mne?] The song, Kryl′ia, appears on the eponymous 1996 album by the group Nautilus Pompilius.

20 This is prefigured in Dmitrii Astrakhan's 1995 film, *The Fourth Planet* [Chetvertaia planeta], in which the astronauts of a Russo-American space flight land on Mars, only

to find themselves in a provincial Soviet town in the Brezhnev period. For more detail on this film see Graffy 1999: 173–5.

21 There are seminal interpretations of this theme in several absorbing articles in the recent *Cinepaternity* volume; see Goscilo and Hashamova 2010.

22 Kuptsov's name comes from the Russian word *kupets*, meaning a merchant.

23 There is also a flying car powered by love in another recent film, Oleg Fliangol′ts's sardonic *Indifference* [Bezrazlichie 1989–2010].

24 For an engaging analysis of the implications of the intricate plot of the film, see Maguire 2010.

25 On the GAZ-21 see, for example, Siegelbaum 2007: 66–7, Thompson 2008: 60–3 and the documentary film *Step on the GAZ. The Story of a Nightmare* [Davit′ na GAZ. Istoriia odnogo koshmara] Edinaia media gruppa, 2007. *Black Lightning* is not the only recent Russian film to have fetishized a Soviet car as a way of evoking the moods and concerns of the Soviet past. The ZiM in Pavel Chukhrai's *Driver for Vera* [Voditel′ dlia Very 2004] and the Zhiguli in Ivan Dykhovichnyi's *The Kopeck* [Kopeika 2002] are treated in similar ways. It is indicative, however, that the most glamorous car in Valerii Todorovskii's *Hipsters* [Stiliagi 2008], which is set during the early Thaw, is not a Soviet vehicle, but a Hudson Custom Eight Convertible Coupe, a reminder that nostalgia for the Soviet past can incorporate a yearning for a time when foreign glamour was rare and difficult to attain.

26 On *Everything Will Be OK*, see Graffy 1999: 167–73; Goscilo 2014: 97–103.

27 The finding of moral value in The Thaw and the time-traveling link between the 1950s and the post-Soviet present are also notable features of *Hipsters*, at the end of which the hero and his wife are instantaneously transported from the 1950s to the post-Soviet present, where they are revealed to have provided the model for all the happily alternative lifestyles now on display on the streets of Moscow.

28 On *A Mirror for the Hero* see Ivanova 1988. On *Prishvin's Paper Eyes* see Shemiakin 1990.

29 On the new films about WWII, see Norris 2007 and Smorodinskaya 2010: 101–6.

30 This sequence, and Nikita Khrushchev's fury at it, have attracted broad commentary. See for example Prokhorov 2010: 32–4.

31 The trope of young Russian men being transported into WWII to find moral value for their lives is repeated in *The Fog* [Tuman 2010, directed by Ivan Shurkhovetskii and Artem Aksenenko], and the success of both films led to their spawning of sequels: *We Are from the Future 2* [My iz budushchego 2 2010, directed by Boris Rostov and Aleksandr Samokhvalov] and *The Fog 2* [Tuman 2 2012, dir. by Ivan Shurkhovetskii]. On the indistinguishability of these films, in which "the past becomes the main project of the future" and "The motherland is one, one for all times" (the slogan of *The Fog*), and on the consequent impossibility of taking them seriously, see Kalinin 2013.

32 On the role of maps in films, see *The Cine Tourist*. On mapping the Soviet Union, see Widdis 2003.

33 The most famous character bearing the name German in Russian culture is the hero of Aleksandr Pushkin's story 'The Queen of Spades.' Like Uchitel′'s hero, he is an outsider, the source of fascinated gossip among the more conventional members of his circle, and is driven by an obsession that leads him to fake a passion for a young woman who, he thinks, can help him to achieve it.

34 For Russian viewers the diminutive nickname Konek ('Horsey') evokes the *Konek-Gorbunok* [Little Humpbacked Horse] of Petr Ershov's children's magic tale of the same name, who flies his peasant master to success and happiness in St. Petersburg. Konek's naïveté and his childish search for answers, as well as the northern setting, also make him the double of Kolia in Dostal′'s *Cloud Heaven.*

35 On how you can "dress yourself Soviet," see Vainshtein 1996, Widdis 2008, and Widdis 2009.

36 On Kabakov's installation see Groys 2006 and the Introduction.

37 On the scale of the celebration of the the Korolev centenary, see Gerovitch 2011: 85–6.

38 On the evasions of *The Taming of Fire*, see Gerovitch 2011: 93–5. On the film more generally, see Wade.

39 For an alternative telling of this story see Siddiqi 2010, especially 121–6, 136–78.

40 Interview with Iurii Kara, *Izvestiia*, 12 October 2007, quoted from Gerovitch 97.

41 On Langemak and Glushko and their arrest see Siddiqi 2010, especially 162–3, 173–8; on the visits to Peenemünde see Ibid. 200–12.

42 The trope in which a son attempts to heal or redeem a broken or dead father is common in recent Russian films about flying, encountered also in *Black Lightning* and *Roads to Koktebel′*.

43 For an analysis of the film see Chapter 8.

44 The use of the trope of illness as a metaphor for the broader moods of an epoch is one of several ways in which *Paper Soldier* echoes the approach of *My Friend Ivan Lapshin* [Moi drug Ivan Lapshin 1979–84], directed by German's father, Aleksei Iur′evich German. I have explored the links between the two films, and German's broader dialogue with films of The Thaw and later Soviet period in *Paper Soldier*, in Graffy 2009.

45 Compare, for example, the following lines: [On peredelat′ mir khotel, / chtob byl schastlivyi kazhdyi, / a sam na nitochke visel: / ved′ byl soldat bumazhnyi].

(He wanted to change the world / So that everyone could be happy, / But he himself was hanging by a thread, / For he was a paper soldier.) For the text of the poem 'Bumazhnyi soldatik', see Okudzhava 2001: 177.

46 Bicycles are associated in *Paper Soldier* both with Dania, who rides one around the Moscow hospital at the beginning of the film and is riding one when he dies, and with the two would-be astronauts, who race each other on bikes just before the flight. Similarly, in *Dreaming of Space*, the bicyclist Konek gives German a ride. The simple pleasure that all these would-be astronauts get from using this most humble means of transport makes them more humanly accessible to viewers, while paradoxically reminding us that the bicycle once represented the highest achievement of transport technology. In H.G. Wells's *The Time Machine*, written in 1898 at the height of the bicycling craze, the time-traveling machine is described in terms reminiscent of a bicycle; for example: "It was at ten o'clock to-day that the first of all Time Machines began its career. I gave it a last tap, tried all the screws again, put one more drop of oil on the quartz rod, and sat myself in the saddle" (chapter 3). The connection of the bicycle and the space rocket as versions of technological innovation is comically present in the numbers performed at the end of Grigorii Aleksandrov's *Circus*. While waiting for Marion Dixon to arrive and perform *Flight to the Stratosphere*, the circus director is forced to reprise an act titled *Three on Wheels, the Technological Wonder of the Twentieth Century!* Also see the Introduction to the volume where we discuss early Soviet artistic projects linking the bycicle and plane.

47 Thus all four films, *Dreaming of Space*, *Korolev*, *'The Cedar' Pierces the Sky*, and *Paper Soldier*, link the Soviet space program to the labor camps.

48 German's reprisal here of the dilemmas of The Thaw also brings to mind the apathy and uncertainty of the protagonist of Mark Osep′ian's late Thaw film, *Three Days of Viktor Chernyshev* [Tri dnia Viktora Chernysheva 1968]. See Prokhorov 2010: 37–41.

49 [A on, sud′bu svoiu kliania, / ne tikhoi zhizni zhazhdal, / i vse prosil: / "Ognia! Ognia!" / Zabyv, chto on bumazhnyi. / V ogon′? Nu chto zh, idi! Idesh′? / I on shagnul odnazhdy / i tam sgorel on ni za grosh: / ved′ byl soldat bumazhnyi.] (And he, cursing his fate, / Did not thirst for a quiet life, / But kept asking for / "Fire! Fire!" / Forgetting that he was made of paper. / Into fire? Well, go, then! Are you going? / And he once stepped forward / And there he burned up for nothing: / For he was a paper soldier.") (Okudzhava 2001: 177.) The association of flight with death in the film is prophesied in the quotation by Dania and Nina in an earlier episode of Aleksandr Blok's June 1910 poem 'Demon,' which ends with the words [I pod bozhestvennoi ulybkoi, / Unichtozhaias′ na letu, / Ty poletish′, kak kamen′ zybkii, / V siiaiushchuiu pustotu . . .] (And under the smile of God, / While being destroyed in flight, / You will fly, like an unsteady stone / Into shining emptiness.) (Blok 1960: 60–1.)

50 German's questioning of Soviet myths and his readiness to stand apart from the cinematic mainstream are equally apparent in his representation of WWII from the point of view of two Germans, stranded in the snows of Russia, in his earlier film *The Last Train* [Poslednii poezd 2003].

51 For a discussion of visual commemoration of these events in contemporary animation, see Chapter 8.

52 For all their shared subversive intent, Fedorchenko's formal approach to the deconstruction of the Soviet space myth is the opposite of that taken by Viktor Pelevin in his *Omon Ra* (1992). If Pelevin's novella contends that Soviet rockets did not go into space at all and that the hoax was perpetrated in the Moscow metro, Fedorchenko's film suggests that there was more space exploration than hitherto admitted and that it took place considerably earlier. See Chapter 10 for discussion of Pelevin's novel.

53 Fedorchenko's enthusiasm for what Serguei Oushakine has called "creative exercises" in the "invention" of the past and its rituals is also evident in his next feature film, *Silent Souls* [Ovsianki 2010]; see Oushakine 2011.

54 "N" 1929, 4. On Soviet sportswomen of the period, see Kobchenko 2011. On the 1930s cinematic reflection of the cult of the female athlete, see Graffy 2013.

55 The association of Soviet achievement with light goes back to Lenin's famous statement of 1920 that "Communism is Soviet Power plus the electrification of the whole country" [Kommunizm est′ Sovetskaia vlast′ plius elektrifikatsiia vsei strany] (Dushenko 2005: 170). See also Chadaga. The link is reiterated in the period in which the film is set in a *Pravda* article of 13 May 1936 proposing that the country go "forward to a still brighter future" [vpered k eshche bolee svetlomu budushchemu] (Dushenko 2005: 265) and in the *Svetlyi put′* [Radiant Path] that transforms the heroine of Grigorii Aleksandrov's 1940 musical of the same name from a simple provincial weaver into a Deputy of the Supreme Soviet. The surname Svetlaia also develops the connection with Gaidai, but does so in a way that subverts the 1930s rhetoric, since in Gaidai's next film, *The Diamond Arm* [Brilliantovaia ruka 1968], the role of the unnamed woman who, in a scene etched in the memory of millions of Russians, involuntarily performs the first strip tease in Soviet cinema, is played by the actress *Svetl*ana *Svetl*ichnaia.

56 This name continues the link to The Thaw since Mikhail Roshchin is also the name of one of the most important playwrights of the 1950s and 1960s. Fedorchenko's Roshchin will also later be revealed to be a writer.

57 See also Chapter 6.

58 The density and care of the film's system of allusions is apparent in the fact that Eisenstein's *Aleksandr Nevskii* was made in 1938, the very year that Fedorchenko chose for the space flight in *First on the Moon*.

59 Sasha Kiselev also places the film in the context of these documentaries. See Kiselev 2005: 38.

60 Asked by Viktor Matizen whether the success of his falsification would dent viewers' confidence in the veracity of other documentary films, Fedorchenko laconically replied, "Probably" (Fedorchenko 2005). But it has not deterred filmmakers from continuing to use the same tropes. Instructive comparisons can be made with the 2007 documentary *The First on Mars. The Unsung Song of Sergei Korolev* [Pervye na Marse. Nespetaia pesnia Sergeia Koroleva, made for the Studiia Roskosmos by Tamara Bocharova]. From the film's very title to the use of an extract from *The Cosmic Voyage*, to interviews with the friends and relations of the participants, to scenes of the destruction of evidence connected to a failed project, detail after detail retrospectively confirms the precision of Fedorchenko's pastiche.

61 In this context, see the interesting engagement with the film by Hashamova 2010.

62 For an analysis of this opening and of *Moscow* in general, see Goscilo 2007.

63 In the lines that Mikhail Aleksandrovich quotes from Pushkin's 1829 poem 'The Drowned Man' [Utoplennik] "black crayfish are sucking his swollen body" [I v raspukhnuvshee telo / Raki chernye vpilis′, Pushkin 1963: 75]. He also recites lines

from two Lermontov poems of 1837, 'Hurrying north from afar' [Spesha na sever iz daleka], which is addressed to Mount Kazbek in the Caucasus, and 'I do not want the world to know' [Ia ne khochu, chtob svet uznal], in which he invokes the image of a granite cliff that the noisy seas cannot destroy.

64 The motif of the Crimean Tatars' exile and the symbolic journey to Crimea are also important in *Days of Eclipse*. At the end of the film the hero's best friend, a scion of a family who experienced that fate, sets sail home to the place of his dreams.

65 For a different reading of this sequence, stressing the boy's "disconnection from the father," see Hashamova 2010: 178.

66 The song "Dorogi" [Roads] was written by Sergei Shnurov, leader of the group Leningrad.

Bibliography

Bergman, J. (1998) "Valerii Chkalov: Soviet pilot as New Soviet Man." *Journal of Contemporary History*, 33, 1: 135–52.

Blok, A. (1960) *Sobranie sochinenii.* Vol. 3. *Stikhotvoreniia i poemy 1907–1921*. Moscow-Leningrad: Gosizkhudlit.

Chadaga, J.B. (2007) "Light in captivity: Spectacular glass and Soviet power in the 1920s and 1930s." *Slavic Review*, 66, 1: 82–105.

The Cine Tourist. (n.d.) Available at http://www.thecinetourist.net/index.html (last accessed 20 January 2012).

Dushenko, K.V. (2002) *Slovar′ sovremennykh tsitat*. Moscow: Eksmo.

—. (2005) *Tsitaty iz russkoi istorii ot prizvaniia variagov do nashikh dnei. Spravochnik.* Moscow: Eksmo.

Fedorchenko, A. (2005) "My ne poliruem vremia" (interviewed by Viktor Matizen). *Novye izvestiia*, 6 June. Available at http://www.newizv.ru/news/2005-06-06/25646/ (last accessed 20 January 2012).

German. A. (2008) "Besplodnye usiliia liubvi k Otechestvu. S Alekseem Germanom-ml. beseduet Liubov′ Arkus." *Seans*, 35–6: 48–52.

Gerovitch, S. (2011) "Memories of space and spaces of memory: Remembering Sergei Korolev." In *Soviet Space Culture. Cosmic Enthusiasm in Socialist Societies*, E. Maurer, J. Richters, *et al.* (eds.) Basingstoke: Palgrave MacMillan. 85–102.

Goscilo, H. (2007) "Re-conceptualizing Moscow (w)holesale." *Slavic and East European Journal* 52, 2: 312–30.

Goscilo, H. and M.B. (2014) *Fade from Red: The Cold War Ex-Enemy in Russian and American Film 1990–2005*. Washington DC: New Academic Publishing.

Goscilo, H. and Y. Hashamova (Eds.). (2010) *Cinepaternity: Fathers and Sons in Soviet and Post-Soviet Film*. Bloomington and Indianapolis, IN: Indiana University Press.

—. (2012) *Embracing Arms: Cultural Representations of Slavic and Balkan Women in War*. Budapest, New York: Central European University Press.

Graffy, J. (1999) "Dmitri Astrakhan: Popular cinema for a time of uncertainty." In *Russia on Reels: The Russian Idea in Post-Soviet Cinema,* Birgit Beumers (Ed.). London and New York: I.B. Tauris. 161–78, 201–5.

—. (2009) "Aleksei German Jr: *The Paper Soldier* (*Bumazhnyi soldat* 2008)," *KinoKultura* 24. Available at http://www.kinokultura.com/2009/24r-bumsoldat.shtml (last accessed 20 January 2012).

—. (2011) "Living and Dying in Sokurov's border zones—*Days of Eclipse*." In *The Cinema of Alexander Sokurov,* B. Beumers and N. Condee (Eds.). London and New York: I.B. Tauris. 74–89.

—. (2012) "'An Unpretentious Picture'?—Igor′ Savchenko's *A Chance Encounter.*" *Studies in Russian and Soviet Cinema*, 6, 3: 301–18.
Groys, B. (2010) *Ilya Kabakov: The Man Who Flew into Space from His Apartment.* London: Afterall Books.
Hashamova, Y. (2010) "Resurrected Fathers and Resuscitated Sons. Homosocial Fantasies in *The Return* and *Koktebel.*" In *Cinepaternity. Fathers and Sons in Soviet and Post-Soviet Film,* H. Goscilo and Y. Hashamova (Eds.). Bloomington and Indianapolis, IN: Indiana UP. 169–90.
Ivanova, N. (1988) "Kto my? Otkuda my? Kuda my idem?" *Iskusstvo kino* 10: 51–56.
Kaganovsky, L. (2008) *How the Soviet Man Was Unmade: Cultural Fantasy and Male Subjectivity under Stalin.* Pittsburgh: University of Pittsburgh Press.
Kalinin, I. (2013) "Future-in-the-past/Past-in-the-future: sovetskoe budushchee postsovetskogo proshlogo." *Seans* blog, 14 May. Available at (last accessed 31 July 2013).
Kiselev, S. (2005) "Kosmicheskaia odisseia, kotoroi ne bylo." *Iskusstvo kino* 9: 36–9.
Kobchenko, K. (2010) "Emancipation within the ruling ideology: Soviet women in fizkul′tura and sport in the 1920s and 1930s." In *Euphoria and Exhaustion: Modern Sport in Soviet Culture and Society.* N. Katzer, S. Budy, *et al.* (Eds.) Frankfurt and New York: Campus Verlag. 251–67.
Kovalov, O. (2006) "Aleksei Fedorchenko: *First on the Moon* (*Pervye na lune*), 2005." *KinoKultura* 11. Available at http://www.kinokultura.com/2006/11r-firstmoon1.shtml (last accessed 20 January 2012).
Maguire, M. (2010) "Dmitrii Kiselev and Aleksandr Voitinskii: *Black Lightning* (Chernaia molniia, 2009)." *KinoKultura* 29. Available at http://www.kinokultura.com/2010/29r-chernaiamolnia.shtml (last accessed 20 January 2012).
Mindadze, A. (2000) "Kosmos kak predchuvstvie." *Iskusstvo kino* 10: 149–64.
"N." (1929) "Krest′ianskaia aktrisa." *Sovetskii ekran*, 10: 4.
Norris, S. (2007) "Guiding stars: The comet-like rise of the war film in Putin's Russia: Recent World War II films and historical memories." *Studies in Russian & Soviet Cinema* 1: 163–89.
Odom, W.E. (1973) *The Soviet Volunteers: Modernization and Bureaucracy in a Public Mass Organisation.* Princeton: Princeton University Press.
Okudzhava, B. (2001) *Stikhotvoreniia.* St Petersburg: Gumanitarnoe agentstvo "Akademicheskii proekt."
Oushakine, S.A. (2011) "Silent Souls (Ovsianki, 2010)." *KinoKultura* 31. Available at http://www.kinokultura.com/2011/31r-buntings.shtml (last accessed 20 January 2012).
Palmer, S.W. (2006) *Dictatorship of the Air: Aviation Culture and the Fate of Modern Russia.* New York: Cambridge University Press.
Prokhorov, A. (2010) "The Myth of the 'Great Family' in Marlen Khutsiev's *Lenin's Guard* and Mark Osep′ian's *Three Days of Viktor Chernyshev.*" In *Cinepaternity: Fathers and Sons in Soviet and Post-Soviet Film.* H. Goscilo and Y. Hashamova (Eds.). Bloomington and Indianapolis, IN: Indiana University Press. 29–50.
Pushkin, A. (1963) *Polnoe sobranie sochinenii* v desiati tomakh. Vol. 3. *Stikhotvoreniia 1827–1836.* Moscow: Izdatel′stvo Akademii nauk SSSR.
Shemiakin, A. (1990) "Pro chto kino?" *Iskusstvo kino* 7: 86–91.
Sibirtseva, E. (2011) "Kino, Gagarin i kosmos," *Sinemateka*, 12 April. Available at http://www.cinematheque.ru/post/144177 (last accessed 20 January 2012).
Siddiqi, A.A. (2010) *The Red Rockets' Glare: Spaceflight and the Soviet Imagination, 1857–1957.* New York: Cambridge University Press.
Siegelbaum, L.H. (2007) *Cars for Comrades: The Life of the Soviet Automobile.* Ithaca, NY: Cornell University Press.

Sirivlia, N. (2005) "Delo v niuansakh," *Iskusstvo kino* 10: 19–23.

Slivkin, E. (2011) "Pobeg v kosmos," *Iskusstvo kino* 2: 63–71.

Smorodinskaia, T. (2010) "The Fathers' War through the Sons' Lens." In *Cinepaternity. Fathers and Sons in Soviet and Post-Soviet Film*. H. Goscilo and Y.Hashamova (Eds.). Bloomington and Indianapolis, IN: Indiana University Press. 89–113.

Thompson, A. (2008) *Cars of the Soviet Union: The Definitive History*. Sparkford: Haynes Publishing.

Vainshtein, O. (1996) "Female fashion, Soviet style: Bodies of ideology." In *Russia-Women-Culture*. H. Goscilo and B. Holmgren (Eds). Bloomington, IN: Indiana University Press. 64–94.

Valerii Chkalov. Montazhnaia zapis' zvukovogo fil'ma. (1941) Moscow: Goskinoizdat.

von Geldern, J. and R. Stites. (Eds.) (1995) *Mass Culture in Soviet Russia: Tales, Poems, Songs, Movies, Plays and Folklore 1917–1953*. Bloomington and Indianapolis, IN: Indiana University Press.

Wade, M. (n.d.) "Taming the Fire." Available at http://www.astronautix.com/articles/tamefire.htm (last accessed 20 January 2012).

Widdis, E. (2003) *Visions of a New Land: Soviet Film from the Revolution to the Second World War*. New Haven and London: Yale University Press.

—. (2008) "Dressing the part: Clothing otherness in Soviet cinema." In *Insiders and Outsiders in Russian Cinema*. S.M. Norris and Z.M. Torlone (Eds.) Bloomington and Indianapolis, IN: Indiana University Press. 48–67.

—. (2009) "Sew yourself Soviet: The pleasures of textile in the machine age." In *Petrified Utopia. Happiness Soviet Style*. M. Balina and E. Dobrenko (Eds.). London and New York: Anthem Press, 115–32, 278–9.

Yurchak, A. (2008) "Post-post-communist sincerity: Pioneers, cosmonauts and other Soviet heroes born today." In *What Is Soviet Now: Identities, Legacies, Memories*. T. Lahusen and P.H. Solomon Jr. (Eds.). Berlin: LIT Verlag. 257–76.

8 Animal aviators

Refashioning Soviet myths in contemporary Russian digital animation

Vlad Strukov

> In English, Disney's moving image is called . . . an animated cartoon. And in this name, both concepts are interwoven: both 'animateness' (anima—soul) and 'mobility' (animation—liveness, mobility). And the drawing actually is 'animated through mobility'.
>
> Sergei Eisenstein, *Eisenstein on Disney*[1]

> Wow! . . . way cooler than in the metro . . . I'm flying!
>
> Venia the Rat, *Belka and Strelka*

Introduction: sacred animals and profane tales—commemorating Soviet modernity

The modern practice of using animals for scientific purposes, including experiments with dogs flying into orbit, fulfills the age-old symbolic function of animal sacrifice in terms of both its ritualistic nature and commemorative imperative. In 2010, excavations in the Egyptian desert revealed the remains of over eight million dogs and jackals buried in the so-called Dog Catacombs that date to 747–30 B.C. (Parry 2011). While Ancient Egyptians routinely sacrificed dogs to please Anibus, the jackal-headed god of the dead, Soviet—and, for that matter, American—scientists brought dogs and other animals to the altar of modernity.[2] The USSR started using dogs for experiments involving rocket-powered flights in the stratosphere as early as 1951. However, it is only on 19 August 1960 that two dogs—Belka [squirrel] and Strelka [little arrow]—in the company of a few mice, rats, plants, and fungi, flew into outer space aboard Soviet Sputnik-56 and returned home alive.[3] Predictably, the Russian full-feature animated film, *Belka and Strelka* (hereafter abbreviated as *B&S*),[4] released on 18 March 2010 (dir. Sviatoslav Ushakov and Inna Evlannikova), celebrates the fiftieth anniversary of the first successful space flight rather than telling the grim history of the early years of Soviet space exploration.[5]

In both the USSR and the USA, aviation and space exploration were subject matter for propaganda cinema, including documentaries and animated films. Making animated films for the US military was the main source of income for the majority of American animation studios in the 1940s. For example, in 1943, Walt Disney Studio completed a six-hour maintenance and repair manual for Beechcraft airplanes. Additionally, "Disney artists designed more than 1,400 logos and insignias for military and civil organizations . . . Mickey Mouse donned a hard hat in posters for the Aircraft Warning Service Volunteer Observers" (Solomon 1998: 147). Disney even considered making a feature-length film based on an unpublished story by Royal Air Force flight lieutenant and future novelist specializing in children's literature, Roald Dahl. It would be a film about gremlins that were responsible for all technical problems RAF pilots encountered during the Battle of Britain. However, the most extraordinary project of the time was *Victory through Air Power* (1943, dir. H.C. Potter, Walt Disney Studios), a topical WWII documentary combining live action and animation and based on the eponymous book by Major Aleksandr de Severskii, a Russian aviation pioneer who emigrated to the USA. In his book, he argued for the creation of an air force as an independent branch of the armed service. The film recounted the development of warfare and included an animated segment, "The History of Aviation" (Lenburg 2009: 239).

In Russia, the artistic connection between animation and aviation was instituted in the imperial period and was more a celebration of technological modernity rather than state propaganda. Born in Moscow in 1882 into a family of Polish origins, Vladislav Starevich[6] is known as the author of the first puppet-animated films that he created in the Khanzhonkov film studios in the 1910s.[7] He fled Russia after the Bolshevik revolution and settled down in France, where he continued to make animated films until his death in 1965. In 1912 Starevich made an animated film titled *Insects' Aviation Week* [Aviatsionnaia nedelia nasekomykh], which celebrated aviation week, held in Saint Petersburg in May 1911; the event resulted in the death of an aviator.[8] Starevich used dead insects for his stop animation: by replacing the beetles' legs with wire, attached with sealing wax to their thorax, he was able to create articulated insect puppets. Puppet animation and computer 3D animation are similar in that they produce a realistic effect of a 3D space through manipulation with a puppet, whether material or virtual, in a defined world. Starevich's animation in the 1910s and 3D animation in the 2010s define the trajectory of Russian animation with particular interest in recreating space as a tangible ludic world and manipulation with perception through stop-motion and vector graphics, respectively. In both cases, aviation provides animators with a context in which to explore the connection between technological modernity and animation, thus signaling a link between the aerial perspective and the 3D experience of flight.[9]

Animation is a form of art that constantly contrasts flatness and depth through manipulation of matter, i.e. metamorphosis. Although always fluent and at times ephemeral, animation is a technical art, in the sense that it is a product of intensive labor used in tandem with machinery. Animation, especially feature-lengths, has always been in dialogue with conventions of realist cinema; from *Snow White and the Seven Dwarfs* (1937, dirs. William Cottrell *et al.*), Disney and other studios reinstituted the laws of perspective and gravity that were challenged only when computer-generated images became available at the end of the twentieth century. As soon as the latter became normalized, they were challenged again by the technology of 3D animation. In this chapter I use the term '3D' to refer to the modern use of digital technologies for production and presentation of full-feature animated films, i.e. digital 3D. I acknowledge that the practice of demonstrating films in cinemas with the effect of 3D goes back to the 1950s: in the USA, 3D became the rage in 1953, and the technique was used in cartoons, too. Walt Disney's *Melody* and *Working for Peanuts* (1953, dirs. Ward Kimball *et al.*, and 1953, dir. Jack Hannah, respectively) were the first cartoons produced and released in 3D. In the USSR, 3D demonstrations became available later on and coincided with the successes of the space program, thus providing a different 'vantage point' in technical and symbolic terms. By choosing the history of Soviet space exploration as the theme of their film, the makers of *B&S* engage with this dialectic of representation in animation, and they achieve their aims not only through allusions to Soviet and American animation of the previous decades, but also through depiction of the evolution of airborne technology as such: they show flight in its various manifestations, culminating in the scene of levitation on board Sputnik-5 when it reaches the orbit.

Employing Walter Benjamin's ruminations on the subject, Esther Lesley defines Disneyworld before WWII as a restitutive utopia (Lesley 2002: 112), a system of representation that offers escape, compensation, and even therapy for damaged lives. I would argue that the function of such a restitutive utopia in the contemporary Russian context lies within the Soviet past, which is simultaneously a site of trauma and of utopian recompense. To complicate the matter, Russian contemporary audiences consume plenty of animation produced in the USA, and thus are involved in the global sphere of the Disneyworld unconscious that, of course, has mutated over decades and yet remains transforming and inspiring. As a result, post-Soviet Russian artistic practice draws simultaneously on two modernities, Western and Soviet, achieving a normalizing effect and engaging in an ironic exploration of the two. In this chapter, I use *B&S* as a case study to explore filmic sites, particularly full-feature animation, scrutinizing and commemorating Soviet and western modernities. I examine the aesthetic potential of animation to represent flight and aviation in both historical and contemporary contexts. However, I am not concerned with an historical overview of the representation of aviation and space exploration in Soviet or American animation, as this would be a task of a completely different order.[10] Similarly, while drawing on the US and Soviet tradition, my study is not comparative. Instead, it aims to interrogate the ways in which contemporary Russian animation engages with historical memory through visual means. I view aviation and animation as particular technologies of vision, with each providing an interpretive framework that is medium-specific. The intersection of aviation and animation as optic modulators enables a broader discussion of technological modernity and its legacy in the post-industrial era, which in our case coincides with the change in Russia's political regime. In this regard, aviation and the linear type of flight it provides function as a metaphor for modernity, with its privileging of technological advancement and social progress—in contrast to the postmodern rhyzomic type of movement, symbolized by instances of levitation. I argue that in the realm of Russian visual culture and specifically in digital animation, multiple cultural references provide opportunities for the occurrence of this type of movement: modernist discourse is suspended in favor of postmodern zigzags and vibrations of meaning. In its cross-medial interpretation, my analysis makes use of theories of dialogism and intertextuality (Bakhtin 1981; Kristeva 1980) and post-structuralist concerns with intercultural and transmedial storytelling (see, for example, Bay-Cheng *et al.* 2010; Bolter and Grusin 2000; Elleström 2010; Grishakova and Ryan 2010; Jenkins 2007; Ryan *et al.* 2004; Szczepanik 2002). My study contributes to these theories of intermedial exchange by exploring mediation as an ideological concern and by examining cultural memory as a visual practice.

Therefore, I explore the process of de-regulation of the animation industry in post-Soviet Russia to reflect on the role of independent studios that construct their own visions of the past by technologizing and visualizing cultural memory. I demonstrate how Russian animation studios are in competition to produce a spectacle of cultural myths and memory which appeals to the audience through ocular participation. I analyze the narrative of *B&S* as an intermedial participatory strategy which assembles and disbands cultural myths of the Soviet era by assimilating and critiquing visual discourses of the period and producing a transmedial visual palimpsest including posters, paintings, film, and other visual displays of the time. In this context the technology of flight emerges as a means to sustain a narrative whereby various intermedial connections reveal gaps in discourse, thus enabling a particular perspective of interpretation and engagement symbolized by the notion of flight. By analyzing cross-cultural connections I emphasize the global dimension of this perspectival strategy of re-tooling memory work for global consumption which finds itself in competition with similar discourses of celebrification and propagandizing. Thus the chapter offers an original insight into the origins, functions, and scope of participatory framework of Russian cultural memory in its reference to animation conceived here as a form of transmedial production.

Tail spin of history: Russian animation in the post-Soviet period

In the USSR, animation played a pivotal role in the children's entertainment and education system, with many studios combining production of educational, factual films featuring animation sequences with the production of animation proper. Soviet animated films normally contained didactic and frequently overt ideological messages,[11] as in, for example, *A Journey to the Land of Giants* [Puteshestvie v stranu velikanov, dir. D. Babichenko, 1947], *Child of the Sun* [Ditia solntsa 1957, dir. A. Ivanov], and *We Are Drawing October* [My risuem Oktiabr' 1977, dirs. E. Gamburg and O. Zakher]. At the same time, literature, art, and films for children enabled writers and artists to criticize the Soviet regime with a lesser risk of prosecution because of the more relaxed attitude of Soviet censors to cultural products for children,[12] such as V. Diagterev's *Stubborn Dough* [Upriamoe testo 1955] which indirectly satirizes Soviet surveillance practices.[13]

After the dissolution of the Soviet Union, the ensuing processes of deregulation and decentralization,[14] and the diminishing role of the state as the sole provider of funds for animation studios, the post-Soviet animation industry went into a sharp decline in the 1990s. In her *State of the Art: Russian Animation Today*, Larisa

Maliukova notes that "during the Soviet era, Soiuzmul'tfil'm alone released up to forty ten-minute films per year," whereas in the mid-1990s the average annual output was six films (Maliukova 2009). The decline in production coincided with Russian audiences' general lack of interest in their national/local culture: the 1990s was a decade that celebrated everything foreign, imported, extraterritorial, untraditional, and non-native. It was a period of intensive engagement with the world (popular) culture that for decades had been unavailable to the average Soviet citizen.[15] This process of reintegration into world culture was facilitated by the widespread illegal duplication and dissemination—initially with the help of video recorders and eventually computers—of audiovisual materials on the growing pirate market. In the early 2000s the pirate emerged as a figure of cultural authority whose role was not only to disseminate audiovisual and computer files but also to advise, instruct, and produce his own artistic content.[16]

These processes resulted in the emergence of independent animators and animation studios that often used computer technologies and networked communication systems for delivery of their products.[17] In 2004, the state reprised its support for production of animated films by distributing funds through Russia's Ministry of Culture. According to Ria-Novosti, in 2006 there were sixty-two animation studios in Russia, making seventy-five animated films, six of which were full-features ("Gosudarstvo finansiruet . . . " 2006).[18] This suggests that animation studios would often be set up to work on a specific project, thus enabling a more fluid, networked system of production. Such studios would also specialize in the re-release of digitally restored Soviet feature-lengths, such as *The Little Hunchbacked Horse* and *Mawgli* [Maugli 1973, dir. R. Davydov]. Since 2005, a number of full-length original animated films have been released, making them a noticeable genre on the Russian film and animation market.[19] These films fall into three main categories: a) literary adaptations: for example, *Rollie and Elf: Amazing Adventures* [Roll i El'f: Neveroiatnye priklucheniia 2007 dir. A. Ignatenko]); b) serialized films: [20] for example, *A Heap of Gems* [Gora samotsvetov, 2004–9 multiple dirs.], and c) films that exploit the Russian and/or Soviet historical legacy.[21] Films in the last category are predominantly products of two studios, *Mel'nitsa* [Windmill] in Saint Petersburg and *Center of National Film* (CNF-Anima) in Moscow. The former specializes in production of films based on Russian history and folklore, such as *Il'ia Muromets and the Nightingale-Robber* [Il'ia Muromets i solovei-razboinik 2007, dir. V. Toropchin],[22] whereas the latter uses the Soviet cultural discourse and artistic legacy in its creative work, as was the case with *B&S*. The competition between the studios is evident not only in the choice of subject matter and artistic style—*Mel'nitsa* produces highly stylized folkloric films promoting Russian ethnic culture, while *CNF-Anima*'s films display features of the Disney-style visual language, with its glossy finish and dynamic plots—but also in their use of technology for production and demonstration. Though both *Mel'nitsa* and *CNF-Anima* use computer technology for film production, *CNF-Anima* was the first Russian animation studio to release a 3D film.[23] Despite some minor technical flaws in the application of 3D technology, *B&S* received a warm welcome from Russian audiences and critics, grossing two million dollars on the weekend of its premiere alone.[24] It is therefore not coincidental that a Russian film which has pioneered 3D animation technology celebrates one of the most important achievements of Soviet technological modernity: flight into space.

Airborne melodramas: canine pedigrees and Soviet legacies

Upon return to Earth, the historical Belka and Strelka continued living at the space institute and gave birth to a few puppies. One of them—Pushinka [Fluff]—was presented to John F. Kennedy's daughter, Caroline, by Nikita Khrushchev in 1961. The canine 'gift,' whether the symbol of a fragile truce between the two superpowers or a scornful mockery, began to procreate at the White House, giving birth to more puppies, so that Pushinka's descendants are still living in the USA. The fictional events of *B&S* begin with a clumsy Soviet puppy called Pushok traveling on a chartered plane to Washington, D.C., where he meets the American president and his canine entourage, and converses with a stunning blonde of some exotic breed. As Pushok attempts to charm her, he retells events from his parents' lives and imagines a lot of improbabilities on the way to impress her. The purpose of the film preamble is not only to provide a historical setting and, by attributing the narrative to a pup, to account for numerous plot digressions, but also to 'translate' the events from the Soviet past to an audience who is largely unfamiliar with the political and technological rivalry of the Cold War and especially with the everyday life in the USSR in the 1950s–60s. Such 'translation' involves replacing political confrontation—there are no references whatsoever to Cold War conflicts—with personal rivalry and adopting a child's perspective on events[25] to present the Soviet historical past as a magical kingdom where most improbable events could take place. Therefore, the film allows for the merger between historical and fantastical, between serious and satirical, and between quotidian and melodramatic. Visually, *B&S* combines two preferred types of shot, the aerial and the low angle, the constant switch between them functioning as a means of transition from one emotional tonality to another, and also as a source of the film's humor. Thus, in its narrative organization, emotional charge, and representation, *B&S* has little to do with the linear progression of Soviet novels and films devoted to aviation and space exploration from the Stalinist period;[26] rather, the movement in the film is more akin to flight as portrayed in the films of the Thaw period; for example, in Larisa Shepit'ko's *Wings* [Kryl'ia 1966], with its zigzags between the past and present, and different camera angles, including aerial vantage points.

Aviation has an important compositional function in *B&S*: it helps divide the film into four main parts. The first tells the story of Pushok's arrival in Washington, D.C.; the second introduces the viewer to Belka and her glamorous life in a Moscow circus; the third follows Belka and Strelka's adventures in the city before their capture by dog exterminators, who send them to what finally proves to be the Baikonur space station in Kazakhstan;[27] and finally, the fourth part focuses on the dogs' training at Baikonur and their flight into open space. In the epilogue, the film returns to the USA and the viewer learns that Pushok is the son of Belka and Kazbek, the cosmonaut trainer dog. Transitions from one part to another are punctuated by the imagery of flight: in the first instance, a bird's flight; then an aircraft's voyage across the Atlantic, followed by Belka's escape from the circus on a toy-like plane; and finally, the journey into space on board Sputnik-5. Each part denotes a stage in Pushok's narrative and provides the film with a cohesion that otherwise would be lacking, for *B&S* includes a large number of secondary characters (the most original probably being a family of fleas who live in Strelka's hair and who unwillingly experience all the discomforts of flying).

In the USA, Pushok's main task is to confirm his status among other dogs living in the White House and to legitimize his pedigree. His mother, Belka (voice: Anna Bol'shova), the viewer learns, leads the privileged life of a circus performer in post-war Moscow: she is a glamorous star of a variety show (or so she imagines herself to be). One day the circus manager fails to find anyone to perform a flying stunt and he pressures Belka into operating a miniature plane. It misfires and crashes in a sandpit in the middle of Moscow, and so Belka is catapulted from the comfort of a 'privileged lifestyle' into the barbarous environment of Moscow's street life. There she meets Strelka (voice: Elena Iakovleva), a homeless dog roaming the streets of Moscow together with her friend, Venia the Rat (voice: Evgenii Mironov). The three unite in a typical Soviet surrogate family, which quickly moves from the simple task of survival to aspirations of becoming the first astronauts. Albeit more mature, Venia is smaller than Belka and Strelka, and he assumes the role of their child. He is a typical trickster,[28] completely devoted to his cause of becoming a millionaire—a reference, of course, to the character of Ostap Bender in Il'ia Il'f and Evgenii Petrov's novel *The Golden Calf* [Zolotoi telenok 1931], adapted for the screen by Sof'ia Mil'kina in 1968. In this regard, Belka and Strelka's adventures are reminiscent of the con masters of "the children of Lieutenant Schmidt" referred to in the novel. The role of the Soviet Trickster is similar to that of the American Eccentric, immortalized in film by Charlie Chaplin, and in animation by Mickey Mouse. They provide a safe release from the dangers and misery of everyday life [byt]: in the USA, the dangers were caused by the Great Depression, and in the USSR, by the new Soviet order that changed the very template of human existence.[29] Thus, Belka and Strelka display an affinity with characters from early Disney films, whereby their mimetic humor, slapstick comedy, and cartoon gags are not merely a matter of style, but a worldview.

While in its visual apparatus *B&S* is similar to American cartoons—an issue I explore in detail in the concluding section of the chapter—the essential function of the characters of Belka and Strelka is to negotiate Soviet ideological discourse. They represent the two competing ideologies, official and unofficial, grand and noble, and mundane and pitiful. The dogs' opposing mottos—"One for all, and all for one" [Odin za vsekh i vse za odnogo] and "All against one, and everyone for himself" [Vse protiv odnogo i kazhdyi sam za sebia]—summarize their worldviews and account for the dogs' position in Soviet society. While Belka is a star in the circus—one of the few art forms approved by Stalin's regime—and enjoys being at the center of a circle of friends and admirers, Strelka is a homeless dog who lives on the margins of Soviet society and relies solely on her dear friend, Venia. Strelka has experienced the many hardships of street life, including hunger, fear, and distress when bullied. By contrast, Belka has enjoyed the privileges of the Soviet upper class, i.e., the official 'artistic' [tvorcheskaia] intelligentsia. Belka and Strelka's first encounter gives no promise of a lasting friendship: they are divided by fundamental differences in upbringing, lifestyle, and aspiration. However, aviation/space exploration functions as a symbolic link between the characters and unites them irrespective of their background. Therefore, *B&S* resurrects Soviet paradigms of social mobility, whereby a small accident might cut short an extremely successful career or, on the contrary, unexpectedly endow an individual with social success. These possibilities are presented in the film as two myths: that of a social outcast and trickster—Strelka, and that of the circus performer loyal to the regime—Belka.[30] The dual nature of the narrative structure and characterization is symptomatic of contemporary Russian cinema;[31] however, *B&S* is an exception in that not only does it employ a binary structure, but it also provides a binate ethical system where the lines between good and evil are clearly drawn, unlike in other film and animation productions of the past decade, for example Pavel Lungin's 2002 *The Tycoon* [Oligarkh], Timur Bekmambetov's 2003 *Night Watch* [Nochnoi dozor], Ivan Vyrypaev's 2006 *Euphoria* [Eiforiia], and so on. This signifies a return of normative discourse—enhanced by allusions to the classics of the early Soviet period, and apparent in films aimed at children—which has replaced the moral entropy of the post-Soviet decades.[32]

Just as in Stalinist comedy, the ideological conflicts/dramatic twists of *B&S* have happy resolutions. The dog exterminators who chase the dogs on Moscow streets prove to be acting on behalf of the space agency, and the dogs are sent to a training center in the vicinity of the Baikonur space station. There, they join a team of other future cosmonauts and enter a competition to select the first animals to fly into space. Their trainer is the laconic dog Kazbek (voice: Sergei Garmash), who subscribes to Spartan principles; his training ethos boils down to the extraordinary notion that "Everything that doesn't kill helps build character," and he persistently dismisses Belka's complaints about the extreme discomforts in the training center by shouting, "This is not a party to celebrate International Women's Day." Kazbek epitomizes bravery, fortitude, and discipline—qualities attributed to Stalin's "bright falcons," aviators who gloriously fulfilled the symbolic mission of conquering new terrains via long-distance flights entrusted to them by the party.[33] The characters of Belka and Kazbek reinforce Soviet gender stereotypes—he is a brutal macho with a tender heart, whereas she appears to be independent, but in fact is unable to achieve her goal without a man's help. So it comes as no surprise that these adversaries develop a romantic relationship, with many comic situations arising on the way.[34] The romance between Belka and Kazbek develops not only on the ground, but also in space.

The main purpose of their flight is to replace the severely discharged battery of a communication sputnik and send it back on its orbit. While Belka is motivated by the promise of individual success and whirlwind romance, Strelka is determined to reach outer space because of a childhood trauma. In a series of flashbacks, the viewer learns that Strelka was not abandoned by her mother, but as a pup got lost in the city and learned how to survive on her own. She lacks a father, however, who, according to Strelka's mother, lives among the stars in the sky. This melodramatic story has sardonic undertones: in the USSR, women in single-parent families often referred to the missing fathers as pilots. The cause of the fathers' absence could have been quite prosaic; alternatively, it was a way to refer discreetly to men imprisoned in Stalin's numerous jails and gulags. Therefore, on one level Strelka's desire to fly into the sky is an attempt to overcome her childhood trauma and reunite with her father, and, on another, it symbolically adverts to the Soviet project of modernization with the ultimate goal of extending beyond the reaches of human life into the sky.

Flight into the filmic stratosphere: Soviet achievements and Stalinist iconography

Much of *B&S* is devoted to depicting action that transpires above ground. This implies an existence of free space where characters develop; cinematically, this free space, void of gravity, enables experimental representation of cultural memory by means of animation. The opening shows a crow performing acrobatic stunts high up in the air, trying to hold onto a generous hunk of cheese. Stylized in the manner of the frivolous bird from the Soviet hit *The Plasticine Crow* [Plastilinovaia vorona 1981, dir. Aleksandr Tatarskii], the happy crow collides with the star on top of the Kremlin's Spasskaia Tower and drops the cheese. The crow's flight provides the film with two recognizable symbols of power—the Kremlin and Red Square—and injects suspense into the film narrative, i.e., the eternal question of whether the crow will finally manage to recover and enjoy the cheese. The crow's flight takes the viewer to an aerodrome, which shows a display of Soviet passenger aircrafts, and later to a Moscow boulevard, where the crow meets Venia, reenacting the pivotal scene from Aesop's fable, i.e. flattering the crow, hoping it would drop the cheese.[35] Whether focusing on the flight of a bird, an aircraft, or a spaceship, *B&S* is concerned with two main themes: the role of technology in Russo-Soviet culture, or aviation and space travel as a Russo-Soviet grand narrative, and flight as a projection of individual aspiration, symbolized by the plane and spaceship.[36]

Most importantly, the crow's flight accounts for the film's dominant aerial views, and the film presents a specific way of reworking visual representations for its own commemorative purposes. For example, the film shows a Moscow street with a red banner stretched across it, reading "Long live the mighty aviation of the socialist country!" It is an allusion to a 1937 poster (author unknown) later released as a postcard (Fig. 8.1).[37] The poster is visually reproduced in the film as one of the shots in Red Square, with the building of the Historical Museum appearing draped in red cloth and the Moskva hotel towering above other buildings in the background. In the film, the red aircraft of the poster is ironically rendered as the red rind of the crow's cheese. The 'cheesy' humor of *B&S* presents the narrative as a playful journey of memory, riddled with inferences, innuendoes, and flashes of sheer inspiration. The film glides over Soviet history—literally, above the Historical Museum—breaking it up into horological syntagms and freely mixing them up, then reconstructing the whole with the help of visual discourses presented as a set of haptic instances. This procedure ensures the viewer's pleasure and authority, as in the sequence under analysis, in which three scenes appear in linear order as dutiful recreations of the poster, including its composition of three layers of representation: the sky, Red Square, and the slogan at the bottom. Thus, *B&S* emphasizes construction as the main principle of the Soviet project, the nature of memory, and the visual aesthetics of the film.

While the main events of the film clearly indicate that the action takes place in Moscow in the late 1950s–early 1960s (the Thaw period), *B&S* additionally

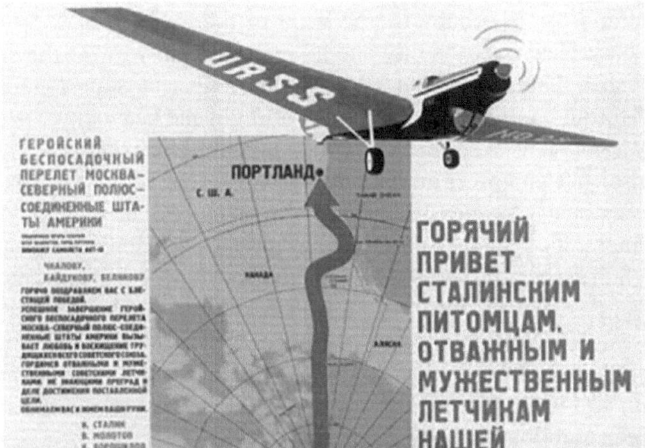

Figure 8.1 Soviet poster (anonymous), representing the planned flight from Moscow, USSR, to Portland, USA, via the North Pole

borrows heavily from Stalinist culture of the 1930s, which is evident not only in the way the aviation poster from 1937 was adapted for filmic representation but also in its usage of Stalinist art and sculpture to present the action and setting, especially street scenes.[38] It is understandable why the main building of Moscow State University (completed in 1953) and the Moscow metro (the film shows the Revolution Square [Ploshchad' revoliutsii] station, which opened in March 1938) appear as the backdrop for much of the film's action, as both are powerful landmarks that define Moscow's skyline as well as its underground world. It is peculiar, however, that Strelka and Venia find refuge on the famous sculpture *The Worker and the Peasant Woman* [Rabochii i kolkhoznitsa, Vera Mukhina 1937], first displayed at the World Exhibition in Paris in 1937, and then moved to decorate the Exhibition of Achievements of the People's Economy (the Russian acronym is VDNKh) in Moscow.[39] Strelka and Venia occupy the pedestal of the tower, which provides them with an amazing view over the city and an unobstructed view of the sky. It is very well established that aviation helped Stalin build his cult of personality by linking him to god through religious connotations of flight and heavenly existence: "Religious associations were even more obvious in phantasmagoric descriptions that ascribed omniscience and other godlike features to Joseph Stalin. . . . Like the resurrected Chris, Stalin was frequently attributed the radiance of an overpowering, brilliant sun" (Palmer 2006: 241). For Strelka, her journey into space is an opportunity to meet the father she has never known. A longing for the paternal figure is a frequent feature of post-Soviet film (Goscilo and Hashamova 2010), and in *B&S* their union is presented in symbolic terms, as Strelka recognizes her father's features in a star constellation. Chasing her father in a run on Saturn's loops, Strelka attains a union with him in a heavenly abode whereby space/heaven is not a site of jubilation but of deep emotional trauma.

An improbable couple beside Mukhina's visionary worker and peasant, Strelka and Venia reenact a popular text/myth of the Soviet period, the nineteenth-century novel by Victor Hugo, *Les Misèrables* (1862). Thanks to its revolutionary message, the work was one of the few Western texts approved for mass consumption in the USSR. Moreover, Soviet authorities would encourage republication of sections from the novel as individual literary texts; a chapter called *Gavroche* about children living inside a sculpture of an elephant was adapted for younger readers and published as a separate text. In 1937, a film with the same title was released (dir. Tatiana Lukashevich) depicting the struggle of young revolutionaries in France.[40]

B&S contains even more subtle allusions to the art of the period. For example, in an unfortunate leap, Venia ends up squashed against the windshield of a passing car. The camera shows his collision from inside the vehicle, with the imagined viewer being positioned behind the driver whose identity remains unknown. The mise-en-scène, the use of the windscreen as an additional frame, and the orchestration of the shot, particularly the position of the camera and the cityscape seen from inside the car, are reminiscent of Iurii Pimenov's painting *New Moscow* [Novaia Moskva 1937]. On one level, the painting celebrates the construction of a new Moscow; on another, it nostalgically registers the passing of the old Moscow architecture—in this case, Okhotnyi riad—and public life in the center of the city. That sentiment would be appreciated by many contemporary Muscovites angered at the architectural jihad bestowed on the city by its former mayor, Yuri Luzhkov.[41] However, *B&S*'s main antecedent text is not Stalinist art and architecture, but rather Stalinist comedy. In Grigorii Aleksandrov's 1936 *Circus*, flight into the stratosphere functions as a symbol of Soviet modernity, of friendship between Soviet and American people (the flight is performed by a Soviet circus performer, Ivan Martynov (Sergei Trofimov), and a visiting star from the USA, Marion Dixon (Liubov' Orlova)), and of the victory of Soviet values. After Marion's lover abandons her for having a black child, the Soviet audience in the circus joyfully welcomes both baby and mother: in a symbolic act of unity and peace, the child falls asleep on the chest of a peasant-looking woman who signifies the all-accepting Russian nation [narod]. In *Circus*, flight into the stratosphere is an aspirational endeavor of two people: Dixon is thrown into the air with the help of a rocket, while Martynov, her Soviet admirer, is able to fly thanks to a complex catapult mechanism. The flight is staged as a performance in the Soviet circus, the magical atmosphere of which enables transformations and transgressions. Dixon's flight is brief and linear, whereas the costumed Martynov, armed with a pair of steel wings patently intended to conjure the myth of Icarus, flies around the circus a few times before landing safely on a podium. The camera captures Dixon shot into the height of the circus dome from below, and it traces Martynov's flight from a single, fixed point of view, so that Martynov appears to be flying around the camera rather than being tracked by it. The elaborate special effects do not obscure an obvious fact: neither Dixon nor Martynov is in charge of her and his flight; they are not aviators, but human charge blasted into the air. In *B&S*, the point of view switches between that of Belka and the circus audience, capturing the flight in the plane from different perspectives and registering Belka's panic as she loses control and ends up flying through the canvas of the circus dome into the open space above Moscow. Her flight—ill-destined to land in a Moscow backstreet—evokes not merely *Circus*, but the whole Soviet project that, as the animated film attests, "went wrong" and left the nation, metaphorically speaking, in unknown terrain. *Circus* presents flight as the ultimate achievement of the Soviet nation; from a narrative point of view, it leads to the film's conclusion, in which the even more Sovietized Dixon—as her Russian accent reveals—takes part in the parade on Red Square. In *B&S*, the flight is a matter of accident, a misguided intention, and it is a moment of transition between parts of the narrative, film locales, and conflicts. The flight helps organize the film's narrative structure and its choice of angles and shots, privileging the so-called overhead 'god shot.'

This vantage point makes mapping the unknown terrain possible. The yellow dotted line marking the route of the airplane shown in the film's beginning is similar to how routes would be marked on Soviet propaganda posters and postcards. The practice is known as 'Stalin's route' and goes back to 1936, when Stalin summoned the three long-distance aviators, Chkalov, Baidukov, and Beliakov, and traced the crew's route on the map in his office. After the meeting the aviators would fly a meandering course that would take them all across the Soviet Union in preparation for a transpolar flight. In her discussion of Soviet film, Emma

Widdis notes that in the 1930s, "drawing the map—real and imaginary—of the new territory was a process of social and political consolidation. It was a means of positioning the citizen and individual within the nation, constructing relationships between centre and periphery, and associatively between private and public space" (Widdis 2003: 3). I argue that in *B&S*, drawing the route is a symbolic act of re-positioning the nation's past in the absence of a coherent national narratrive in the present. The device betrays the lack of a projected imagery in the popular imagination. It is as if, in spatial terms, the contraction of the state's boundaries in 1991 had caused, in temporal terms, an expansion of imagination and memory.[42] Widdis claims that in Soviet discourse aviation played a crucial role in the process of "osvoenie" [making one's own] and "zavoevanie" [conquest through battle] (Widdis 2003: 7) of new territories, whether real or imaginary. I argue that by staging flight as a backward movement in time, *B&S* demonstrates that aviation/space exploration is central to post-Soviet 'ozhivlenie' [reinvigoration/animation] of the object world and its transformation through 'obygryvanie' [play, simulation].

Ironically, *B&S* emerges as an encyclopedia of both Stalinism and Khrushchev's Thaw, a fact indicating that Soviet history has acquired the status of a mythological past, in which boundaries between epochs are blurred. In contemporary Russia, the process of mythologization continues, with a steady stream of publications that—albeit adding new details to existing accounts— prolong the established narrative of Soviet achievements under Stalin. A good example would be Iurii Kaminskii's *Kremlin Flights* [Kremlevskie perelety 1998], which, on the one hand, attempts to re-examine Stalin's role in the development of Soviet aviation in the 1930s, and, on the other, glorifies the achievements of Soviet aviators during the same period by including documents and memoirs of such revered pilots as M. Gromov. In addition, *B&*S depicts the Soviet period according to the notions preserved by cultural tradition, thus segmenting the post-Soviet period as a type of modernity in its own realm, derived simultaneously from Soviet and western sources, hence symptomatic of normalization.

Flying West: Russian animation at the crossroads of global enterprise

Gromov's and Chkalov's record-breaking long-distance flights to the USA were not the only cross-over of Soviet and American cultures in the 1930s. Perhaps the most high-profile encounter of Soviet filmic culture and American animation industry was Sergei Eisenstein's visit to the USA and his work with Walt Disney. They met in 1930, when Eisenstein was on a tour of the United States, officially exploring the possibilities of sound technology in film, and covertly trying to escape Stalin's prosecution. In 1935, when Eisenstein was a member of the jury of the First Moscow International Film Festival, Disney animation was showcased and the Soviet filmmaker vigorously backed Disney's productions; however, the first prize understandably went to *Chapaev* (1934, dirs. the Vasil'ev brothers. Both Disney and Eisenstein used storyboards for their work, and Eisenstein was a passionate cartoonist. He was mesmerized by the ability of animation to (re)produce nature in its dialectic forms (nature versus anti-nature, etc.), a theme central to his theory of montage (see, for example, Eisenstein 2000). By alluding to Eisenstein I wish to suggest that the link between the two cinematic cultures was not only technological but also aesthetic, and *B&S* represents this connection not only on the level of plot (Pushok's arrival in Washington, D.C.) (Fig. 8.2), but also on the more profound level of cultural appropriation and intermedial borrowings. In the remaining part of the chapter I aim to showcase such an alliance by analyzing a few examples, though unfortunately, I will not be able to address fully here the many other interconnections.

In an interview,[43] Vadim Sotskov, producer of *B&S*, complained about his difficulties finding a team of scriptwriters that would produce a high-quality scenario for the film.[44] Having tried many Russian authors, Sotskov eventually turned to his American colleagues, contracting John Chua[45] to finish the script. This is one of the reasons why, in its very visual quality, *B&S* pays tribute to early full-feature animation: it looks like animated cartoons of the 1930s and 1940s in that it presents the quality of the initial stage of cell setup rather than full-blown implementation. The artful simplicity of the organization of the shot in *B&S* is similar to that in Disney's *Dumbo* (1941, dir. Ben Sharpsteen), a film that stands out from Disney's experiments of the time (Schickel 1968: 178; 234–6). It features a baby elephant, Dumbo, who is like no other because he has ears as big as sails. His imperfection becomes an asset when he discovers that he can use his ears to fly in the circus where his mother works. Both *Dumbo* and *B&S* accentuate the child–parent relationship signposting the absence of the father. If *Dumbo* tells of the elephant's maturation and eventual success in the circus, *B&S* divides the conflict and storyline between

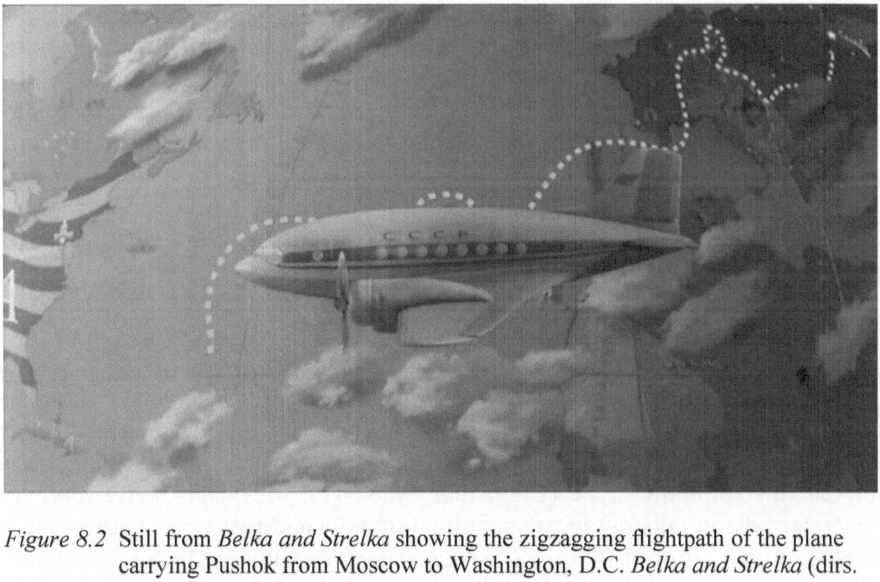

Figure 8.2 Still from *Belka and Strelka* showing the zigzagging flightpath of the plane carrying Pushok from Moscow to Washington, D.C. *Belka and Strelka* (dirs. Sviatoslav Ushakov and Inna Evlannikova, 2010)

the two dogs, i.e., each of them is a reincarnation of Dumbo at different stages of his performing career. The conflict in each of the films is grounded in the tensions of 'compelled action': Dumbo and Belka are forced to perform stunts that neither of them is equipped to do and that endanger not only their careers but their very lives; their predicament is eventually resolved to maximize the characters' success and appeal. The main characters of *B&S* and *Dumbo* achieve success thanks to the good-humored guidance provided by Venia and Dumbo's mouse-friend, respectively. The secondary characters of *B&S,* such as the crow and the elephant performer in the circus, are based on those from the American film, testifying to a sophisticated visual crosspollination. Even the music that accompanies Belka's flight on a plane in the circus emulates the musical score of the opening scenes of *Dumbo* in its melody, tempo, and orchestration. Finally, the films share a preoccupation with flight, aerial shots, and playful spatial schemas. Like *B&S, Dumbo* is concerned not just with flight, but specifically with aviation. Dumbo rises to international fame after his victorious flight in the circus, and the film shows reprints of newspaper publications that glorify Dumbo's achievement and emphasize its sporting and military significance. In other words, *Dumbo* captures America's fascination with aviation, particularly at the start of WWII, when the film was in production. Thanks to his ability to fly, Dumbo achieves a celebrity status comparable to the cult of the aviator in Stalin's Russia. That status is evoked in the 2010 film's full Russian title, *Zvezdnye sobaki,* meaning 'star dogs.'

In the USSR, the aviation industry depended heavily on the acquisition and adaptation of Western technology: "The country's most important advances continued to be made possible by aircraft and expertise obtained from abroad" (Palmer 2006: 278).[46] Similarly, Soviet animation relied on technologies adopted from the West. In *B&S,* the producers pay homage to the history not only of Soviet aviation and the space program, but also of animation by engaging with the 3D technology pioneered for animated films in the West and by borrowing imagery from American films. Disney films of the 1930s and 1940s serve as a historical source of inspiration for the makers of *B&S,* who also acknowledge the influence of contemporary cinema by deriving their characters and mise-en-scène from various productions. For example, the character of Venia is heavily indebted to the rat in *Ratatouille* (2007, dirs. Brad Bird and Jan Pinkava, Pixar studios), the only striking departure being Venia's ability to speak—and it is the amazing voice of Evgenii Mironov and his hilarious gags that ensure Venia's distinctive qualities.

A significant difference between *Dumbo* and *Ratatouille,* on the one hand, and *B&S,* on the other, lies in their employing different types of animation. Eisenstein developed the idea of totemism, i.e. "the immersion of *self* in nature and animals," which is evident in the two Western films. The animal characters in *B&S* are not juxtaposed with people and therefore do not appear as what Eisenstein called "humanized," or "synthetic" characters, but rather as those of the "metamorphic" type (Eisenstein 1986: 41–2).[47] In this sense, *B&S* is cinematized and falls out of the paradigm of animated cartoons. It even incorporates footage from the Soviet screen classic *The White Sun of the Desert* [Beloe solntse pustyni 1970, dir. Vladimir Motyl'], which, legend maintains, is viewed by all Soviet/Russian cosmonauts before their journey into space. The effects of "cinematization" are particularly evident in the ways in which the film uses mise-en-scène and visual orchestration of the sort familiar from recent Hollywood blockbusters. For instance, when a fire breaks out in the spaceship, Strelka and Kazbek begin to extinguish it using hoses provided to dispense food; they stand back-to-back to cover the largest possible area. Their stance evokes a pose typically adopted by characters in blockbusters when engaged in a violent fight (as, for example, the eponymous Mr. and Mrs. Smith defending themselves against the police in a shopping mall in Doug Liman's 2005 film). Thus, *B&S* evidences parodic representation as the main feature of the visual discourse surrounding the Soviet past. The mix of historical facts and particularly of (previous) pictorial styles provides the film with the necessary environment in which to examine the nation's past. It also indicates the trajectory of such an examination by signaling that while technologies traditionally move from the West to the East, aesthetics and theories transcend the East–West divide.

Touchdown: mimetic humor and the glory and trauma of the past

In his study of Soviet aviation Scott Palmer maintains, "for Western European and American audiences, the airplane symbolized power and progress in the present, for Russians, it represented a portent of the future" (Palmer 2006: 282) (Fig. 8.3). Palmer's statement, of course, refers to Soviets; nevertheless, as *B&S* demonstrates, for contemporary Russian society, aviation is no longer a reliable icon of an ideal future but rather a means of reconciliation with its often victorious and frequently troublesome past. *B&S* marks a peculiar tendency of contemporary Russian culture—its infatuation with the past and (government-sponsored) revision of Soviet and, for that matter, Tsarist imperial grandeur. It is not always that the Russian state succeeds in appropriating the 'great' past for constructing its symbolic power in the present. Whereas the remains in Egyptian pyramids may stay intact for centuries, not all elements of Russian history have been equally carefully mummified. In 2011, Putin's government initiated a campaign for exhuming the remains of Russian soldiers who lost their lives to Napoleon's army at Borodino in 1812 to move them to a new commemorative site in Moscow. Despite the government's zeal to construct yet another monument celebrating Russia's past and thus build a revised national lineage, Russian scientists came across an unsolvable problem: in 1812 the bodies of Russian soldiers were cremated and their ashes scattered. To trace their remains at Borodino is therefore impossible ("Prazdnovanie 200-letia . . . "). However, this nuisance will not prevent filmmakers from commemorating the event cinematically. In fact, as the producer of *B&S*, Vadim Sotskov, revealed during my visit to the studio in March 2011, CNF-Anima was considering making a feature-length film that would celebrate the events of Borodino. These films illustrate the manner in which Russian contemporary culture operates. Creating cultural products that celebrate yet another anniversary is not only a successful business model—memory of the distant past seems to encourage Russian citizens to open their wallets—but is also a marker of specific cultural memory that is quite selective in the use of its commemorative apparatus.[48]

Such a form of memorialization is part and parcel of Russian celebrity culture, which, in the case of *B&S*, is grounded in the longstanding tradition of celebrating aviation heroes as stars who have attained a high position physically by going into space but also metaphorically by elevating themselves politically and socially. If in the USSR the focus was on achieved celebrities, to use Chris Rojek's classification (Rojek 2001), with *B&S* we notice a shift in focus to ascribed celebrities, i.e., those whose stardom is based on lineage, not on an individual's achievements.

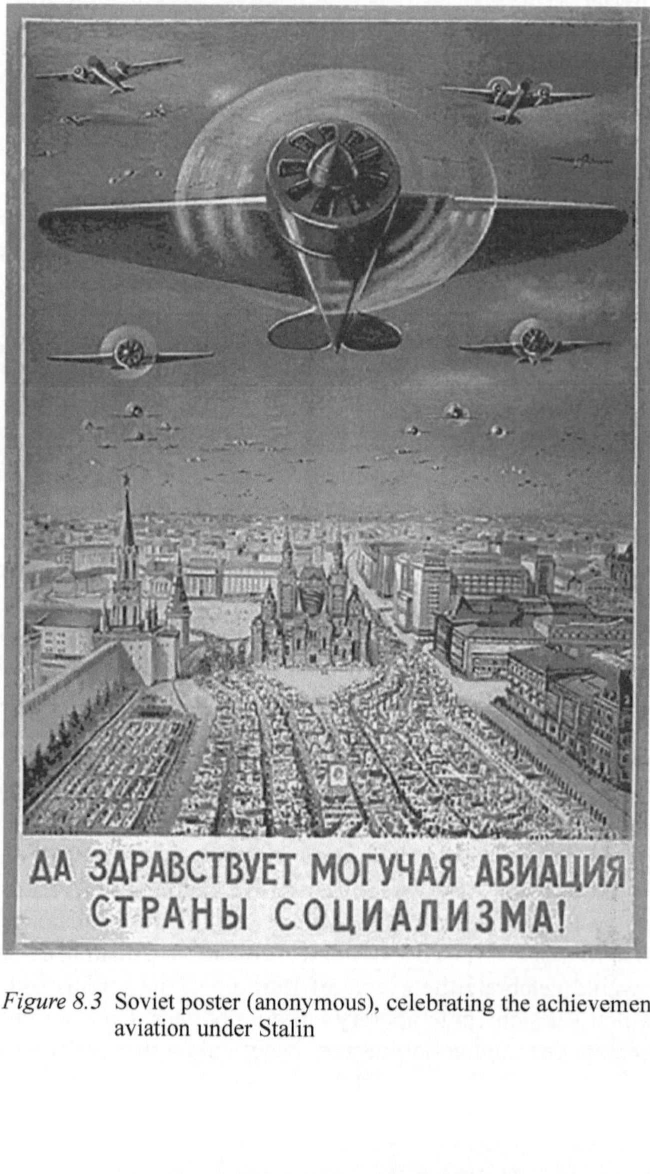

Figure 8.3 Soviet poster (anonymous), celebrating the achievements of Soviet aviation under Stalin

By using iconography of the 1930s Soviet propaganda machine, *B&S* legitimizes historical discourse over Stalin and his period. In spite of its active use of Stalinist iconography, *B&S* is anything but a glorification of the period and its gruesome politics. The first animal cosmonauts are the opposite of what aviators were meant to symbolize under Stalin. They are neither "proud" [gordye] nor "bold" [smelye, besstrashnye], but quite humble and fearful. They are not a supreme expression of socialist progressive powers as symbolized by aviators, but, rather, Soviet tricksters who depend entirely on luck and emotion. Though the film contains elements of the making of a hero—the trials and the tribulations, in which ordinary people transform into supermen— these heroes are not socialist ones, for they lack determination and are driven initially by fright and later by individualist ambition rather than by loyalty to the party. Their motivation is not grounded in collectivism and ideals of communism; on the contrary, the film repeatedly emphasizes personal glory and enjoyment of star status.

In its revisionist strategies, *B&S* reconstructs not only the period of the 1950s and 1960s that would be appropriate for the story, but also the 1930s. In other words, its prime objective is to provide an account of the history of Soviet aviation and space exploration during its two most successful periods. The combined, aesthetically integrated exploration of these historical periods is a novelty in Russian cinema, thus signaling a new reading of the Soviet past and its traumatic legacy. Finally, the film's strategy is to situate the narrative within the visual experience of the spectators, those keen on Hollywood productions and those familiar with Soviet cinematic culture, drawing on two types of modernity, brought together by the notion of flight and human endeavor (Figs. 8.4 8.5, 8.6, 8.7, 8.8).

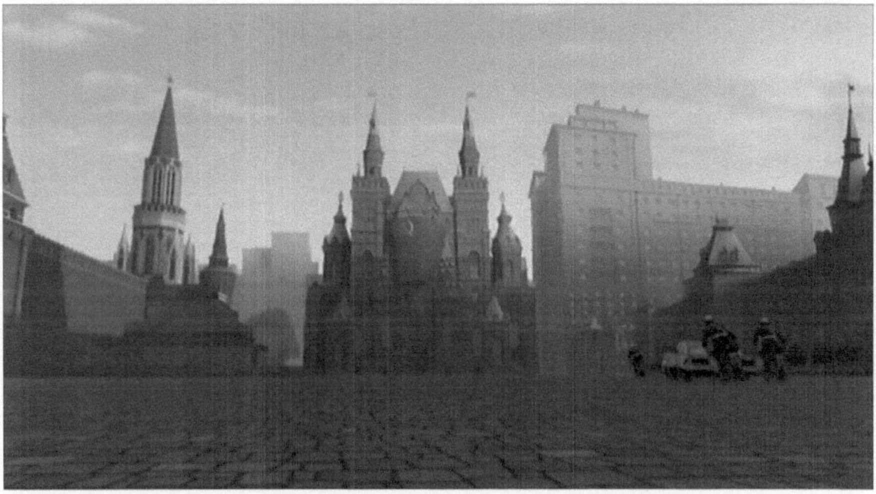

Figure 8.4 Still from *Belka and Strelka* showing Red Square from the bird's-eye perspective. *Belka and Strelka* (dirs. Sviatoslav Ushakov and Inna Evlannikova, 2010)

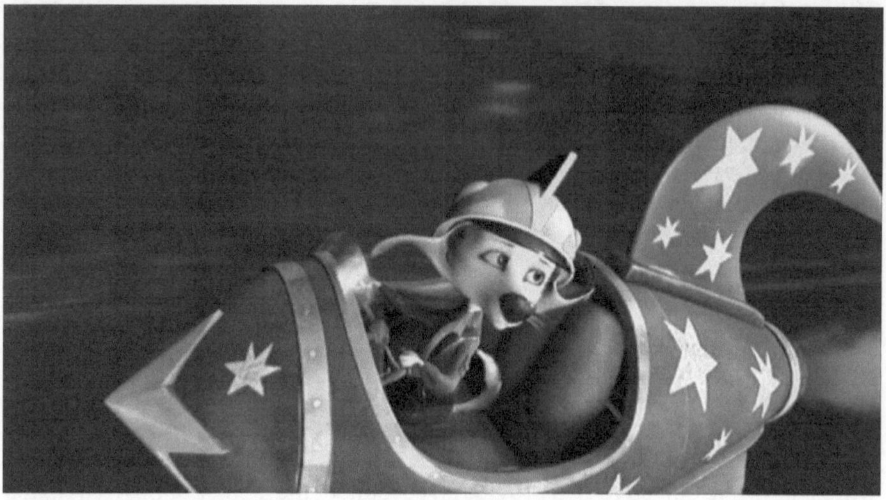

Figure 8.5 Belka flying the stunt plane in the Moscow circus. *Belka and Strelka* (dirs. Sviatoslav Ushakov and Inna Evlannikova, 2010)

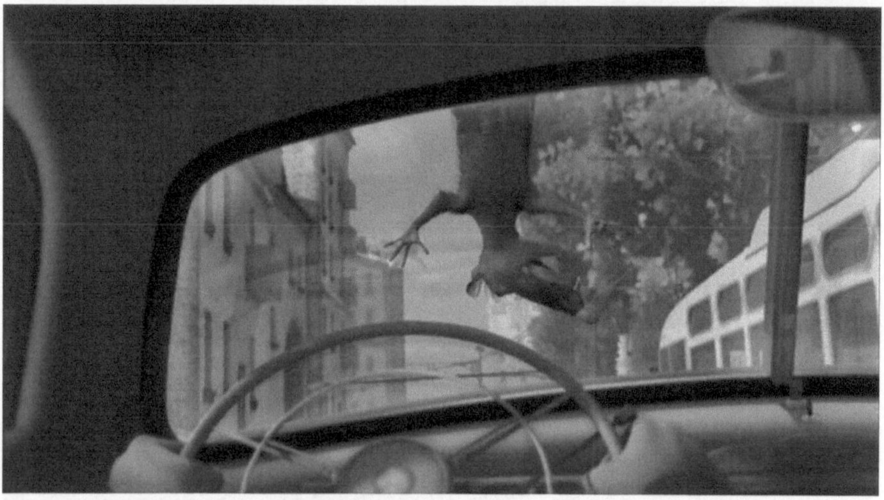

Figure 8.6 Venia landing on the windshield of the passing car. *Belka and Strelka* (dirs. Sviatoslav Ushakov and Inna Evlannikova, 2010)

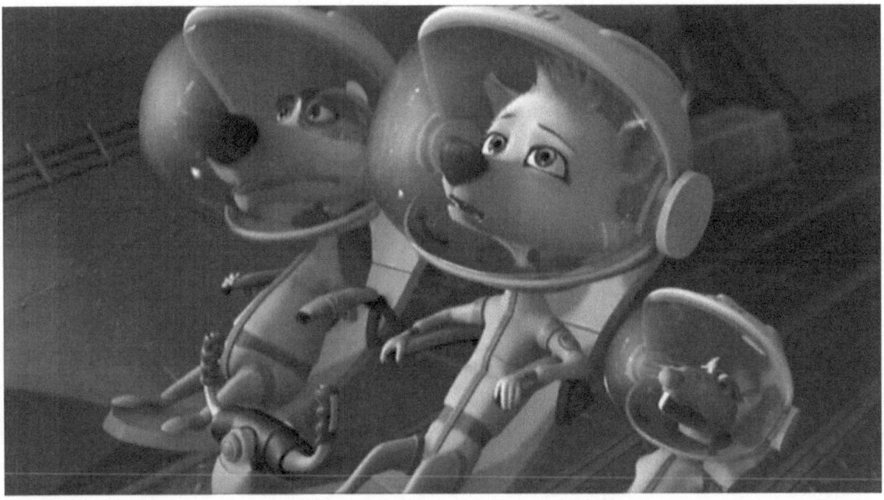

Figure 8.7 Strelka, Belka, and Venia on their way into outer space. *Belka and Strelka* (dirs. Sviatoslav Ushakov and Inna Evlannikova, 2010)

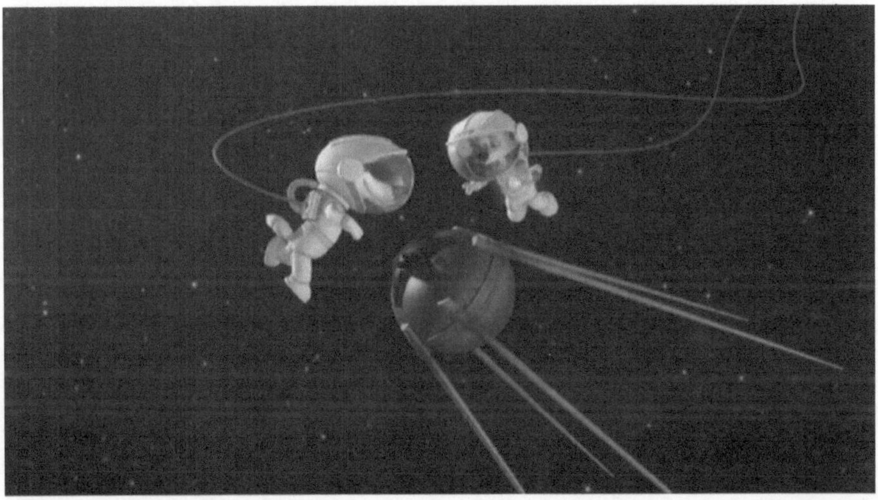

Figure 8.8 Belka and Strelka repairing the first human-made satellite. *Belka and Strelka* (dirs. Sviatoslav Ushakov and Inna Evlannikova, 2010)

Notes

1 Eisenstein 1986: 54.

2 In a wider theoretical context, this association with the ancient culture of Egypt enables me to suggest that animation is one of the oldest forms of art invented by humankind: images drawn millennia ago should be read as attempts at animating static objects, animals, and people, with a vantage point providing an illusion of movement. An example of such a technique would be images drawn on the ceiling of the cave in Altamira, Spain, which correspond not only to the uneven surface of the ceiling but also create an illusion of movement if observed from a specific point of view.

3 Olesya Turkina supplies a cultural history of exploitation of dogs in Soviet space industry in her 2014 *Soviet Space Dogs*. Unlike preceding studies, such as Burgess 1991 and Harvey 1993, with their focus on bio-technologies of the Soviet space program, Turkina traces the ways in which the canine myth was re-mediated in Soviet popular culture.

4 *Zvezdnye sobaki: Belka i Strelka* [Star Dogs: Belka and Strelka]; the English title used for release abroad is *Space Dogs*; scriptwriters—John Chua, Sviatoslav Ushakov; soundtrack—Uma2rman, production design/animation— Aleksandr Khramtsov, Stepan Grudinin, Sviatoslav Ushakov; producers—Sergei Zernov, Vadim Sotskov; production Center of National Film, CNF-Anima, with support from the Federal Agency for Culture and Cinematography.

5 A reference to the vanished animals is made when Belka and Strelka, the film characters, walk through the Baikonur building, decorated with sculptures of "animal heroes" who perished in the Soviet space experiment.

6 Different spellings of his name occur in literature on animation; I use the transliterated form of the Russian spelling.

7 See Chapter 6 for a discussion of his *Flight to the Moon.*

8 Aleksandr Blok reflects on the accident in his poem "The Aviator" (1912). Irina Gutkin notes that "the poet admires the airplane's flight as a bold ascent from ordinary reality towards what seems to be unbounded freedom. However, the pilot perishes in a horrifying crash, and the poet is left pondering the bewildering symbolic meaning of this daring act and its fatal end: Was it driven by a search for glory or a death wish, or was it a premonition of future wars and destruction?" (Gutkin 1999: 199).

9 In Chapter 10 I explore computer flight simulators and their ludic applications as another form of intersection of aviation and animation.

10 American feature-lengths often show various forms of flights, typically assisted by the unusual characteristics of the subject rather than any special machinery. Early American full-length animation includes the following examples: *Fantasia* (1940) famously features animated images, primarily of animals, intended to translate into visual form the classical music conducted by Leopold Stokowski on the soundtrack; *Mr Bug Goes to Town* (1941, dir. Dave Fleischer, Fleischer Studios productions) chronicles an endless battle between flying insects and the human race in Manhattan. In *Johnny the Giant Killer* (1953, dir. Jean Image and Charles Frank, Lippert Pictures productions), Johnny and his friends are able to fly because a giant reduces them to the size of a bee and, forming an alliance with the bees, they attack the giant. Soviet examples include: *The Humpbacked Horse* [Konek Gorbunok, 1947, dir. Ivan Ivanov-Vano, Soiuzmul'tfil'm] based on a literary fairy tale authored by Petr Ershov in 1834 about a horse able to fly; *Flight to the Moon* [Polet na lunu, dir. Valentin Brumberg and Valentina Brumberg, 1953, Soiuzmul'tfil'm] about a young Pioneer who takes part in a scientific expedition to the moon.

11 American scholarship has argued that Disney conveyed his own, profoundly American ideology; see, for example, Zipes 1994.

12 For a discussion of such uses in relation to Soviet animation, see Pontieri 2012.

13 Animation was one of the 'marginal' artistic forms that the state treated with relative tolerance. A well-known example of such practice is Il'a Kabakov's work as an illustrator of children's book early in his career.

14 Polet Studio, founded by Aleksandr Tatarskii, Igor' Kovalev, and Anatolii Prokhorov in 1988, emerged as a result of these processes.
15 For documentation and discussion of such celebration and engagement, see Cherednichenko 1999.
16 This has been manifested in the work of Goblin (Dmitrii Iur'evich Puchkov), who started as a video pirate and translator of unauthorized copies of Hollywood films and eventually emerged as one of the most original players on Russian cinematic and television markets. For an analysis of his artistic strategies, see Rulyova 2005; Strukov 2011. Nancy Condee identifies the profound effect of piracy not only on the Russian film market but also on the very concept of post-Soviet modernity (Condee 2008).
17 For an analysis of new animation studios and artistic strategies, see Strukov 2007.
18 In the period 2009–10 the revenue from the screening of animated films in Russian cinemas grew by 66.8%, from 3.62 billion rubles in 2009 to 6.03 billion in 2010. The revenue of Russian animated films grew from 50.8 million rubles to 721.9 million. The share of animated films in cinema screenings was 20.3% (Belavin 2011).
19 It is possible to argue that, as a genre, full-feature animated films were poorly developed in the USSR. That issue, however, lies outside the scope of this chapter.
20 In fact, a different animation studio, Toonbox, produced an animated serial that tells the story of Belka and Strelka in 2007–8, that is a long time before the release of Ushakov's *B&S*. The serial, produced by Pavel Muntian, is aimed at teenagers and is distributed online (http://www.belka-i-strelka.com/), as well as over satellite and cable television. The studio produced sixteen 10-minute episodes that tell the 'real' story of Belka and Strelka's adventures in space.
21 It is worth mentioning that *B&S* eventually became a subject of serialization strategies: Center of National Film, directed by Vadim Sotskov, released a sequel to *B&S* in 2014 entitled *Belka and Strelka: Adventures on the Moon* [Belka i Strelka: Prikluchenlia na lune]. In 2011–12 Studio KinoAti made fifty episodes of *Belka and Strelka: A Mischievous Family* [Belka i Strelka: Ozornaia semeika, dir. by Aleksei Nekrasov, Vladislav Bairamgulov, etc.] that retell the story of the original film; the series was produced for Russian television. These examples show how films produced in one category move to a different one, thus highlighting the arbitrariness of the classification, which should be viewed as an indication of current trends. The discussion of the extensions of *B&S* and their transmedial potential (Jenkins 2007) is outside the scope of this study.
22 See my review of this film in *Kinokultura* (Strukov 2008).
23 As many critics have noted (see, for example, Obzorkin 2010), *B&S* was actually reworked from a 2D into a 3D film: as a result, only central characters and some objects appear in 3D space, while the background is often static or non-3D.
24 The budget of the film is estimated at $3.5 million and its revenue exceeded $6.2 million. The film was released internationally, with, for example, one hundred and twelve copies of the film circulating on the Polish cinema market alone ("Belka i strelka obognali . . . " 2010).
25 *Belka and Strelka* is exceptional in contemporary Russian film and animation because the film's narrative authority is delegated to a child rather than to one of the parents, as regularly occurs in a great number of films produced since 2000. For a detailed discussion of the parent–child relationship in Soviet and Russian cinema, see Goscilo and Hashamova 2010. Even in *Aziris Nuna: Akuna Matata* [2006, dir. Oleg Kompasov], a science fiction film about the adventures of two brothers boarding a spaceship and traveling to a fantastic world that combines elements of Ancient Egypt and the distant future (i.e., like *Belka and Strelka*, appealing to children and employing a complex temporal organization), the main focus of the film is on the filial link and on paternal figures that are there not to ensure a child's transition to adulthood, but rather to explore dilemmas of nationhood and personal dramas.
26 For example, Boris Polevoi's *Tale about a Real Man* (1946) and *Valerii Chkalov* (1941), dir. M. Kalatozov and Semen Derevianskii, respectively.

27 This twist in the narrative may be read as a reference to the Stalinist practice of expunging individuals and sending whole peoples to Central Asia—Sergei Korolev, the 'father' of Soviet rocket technology, was arrested on Stalin's orders and exiled to a Siberian gulag in 1938. However, it is possible that the film creators' intention was to parody an anecdote, commonly cited in the Soviet history of aviation: in 1848 *Moskovskie gubernskie novosti* reported, "Mr. Nikifor Nikitin should be exiled to Baikonur for uttering seditious words about a flight to the moon" (Pervushin 2005: 28).

28 For more on the function of tricksters in Soviet culture, see Lipovetsky 2009.

29 The duality of *Belka and Strelka*, i.e., simultaneous celebration and fear of the Soviet past (hence, comedy and thriller as the film's main genre affiliations), recalls early Soviet expressions of anxiety as evidenced in Sergei Eisenstein's *Battleship Potemkin* [Bronenosets Potemkin 1925], in which modern technology (film) and modern philosophy (montage cinema) are used to show fear of another type of modernity—the weaponry of the Russian army, which threatens the existence of the sailors and citizens of Odessa.

30 Belka's performance is stylized in the manner of the Stalinist musical comedy *Circus* [Tsirk 1936, dir. Grigorii Aleksandrov] and Belka herself has the striking looks of the immensely popular Liubov' Orlova. I discuss the link between the two films later in the chapter.

31 The following three examples represent the full range—blockbuster, art-house, and popular genre film, respectively: Timur Bekmambetov's *Night Watch* [Nochnoi dozor 2004], Aleksei Popogrebskii *How I Ended This Summer* [Kak ia provel etim letom 2010], and Dunia Smirnova's *Ko-Ko-Ko* (2012).

32 One indicator of such moral entropy is the cinematic style of 'dark cinema' [chernukha] that prevailed in Russia in the 1980s and 1990s (see, for example, Beumers 1999; Graham 2000; Larsen 2003).

33 See, for example, Palmer 2006: 227.

34 A good point of comparison is Iurii Chekiukin and Lev Indenbom's comedy *Gals* [Devchata 1961], which captures the optimistic spirit of Khrushchev's Russia.

35 The fable is called *The Fox and the Crow*, and it is used as a warning against listening to flattery. The earliest surviving versions of the fable, in both Greek and Latin, date from the 1st century AD. In Russian culture, the fable is known thanks to creative translations/redactions by Ivan Krylov, published in 1808.

36 This link is made explicit in the dream sequence, in which Belka orbits the earth in her circus plane.

37 An attentive viewer will notice that *B&S* is preoccupied with the events of the year 1937. Virtually all evoked cultural elements go back to 1937, including a visual reference to the first Soviet transpolar flight: on 18–20 June 1937, Chkalov, Baidukov, and Beliakov flew from Moscow to Vancouver over the North Pole in an Ant-25. On sculptures of those pilots, see Chapter 1.

38 In this regard, Pushok's arrival at the White House is accurate, not only in terms of the actual story but also as a representation of the Stalinist practice of posting Soviet engineers—and with them, spies—to American aircraft building companies, as Scott W. Palmer attests: in the 1930s "the number of Soviet aviation delegations to the United States increased dramatically . . . . Initially confined to short tours of workshops and production plants, Soviet officials exploited American liberal free-trade policies and few controls over technology transfer to gain increasing access to specific construction techniques and to familiarize themselves with engine systems licensed for production abroad" (Palmer 2006: 199).

39 The sculpture became the symbol of the Mosfil'm studio in 1947. Therefore, in *B&S* it is evoked not only as a monument created during the Stalinist period, but also as a monument to Soviet cinema and its cultural legacy.

40 In post-Soviet Russia, Gavroche is perceived as a French trickster rather than a revolutionary; he is celebrated in Russia's conspicuous consumer culture, for example, in the name of a restaurant in Moscow that ironically replicates the very bourgeois style the nineteenth-century Gavroche fought against.

41 For a full analysis of Moscow's transformation, see Harris 2007.
42 In their function such spatio-temporal compressions evoke Mikhail Bakhtin's notion of the chronotope (1981), which suggests a creation of a new cultural configuration in contemporary Russia.
43 I am grateful to Mr. Sotskov for agreeing to speak to me personally. The interview took place on 11 March 2011 on the studio premises.
44 For an insightful overview of the crisis situation with scripts on Russian cinematic market, see Abdulaeva *et al.* 2011.
45 Chua is known as the writer of *Something's Wrong in Kansas* (2008, dir. Louis Paul Tochhet), a hybrid of thriller and horror aimed at young people, as well as a number of documentary films produced both for video and television release. *B&S* bears evidence of the scriptwriter's background in comparative studies as well as his experience in producing thrillers and documentaries in the construction of the story and historical accuracy.
46 This part of Soviet history tends to be omitted in Russian contemporary historical accounts of Soviet aviation. See, for example, Baushev 2007, published by the Russian State Library. This omission indicates that aviation and space exploration still occupy a sacred place in Russia's cultural imaginary.
47 Aleksei Kharitidi's animated film *Gagarin* (1994) achieves 'metamorphosis' in a completely different way. The cartoon laconically depicts human desire for, and fear of, flight: a dragonfly and a helicopter inspire a little caterpillar to crawl into an idle shuttlecock and to experience flight as children pick up the shuttlecock and resume playing badminton. Here metamorphosis is charged with metaphysical connotations as the caterpillar eventually turns into a butterfly, in contrast to machines that enable the metamorphosis of humans and facilitate their flight. The caterpillar, however, is unable to fly, as it now experiences existential anxiety and unease.
48 Most recently this tendency has mutated to include historical biopic films such as Petr Buslov's *Vysotskii* [Vysotskii: Spasibo, chto zhivoi, 2011] and Nikolai Lebedev's *Legend 17* (2012).

Bibliography

Abdulaeva, Z., K. Matvienko, and A. Medvedev. (2011) "Katarsisa ne budet. Konkurs stsenariev 'Lichnoe delo-2010'. Itogi." *Iskusstvo kino* 1: 47.

Bakhtin, M. (1981) *The Dialogic Imagination: Four Essays by M.M. Bakhtin*. Trans. C. Emerson and M. Holquist. Austin: University of Texas Press.

Baushev, I.S. (2007) *Sozdateli otechestvennoi voennoi aviatsii*. Moscow: Russian State Library/Pashkov dom.

Bay-Cheng, S., Ch. Kattenbelt, and A. Lavender. (2010) *Mapping Intermediality in Performance*. Amsterdam: University of Amsterdam Press.

Belavin, P. (2011) "Mul'tfil'my sobrali kassu." *Kommersant*. 17 February. Available at: http://www.kommersant.ru/doc/1586791 (last accessed 17 May 2011).

"'Belka i strelka' obognali zombie i Milu Iovovich v pol'skom prokate" (2010) lenta.ru. Available at: tp://lenta.ru/news/2010/09/22/better/ (last accessed 12 May 2011).

Beumers, B. (1999) "Cinemarket, or the Russian Film Industry in 'Mission Possible.'" *Europe-Asia Studies*51.5: 871–96.

Bolter, J.D. and R. Grusin. (2000) *Remediation: Understanding New Media*. Cambridge, MA: MIT Press.

Burgess, C. (1996) "Dogs Who Rode Rockets: Early Soviet Space Flights with Dogs and their Role in Paving the Way for Manned Space Flight." *Spaceflight* 38.12: 421–23.

Cherednichenko, T. (1999) *Rossiia 90-ykh: Aktyal'nyi leksikon kul'tury*. Moscow: Novoe Literaturnoe Obozrenie.

Condee, N. (2008) "From Emigration to E-migration: Contemporaneity and the Former Second World." In *Antinomies of Art and Culture: Modernity, Postmodernity, Contemporaneity*. Ed. T. Smith, O. Enwezor, and N. Condee. Durham: Duke University Press. 235–49.

Eisenstein, S. (1986) *Eisenstein on Disney*. Ed. J. Leyda, trans. A. Upchurch. London: Methuen.

Eisenstein, S. (2000) *Montazh*. Moscow: Muzei kino.

Ellestrôm, L. (2010) *Media Borders, Multimodality and Intermediality*. London: Palgrave Macmillan.

Goscilo, H. and Y. Hashamova. (2010) *Cinepaternity: Fathers and Sons in Soviet and Post-Soviet Film*. Bloomington: Indiana University Press.

"Gosudarstvo finansiruet s'emki 150 rossiiskih polnometrazhnylh fil'mov" (2006). Available at: http://ruskino.ru/news/349 (last accessed 21 April 2011).

Graham, S. (2000) "Chernukha and Russian Film." *Studies in Slavic Cultures* 1: 9–27.

Grishakova, M. and M. Ryan. (2010) *Intermediality and Storytelling*. Berlin: Walter de Gruyter.

Gutkin, I. (1999) *The Cultural Origins of the Socialist Realist Aesthetic: 1890–1934*. Evanston, IL: Northwestern University Press.

Harris, E. (2007) *Moscow Heritage at Crisis Point*. Moscow: Moscow Architecture Preservation Society and Save Europe's Heritage.

Harvey, B. (2007) "Biosputniks: The Use by the Soviet Union and Russia of Dogs, Monkeys and Other Animals in the Exploration of Space 1949–93." *Journal of the British Interplanetary Society* 46(10): 381–97.

Jenkins, H. (2007) "Transmedial Storytelling" [Blog posting]. Available at: http://henryjenkins.org/2007/03/transmedia_storytelling_101.html (last accessed 20 July 2013).

Kaminskii, Iu. (1998) *Kremlevskie perelety*. Moscow: Glasnost'.

Kristeva, J. (1980) *Desire in Language: A Semiotic Approach to Literature and Art*. Oxford: Blackwell.

Larsen, S. (2003) "National Identity, Cultural Authority, and the Post-Soviet Blockbuster: Nikita Mikhalkov and Aleksei Balabanov." *Slavic Review* 2: 491–511.

Lenburg, J. (2009) *The Encyclopedia of Animated Cartoons*. New York: Facts On File.

Leslie, E. (2002) *Hollywood Flatlands: Animation, Critical Theory and the Avant-Garde*. London, New York: Verso.

Lipovestsky, M. (2009) "Trikster i 'zakrytoe obshchestvo.'" *Novoe literaturnoe obozrenie*, 100. Available at: http://magazines.russ.ru/nlo/2009/100/li19.html (last accessed 12 May 2011).

Maliukova, L. (2009) "The State of the Art: Russian Animation Today." *Kinokultura* 23. Available at: http://www.kinokultura.com/2009/23-maliukova.shtml (last accessed 20 June 2009).

Obzorkin, F. (2010) "Proryv otechestvennoi animatsii v vysshuiu ligu: 'Zvezdnye sobaki: Belka i Strelka.'" *Afisha.ru*. Available at: http://www.afisha.ru/movie/199150/review/319492/ (last accessed 21 April 2011).

Palmer, S.W. (2006) *Dictatorship of the Air: Aviation Culture and the Fate of Modern Russia*. Cambridge: Cambridge University Press.

Parry, W. (2011) "Millions of Mummy Puppies Revealed at Egyptian Catacombs." *Live Science*. Available at: http://www.livescience.com/13473-mummified-puppies-egyptian-dog-catacombs.html (last accessed 6 April 2011).

Pervushin, A. (2005) *Kosmonavty Stalina: Mezhplanetnyi proryv Sovetskoi imperii*. Moscow: Eksmo, Iauza.

Pontieri, L. (2012) *Soviet Animation and the Thaw of the 1960s: Not Only for Children*. London: John Libbey Publishing Ltd.

"Prazdnovanie 200-letia pobedy pod Borodino natolkulos' na neozhidannuiu problemu." (2011) *RBK Daily*. Available at: http://www.rbcdaily.ru/2011/05/04/focus/562949980194838 (last accessed 4 May 2011).

Ryan, M., J. Ruppert, and J.W. Bernet. (2004) *Narrative Across Media: The Languages of Storytelling.* Lincoln, NE: Nebraska University Press.

Rojek, C. (2001) *Celebrity.* London: Reaktion Books.

Rulyova, N. (2005) "Piracy and Narrative Games: Dmitry Puchkov's Translations of 'The Lord of the Rings.'" *Slavic and East European Journal* 49.4: 625–38.

Schickel, R. (1968) *Walt Disney.* London: Weidenfeld and Nicolson.

Solomon, C. (1998) "The Disney Studio at War." *Animation: Art and Industry.* Ed. M. Furniss. New Barnet, UK: John Libbey Publishing Ltd.

Strukov, V. (2007) "Video Anekdot: Auteurs and Voyeurs of Russian Flash Animation." *Animation: An Interdisciplinary Journal* 2.2: 129–51.

—. (2008) "Vladimir Toropchin: *Il'ia Muromets and the Nightingale*-Robber." *Kinokultura* 22: http://www.kinokultura.com/2008/22r-muromets.shtml (last accessed 22 April 2011).

—. (2011) "Translated by Goblin: Global Challenge and Local Response in Post-Soviet Translations of Hollywood Films." *Contexts, Subtexts and Pretexts: Literary Translation in Eastern Europe and Russia.* Ed. B. Baer. Amsterdam: John Benjamins Publishing Company. 235–48.

Szczepanik, P. (2002) "Intermediality and (Inter)media Reflexivity in Contemporary." *Convergence: The Journal of Research into New Media Technology* 8.4: 29–36.

Turkina, O. (2014) *Soviet Space Dogs*. Trans. I. Cannon and L. Wasserman. London: Fuel.

Widdis, E. (2003) *Visions of a New Land: Soviet Film from the Revolution to the Second World War*. New Haven and London: Yale University Press.

Zipes, J. (1994) *Fairy Tale As Myth / Myth As Fairy Tale*. Lexington, KY: Kentucky University Press.

9 Screening aviation, mediating Afghanistan

Andrei Kavun's *Kandahar*

Anindita Banerjee

We may say that in the age of aerial bombing, the world has also been transformed into—is essentially conceived and grasped as—a target.

Rey Chow, *The Age of the World Target* (Chow 2006: 31)

Down from the heights that once we commanded
With burning feet we descend to the ground
Bombarded with calumny, slander and lies,
We're leaving, we're leaving, we're leaving.
. . .
Farewell, bright world, Afghanistan,
Perhaps we should forget you now,
But sadness grips us as we go:
We're leaving, we're leaving, we're leaving.

Igor Morozov, 1989
(cited in Braithwaite 2011: 110)

Introduction

A screen is first and foremost a surface on which something can be projected. Simultaneously, it also protects, shields, and hides from view, and in this function it had been in use for millennia. In military usage, aviation fulfills similarly contradictory functions, visualizing the target, on the one hand, and distracting the enemy, on the other. Airplanes, which began to be used for surveillance as early as the Italo-Turkish War (1911–12), and then more frequently during WWI (Banerjee 2013: 49–52), later evolved into the principal instruments for staging screening attacks, which may include non-lethal or lethal bombardment to facilitate the movement of ground troops. Bodies of men serve as human shields for the same purpose. An uncanny convergence may be observed, moreover, between the visual functions of the screen, its instrumental linkages with aviation, and the psychological valences the term has acquired over the last century. "Screen memory," a term coined by Sigmund Freud, found renewed resonance in the contemporary context of global warfare, with its associated condition of post-traumatic stress disorders. According to Freud, screen memory is a recollection whose actual function is to hide highly emotional material; thus the memory offered up for analysis deliberately seeks to substitute itself for the unspeakable depths that it conceals (Freud 1996: 46). In Cathy Caruth's formulation, screen memories constitute the predominant modality of negotiating the disjunctures among the subject, the state, and history in which veterans of late modern wars often find themselves (Caruth 1996: 28, 63).

It is not surprising, then, that the big screen has emerged as an important site for collective reckoning with conflicts that confound historical sensibilities yet leave indelible marks on public memory. Like Gillo Pontecorvo's *Battle of Algiers* (1966) in France or Francis Ford Coppola's *Apocalypse Now* (1979) in the United States, cinema in post-Soviet Russia provides a crucial medium for recuperating the 'forgotten war' waged by the Soviet Union in Afghanistan from 1979 to 1989—a war doubly repressed by virtue of its inexorable association with the dissolution of the USSR on the one hand and its resonance with the ongoing conflict in Chechnya on the other.[1] Since 1991, a small but significant body of films—including *The Leg* [Noga 1991, dir. Nikita Tiagunov], *Peshawar Waltz* [Peshavarskii val's 1994, dir. Timur Bekmambetov], and *The Muslim* (Musul'manin 1995, dir. Vladimir Khotinenko]—have sought to narrate the Soviet experience in Afghanistan by adopting the perspective of ordinary soldiers on the ground caught up in an unconventional war.[2]

The most successful example of the Afghanistan subgenre in post-Soviet cinema is undoubtedly Fedor Bondarchuk's *9th Company* [9-aia rota, 2005], a film about a group of infantrymen abandoned during the massive Soviet withdrawal. Released the same year in which Vladimir Putin declared that the fall of the Soviet Union was "the biggest geopolitical catastrophe of the twentieth century" (Putin 2005), *9th Company* was named "the biggest Russian blockbuster since 1991" by the *New York Times* (Kishkovsky 2005).[3] The coincidence did not go unnoticed. After a special screening of the film at the Kremlin, one of the journalists invited to the event commented, "This film has already been called our own *Apocalypse Now*. It seems, however, that it will be our *Apocalypse Yesterday, Now, and Forever*" (anonymous, cited by Oushakine 2009: 156). In contrast with the critique implicit in this comment, officials interpreted the depiction of individual endeavor and endurance on the screen as an invitation to foreclose any discussion of why remembering Afghanistan should matter in Russia now. At the same event, President Putin hailed Bondarchuk's project as a call for "putting to rest once and for all this political noise about Afghanistan" (cited by Oushakine 2009: 156). Recounting these contradictory responses, Serguei Oushakine suggests that while reconstructing Afghanistan on the cinematic screen may compensate for the "symbolic absence" of Chechnya in post-Sovivet public discourse, it might also preempt rather than prompt a broader dialogue about the deterritorialized nature of contemporary warfare (Oushakine 2009: 158–63). This chapter focuses on a recent Afghanistan film that mobilizes aviation—the third element in the semiotic field of the screen—to stage a remarkable intervention in the intertwined politics of visualization and memorialization.

Despite its direct invocation of Afghanistan's topology, *Kandahar*, a film by Andrei Kavun released in 2010, self-consciously unmoors itself from the conventional timeframe, setting, characters, and perspective of its predecessors. It is not set in the historical period of the Soviet occupation, nor does it purport to recreate the viewpoint of the iconic infantryman reporting from the ground as in *Apocalypse Now* or *9th Company*. Instead of being bracketed by the landmark dates of 1979 and 1989, the film's point of departure is the interstitial year of 1995—exactly midway in the decade between the fall of the Soviet Union and 9/11, but also at the peak of the first Chechen war that began in 1994. *Kandahar* attempts to narrate the experience of a group of Russian airmen. Unlike the protagonists of previous Afghanistan movies, they do not represent the military arm of the state. Ostensibly engaged in a civilian humanitarian mission to Kabul on a repurposed Ilyushin-76 military transport plane, the pilots are abducted by the Taliban and held in captivity for more than a year before they manage to escape in the same aircraft.

As such, unlike other Afghanistan epics, *Kandahar* seems to make no overt claim for official recognition or public commemoration of the decade-long conflict immediately preceding the demise of the Soviet Union. This chapter contends, however, that instead of merely valorizing the distant field of a forgotten war, *Kandahar* opens up the Soviet experience to a contemporary global dialogue on visuality and subjectivity in which Afghanistan plays a particularly important role. The film enmeshes the past and the present through the intersections between the least tangible yet most symbolic instrument of the Soviet war in Afghanistan and the most prevalent modality of imagining conflict today: aviation and technologically mediated vision. The former, thematized in the film as the IL-76, became a ubiquitous icon not only of the contemporaneous experience of the Soviet war, but also of post-Soviet memorial practices. As estimated in historical accounts, between 1979 and 1989 IL-76 transport planes carried 89% of the troops deployed and 70% of the armaments to Afghanistan (Westermann 1999: 5). They also conveyed a much more bleak cargo, the bodies of the fallen, on the return journey, earning them the euphemistic nickname of Black Tulip [chernyi tiul'pan]. The Black Tulip appears in practically every physical site commemorating Afghanistan in post-Soviet Russia.[4] The airplane's material presence in the film, however, acquires far greater significance when considered in conjunction with the screen on which we see it.

As Paul Virilio (1996) and Rey Chow (2006), among others, have theorized, militarized aerial machines have increasingly come to determine not only the perspective from which continuously evolving visual media—ranging from film to television and satellite pictures to Google Earth—represent the world, but also the very apparatus and aesthetics of the cinematic, televisual, and electronic screens that mediate our relation to it. Afghanistan, an epicenter of global geopolitical struggle for at least two centuries, in the last decade has also become the ultimate case study of what might be termed aero-visuality: a technologically mediated economy of perception from real or simulated heights (Parks 2013). In the post-9/11 "age of the world target," as Chow eloquently puts it (Chow 2006: 31), the most extreme manifestation of aero-visuality may indeed be the drones deployed over the frontier between Afghanistan and Pakistan, whose targets are determined on the basis of data transmitted from satellites. Derek Gregory summarizes the characteristic features of aero-visuality as follows: "the spatialities of targeting and the virtualization of violence," "linking and de-linking 'sight' and 'site'," and "the visual performance of the social field rather than the social construction of the visual field" (Gregory 2013: 182).

Following Gregory's terminology, this chapter argues that whereas historical heroic dramas such as *9th Company* attempt the social construction of an un-representable field of war, *Kandahar* stages a unique performance of Afghanistan that actively meditates upon the aero-visual economy of its own material medium, the screen. By foregrounding and narrativizing the screen itself as both the inscriptional surface and the symbolic repository of memory, it simultaneously exposes and disrupts the continuum between Soviet and contemporary global imaginations of Afghanistan. Far from relegating the 'forgotten war' to a glorious chapter of the Soviet past, the film transforms it into an integral part of ongoing debates about violence, identity, and history within and beyond Russia's borders. The following sections demonstrate how *Kandahar* attains this objective by actively engaging the multifaceted associations of the screen with aviation, visualization, and memorialization.

The Icarus effect: Afghanistan through a fractured screen

Visualizing the world through overhead positioning systems, as Chow suggests, is intimately related to practices of power and the systematization of knowledge (Chow 2006: 31). By flattening out the teeming practices of everyday lives, aerial fields of vision create an illusion of easy navigation through space and time. As sight reduces places to mere sites, Gregory argues, pixelated grids begin to stand in for a country, a city, or a neighborhood (Gregory 2013: 183). One might add to this trenchant observation that as aero-visuality performs its own screening of space—in terms of both projecting and obscuring—it also de-texturizes the complexity of time. Sites have lives that can be summarized in a few bullet points of landmark events; sighting them as embodied entities, however, catalyzes far more complicated interactions between memory and the lived present. It is through precisely such a subversive engagement with the aerial logic of visualization that we can begin to approach *Kandahar*'s resistance to the existing conventions of Afghanistan movies. The fissure first makes itself visible in the disjuncture between the packaging and marketing of the film and its actual field of vision.

Probably because of the huge success of Bondarchuk's Afghanistan epic, *Kandahar* is packaged as an action thriller of personal redemption. The DVD cover proclaims "Survive and Return" [Zhivi, chtoby vernut'sia] below a portrait of uniformed pilots, heroically arrayed in front of their craft, invoking not just the romantic images of airmen from the Great Patriotic War (Palmer 2006: 261–74), but also the end title of *9th Company*: "We were forgotten, yet we won our own war" (Fig. 9.1).

It is the background of what the viewer imagines to be Afghanistan, however, which makes the cover image recognizable to a wide spectrum of domestic and international audiences: overhead views of flattened-out desert landscapes

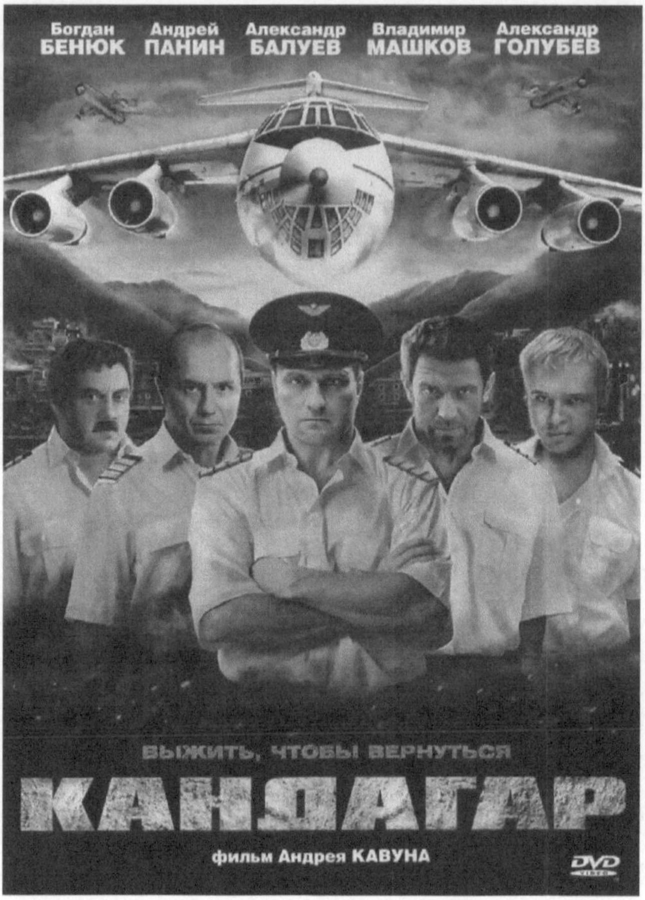

Figure 9.1 DVD cover of Kandahar. Anindita Banerjee

in which any remaining traces of human habitation are continuously exploding and realigning themselves into even more featureless rubble. The elaborately designed internet site for the movie (www.kandagarfilm.com) also privileges this modality of visualization. It places special emphasis on the director's masterful deployment of computer-generated special effects, relying precisely on the kind of aerially targeted imaging that has become so ubiquitous in the post-9/11 world as to be almost unremarkable. The site, itself a deterritorialized extension of the film on the internet, becomes a virtual repository of anonymous pixelated grids, propagated endlessly across multiple media ranging from the news to computer games and global positioning systems, which have completely substituted any real place called Afghanistan (or Iraq, or any topographies of today's wars). Unlike the film itself, the DVD cover and internet site block from view the embodied "practices of everyday life," as Michel de Certeau famously put it, that constitute

Kandahar and by extension Afghanistan (de Certeau 1984: 93). Transforming the sight of a place to an abstract site in this fashion visually mimics the operation of screen memory as described by Freud (Freud 1996: 46), in which a formulaic response obscures the lived experience of trauma.

What the viewer sees once the film starts, however, stands in stark contrast to both its physical and online packaging. Technologically mediated views occupy very little space on the screen. Instead, the dramatic force of *Kandahar* derives from the triangulated embodiment of three elements that aero-visuality elides: the materiality of the aircraft, the subjectivity of the humans operating it, and the phenomenology of life suppressed by the visual abstraction of targeting. Through three sets of complex relationships—between the aircraft and the aviators, the aviators and their captors, and the aviators themselves—the film transforms Kandahar from an aerially mediated site into a deeply textured, complex sight that refuses to submit to any totalizing view of geography or history.

The IL-76 portrayed in the movie is itself an ambiguous visual entity, a recalcitrant painted-over relic of a past war engaged in a dubious mission that is called humanitarian but whose real purpose is never revealed. Unlike the mythic Icarus—whose fall provided Igor Morozov, the bard cited in the epigraph, with the paradigmatic metaphor for the Soviet Union's shameful exit from Afghanistan—the pilots of this plane ignominiously follow the orders of a Talib aviator, flying a Soviet-era military jet, to land on Kandahar's dilapidated airstrip "without any heroics." The subsequent narrative unfolds at ground level in the ruined city, besieged by the rival insurgencies of the Taliban and the Northern Alliance in the chaos that followed the Soviet withdrawal. Kavun's *Kandahar*, consequently, evolves into a visual account of conflict as it unfolds in lived, embodied space. Locked up in walled stone compounds of a *terra incognita* pockmarked by incredible brutality, the pilots navigate the city of Kandahar in all its material manifestations alongside and together with their captors. The film's challenge to aero-visuality becomes even more apparent when considered in light of the pilots' status in the eyes of the Russian public and the Russian state. Instead of Kandahar, a place predominantly imagined as an abstract target, it is the Russian protagonists who become invisible: no attempt is made by the government to locate them or negotiate their release. Only after the airmen engineer their own escape on the same hulking craft do they enter the public eye as officially recognized Heroes of Russia. The film appropriately relegates this anticlimactic moment of canonization to the background images of the closing credits.

The focal point of *Kandahar* is the endless repetition of small-scale violence perpetrated by individuals upon each other, transforming the heroic tale of redemption into a meditation on interiority and inter-subjectivity. Eschewing the aerial gaze, it trains a minute eye on claustrophobic enclosures that dominate the screen: the walled compound, the dilapidated rooms where the captives spend their days, and the rusty Soviet-era truck in which they are periodically moved from house to house past narrow lanes and courtyards. Even in the scenes where the aviators are pictured flying, the camera captures only the cramped interior of the cockpit or the microscopic details of the controls.

This visual performance of spatial interstitiality, which translates Afghanistan from an aerially disembodied site into a complex sight of lived places and practices, finds its counterpart in the film's engagement with temporality—and by extension the question of memory. *Kandahar* challenges the very claim of historical reconstruction that underlies heroic cinematic epics such as *9th Company* through another crucial element. In bold red letters, the opening titles proclaim that the film is "based on real events." As Kavun elaborated in an interview to the *Los Angeles Times*, both his motivation and source material came from personal notes that Vladimir Sharpatov, the captain of the kidnapped crew, kept during his captivity (Loiko and Stack 2010). The citation of Sharpatov's unpublished document, however, reflexively infects the film's reception with the problematics of screen memory. How do Sharpatov's notes align with his reconstructed character on the screen? What do they elide in the process of speaking for all the individuals sharing the experience? Is the witness's voice not already shaped by the grand narratives of retrospective heroism that Afghanistan, and its principal metonym of the Black Tulip, has come to emblematize in the post-Soviet culture of commemoration?

The screen in *Kandahar* thus simultaneously negotiates multiple fronts of conflict, multiple codes of commemoration, and multiple layers of trauma. The protagonists, a group of five aviators ranging from the generation that would have served as Soviet combatants in Afghanistan to a young radioman barely twenty years old, literally embody the ways in which the not-so-distant past collides with and penetrates the unfolding present and imminent future. The saga of dubious civilian heroes whom the state abandons in a no-man's land in 1995 consequently becomes a unique conduit for compelling the post-Soviet audience to confront the fissures in an aerial, uncomplicated view of the relationship between remembering and belonging. Instead of remaining confined to a clearly demarcated, potentially redeemable point in time, Apocalypse indeed flows through the cracks into the present and the future imperfect.

Heights of memory: aviation and the screen of commemoration

It was not accidental that the journalist at the Kremlin screening of Bondarchuk's *9th Company*, cited in the introduction to this chapter, called the film "our *Apocalypse Now*." The American experience in Vietnam provided a potent historical analogy for the Soviet experience in Afghanistan (Oushakine 2009: 158). Anatolii Adamishin, a Soviet officer, wrote in his personal diary in 1979:

> Did we not learn anything from Vietnam? . . . The action in Afghanistan is the quintessence of our own internal affairs. The economic disorganization, the fear of the Central Asian republics, the approaching Congress, the habit of deciding problems by force, the ideological dogmatism. . . . There is the same obscurity everywhere.
>
> (cited by Braithwaite 2011: 11)

The primary object of Adamishin's critique, the "obscurity" surrounding Afghanistan, found its counterpart in American counternarratives to Vietnam. In a chapter evocatively titled "Commemorating the Unwon War: On *Not* Remembering Vietnam," the journalist Peter Ehrenhaus noted that "the tradition of U.S. public discourse in the wake of war is founded upon the premises of clarity of purpose and success; when such presumptions must account for division, equivocation, and failure, commemoration seems somehow inappropriate" (Ehrenhaus 1989: 96).

The illogic that Ehrenhaus points out and Adamishin rails against concerns precisely the rhetorical and psychological operations of screening that shaped the Soviet discourse of Afghanistan and continues to pervade its global image in our time: substituting deep internal anxieties with forceful antipathy toward an externalized Other, demanding moral clarity from shaky ideological positions that result in tremendous human loss, and ultimately, the doomed attempt of making heroes out of reluctant victims. The ambiguous semiotics of the IL-76 featured in *Kandahar* is a case in point. As mentioned earlier, the Black Tulip has emerged as the quintessential symbol of commemoration after 1991. It features as the name of monuments erected to the fallen and serves as the emblem of numerous informal associations through which the veterans of both the Afghan and Chechen wars have tried to organize intergenerational campaigns of recognition and restitution (Oushakine 2009: 158–9). The sublimation of the Black Tulip into a monumental symbol of heroism, however, obscures the material function of the IL-76 as a machine of apocalypse in Afghanistan.

Of all the elements of airpower deployed between 1979 and 1989, the IL-76 played the least glamorous but arguably the most important role. In addition to transporting material and personnel, it was used for direct attacks as well as to screen the movements of ground troops. It was an IL-76 that staged the initial landing in the dead of night at the Bagram Air Base—which, ironically, was established in 1950 with American support and has been reclaimed since 2001 as the main hub of the coalition forces. In an eerie foreshadowing of the 'shock and awe' attack on Baghdad more than two decades later, as the initial takeover of Kabul evolved not into a brief regime-changing foray but a protracted, increasingly complex engagement on multiple fronts, the role of aviation also changed drastically. Edward Westermann, a military historian and former air force officer, argues that air power, far from remaining distant from day-to-day counterinsurgency operations, became an integral part of the action on the ground. The aircraft "became a major instrument for separating the *mujahideen* from the populace and coercing them into abandoning their fight."

> According to Soviet doctrine, aviation constituted a critical 'force adjunct' for the support of ground operations. However, in a relatively short period, the VVS [acronym for the Soviet Air Force: AB] experienced a fundamental transformation in character from a 'force adjunct' to a 'force substitute.' As the war in Afghanistan became a prolonged conflict, the VVS became increasingly important as a force substitute employed to minimize Soviet

> casualties and to compensate for the comparatively small Soviet ground force. In addition, VVS operations in Afghanistan rapidly expanded from a primarily combined arms emphasis to encompass the routine employment of Soviet aviation assets as instruments for punishment and terror. During the nine-year occupation, the Soviets embarked upon a strategy centering on the use of airpower as a, if not the, primary instrument with which to eradicate the growing Muslim insurgency and cow the indigenous population.
>
> (Westermann 1999: 2)

As Danny Nelson, a lieutenant colonel of the U.S. Air Force, noted as early as 1985, transport airplanes were actively involved in the strategy of punitive coercion against the local populace, which included bombardment, chemical attacks, and aerial mining. "Ironically," he comments, "the Soviets may be copying U.S. transport tactics from Vietnam" by rolling out bombs down and off the tail ramp while in flight in an effort to clear helicopter assault zones closer to the ground (Nelson 1985).

It is not coincidental, therefore, that the aircraft that becomes both the instrument of captivity and the medium of escape in *Kandahar* is an IL-76. Despite the ostensibly civilian mission that it undertakes in 1995, neither the plane nor the crew whose fate is inexorably bound to it can be read as innocent signifiers in the misadventures of a few individuals who just happened to fall into the wrong hands at the wrong place and time. Even before the aircraft, whose surface has been literally painted over in the colors of a civilian transportation company, takes off from Istanbul—a location that has long served as the historical and symbolic epicenter of Russia's conflict with the Islamic world, from the fall of the Byzantine empire in 1453 to imperial contestation with the Ottomans through the nineteenth century—the IL-76 is freighted down with a correspondingly insurmountable burden of time. Through the imperfect paint job that does not quite manage to obliterate the Soviet military insignia on its sides, the physical body of the plane becomes a palimpsest through which the past insistently forces its way into the present. By virtue of its dual role between 1979 and 1989 as both a screen and a weapon, force adjunct and force substitute, troop carrier and mobile cemetery, the aircraft is already indelibly marked as a fractured screen that cannot reflect the clarity of aerial vision.

The first invocation of the airplane's burden of memory is itself mediated by a whole wall of screens in the control room of the Turkish airfield from which the IL-76 is due to take off. *Kandahar* opens with a close-up of the flight navigator Aleksandr Zdora (Andrei Panin) against a wall of small screens, all tuned to a television broadcast of the unspeakably brutal forms of public justice that became the hallmark of the Taliban. The airman gazes directly at the audience and remonstrates: "Do you have any idea of what it looks like over there?" (Fig. 9.2.)

The impact of this establishing shot is only intensified when the camera finally reveals the addressee, a young representative of the cargo company. Zdora's rhetorical question, directed simultaneously to his youthful countryman and the film audience, opens up a vast abyss of the past behind the miniaturized images of torture and execution being broadcast from a little-known place halfway across

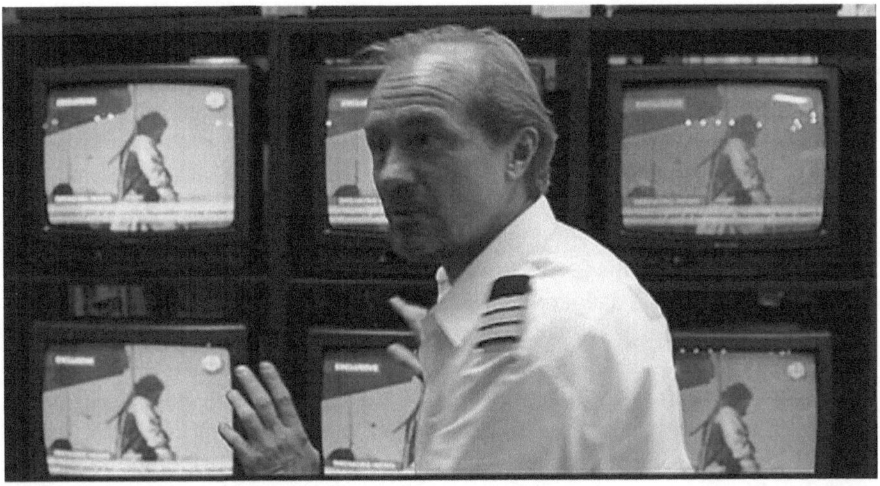

Figure 9.2 Wall of screens. Source: Film still from Andrei Kavun's *Kandahar* (2010)

the world in 1995. In asking—or, indeed, commanding—his on- and off-screen interlocutors to envision "what it looks like over there," the airman places himself in a uniquely privileged position of seeing. The authority of his voice derives from the implication that he himself has greater access to what the "there" looks like. Although Zdora's young companion never voices the questions that this question inevitably calls forth in response—How is the airman able to imagine "what it looks like over there"? Has he been there? What did he actually witness?—the opening of the film superimposes an altogether different field of vision on the unnamed Afghan street projected on the television screen six years after the Soviets have departed: the perspective of a former Soviet combatant who might have once gazed upon equally unspeakable violence in the same or similar places.

Re/visioning aviaton: air war and screen memory

In keeping with the screen's simultaneous ability to display and block, the narrative that follows fails to reveal the exact nature of witnessing that Zdora's questions imply. Rather than using an extended flashback to respond to them, *Kandahar* bypasses the principal driving force of both Coppola's *Apocalypse Now* and Bondarchuk's *9th Company*. It refuses to reconstruct visual memories, not just of the navigator but also of the other three senior crew members who belong to his generation and might possibly have served in the Soviet Air Force in Afghanistan: the captain, Vladimir Sharpatov (Aleksandr Baluev), the co-pilot, Gazinur Hairullin (Vladimir Mashkov), and the flight engineer, Ashkat Abbyazov (Bogdan Benyuk). What the opening scene does instead is to cast an unshakable aura of anteriority over both the craft and the crew about to embark on a journey to the same airfield where the first IL-76 landed in 1979: "We are headed for Bagram. Do you read me? Bagram, in Afghanistan." It is characteristic of the film's complex engagement with both spatiality and temporality that these emphatic iterations of the destination, itself loaded with physical remnants of the Soviet era that are even now being discovered and repurposed by the coalition forces, comes from the mouth of the one crew member who could not possibly have served in Afghanistan and who belongs to the same post-Soviet generation as Zdora's bemused companion in the first scene. Yuri Vshivtsev (Aleksandr Golubev), a carefree young radioman, pronounces the names of these historically freighted locations with absolutely no awareness of their significance. For him, as he explains in a call to his family once the plane is airborne, the trip is going to be "a great adventure to a dangerous place."

A voice speaking heavily accented Russian over the same radio is the first element to break through Vshivstev's—and simultaneously the putative audience's—thrall to the blinding logic of aero-visuality, in which Bagram is merely an abstract "dangerous place" for the younger generation. A Soviet-era bomber looms into view, breaking through the wall of collective amnesia that has rendered Afghanistan a closed chapter of Soviet history. The sight of the Talib pilot who mans the airplane intensifies the uncanny sense that the IL-76's mission all along was a reenactment of countless other missions to destroy precisely such enemies. An instrumental reversal of perspective has occurred during the intervening decade. Whereas the Afghan *mujahideen* of the past were confined to the ground, invisible to the Soviet airmen who bombarded them from the sky, this time the former adversary has risen up, literally resurrected as the monstrous alter ego of a fellow aviator, to force the crew to descend to its own level. Even more unnerving than the embodiment of this formerly invisible victim, however, is the shared language in which he claims intimacy with the captain. The bearded insurgent addresses Sharpatov in Russian, using the diminutive of his first name, "Volodia," and asks after his wife and daughter by name.

The unexpected communion between the two men is far more significant than what it might imply about their personal history. Sharpatov's exchange with the Talib pilot represents a radical subversion of the ways in which aviation itself has historically mediated the poetics and politics of commemoration. In a lecture delivered in 1997 titled "Air War and Literature," the German author W.G. Sebald suggested that in spite of extensive visual records that have been preserved of places utterly destroyed by aerial bombardment, the very nature of this mode of warfare precludes its understanding as an "experience capable of public decipherment" (Sebald 2003: 4). Although Sebald's subject is the silence surrounding the firebombing of German cities toward the end of WWII, his observations are particularly useful for understanding why ground-level combat serves as the privileged lens of remembering war in the aero-visually flattened sites of both the past and the present.

The crucial difference, Sebald contends, is a matter of visualization. From the heights of flying bombers, the landscape below can hardly be conceived as real, let alone teeming with human life; thus for a journalist embedded with a Royal Air Force unit, the complete destruction of a densely populated city remains "the most gigantic display of soundless fireworks in the world" (Sebald 2003: 21). Conversely, eyewitness accounts from the ground, unable to convey the sublimity of the horror unleashed by invisible agents from the sky, can only resort to clichés readily identifiable as Freud's screen memories. Examining phrases such as "on that dreadful day when our beautiful city was razed to the ground," the author concludes that they are "really no more than a gesture sketched to banish memory" (Sebald 2003: 25). Sebald's meditations on the limits of commemorative language seem even more relevant in conjunction with the newer media of what Chow calls "the age of the world target" (Chow 2006: 31). Electronic images of war-torn Afghanistan and Iraq demonstrate that the visual terms of recording and remembering air war have not changed much in spite of the profound transformations in technologies of sighting and targeting between the mid-twentieth century and our contemporary present.[5]

The uncanny similarity of aerial images from WWII to our modes of seeing the ongoing global war on terror reveals why it is so difficult to produce authentic testimonies from either the air or the ground. What dehumanizes both the perpetrators and the victims of air war is the fact that they never actually 'see' each other. It is therefore through the lens of inter-subjective intimacy that *Kandahar* addresses the silence surrounding the largest element of the Soviet military presence in Afghanistan: aviation. The intensely personal interpellation that brings Captain Sharpatov eye-to-eye with a representative of the enemy from the past perversely humanizes the entire crew of the IL-76. Coming face to face with an adversary who would have remained completely invisible to the Soviet-era aviator literally and symbolically collapses the two kinds of eyewitnesses—those in the air and those on the ground—who, in Sebald's analysis, remain isolated from each other.

Nowhere is the conjuncture of the two perspectives—that of the pilot invisible from the ground and of the bombarded populace that remains a mere abstraction from the air—more apparent than in the only sequence in Kavun's film that engages directly with aerial warfare. One night the captives are jolted awake by what seems to be the beginning of Apocalypse—a carpet-bombing by government forces against the Taliban-occupied city. Unlike Soviet times, however, the airmen are now trapped in the same space as the victims, looking through the latters' eyes at the total destruction wrought from above. Instead of a warning built into the plot, the camera's sudden ascent to the aerial perspective signals the opening of this crucial sequence. The screen fills with exactly the kind of image showing anonymous bombarded desert landscapes to which we have by now become completely inured. This is also the only part of the film in which the director makes extensive use of computer-generated graphics, which lends a further aura of virtuality to the scene. Its visual engagement with air war approximates the many combat games consumed on video or computers around the globe today; significantly, many forms of such games are also used for training military personnel for unconventional warfare in places like Afghanistan (Chow 2006: 35) (Fig. 9.3).[6]

Although the viewer is well aware that the protagonists could not have accessed an aerial view of Kandahar from their confined position on the ground, the bombardment sequence, like the Freudian screen memory, could very easily stand in

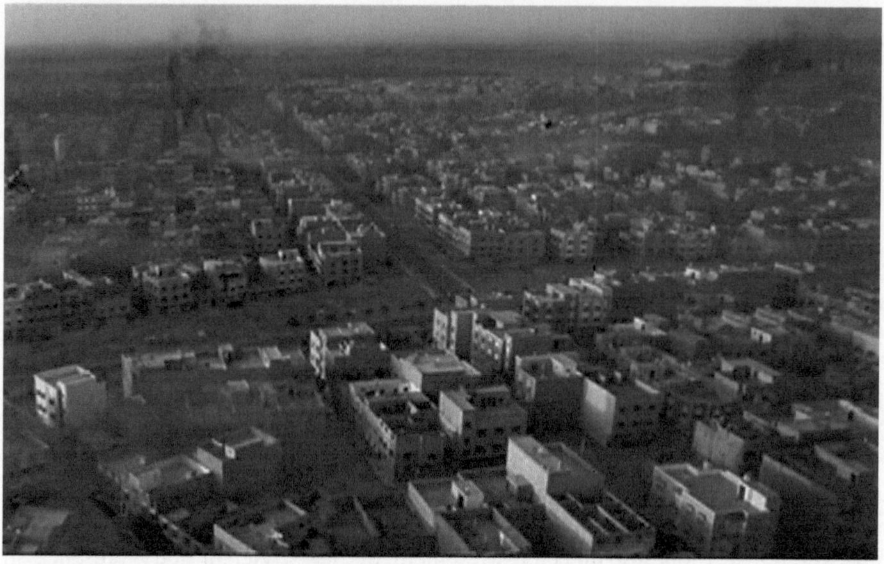

Figure 9.3 Air war simulation. Film still from Andrei Kavun's *Kandahar* (2010)

for another kind of visual witnessing from the not-too-distant past. The way of seeing depicted on the screen relates directly to the transformation of Soviet airplanes from a 'force adjunct' in the background of Afghanistan into a 'force substitute' engaged intimately with ground-level operations. From the few vivid accounts that have emerged in recent years, the strategy of total obliteration unleashed upon the cinematic Kandahar in 1995 appears identical to those used by Soviet pilots against countless Afghan settlements. One such description by Aleksandr Prokhanov, who accompanied air squadrons as a journalist in Afghanistan, fits perfectly into the peculiarly abstracted yet strangely moving category of testimonies recorded by Sebald:

> The aircraft flew over the green zone, dropped bombs, flattened the garden and the walls, reached down to destroy the roots of the plants, diverted and blocked up the underground arteries of the irrigation system, smashed the *kishlaks* [Soviet military slang for Afghan settlements: AB] to dust, burned up the very oxygen with the heat of their explosions, and turned the valley into a lunar landscape, grey, friable, where the insects, the seeds, bacteria, the pollen of the flowers, were dying in agony.
>
> (Prokhanov 1991: 61)

How would the 'eye/I' recuperate from this apocalyptic scene if it were not safely ensconced in the shell of an aircraft, floating far above the spaces and bodies of everyday life being blown to smithereens? This is the question that *Kandahar* seems to ask in the only frenetically paced section of the film. Forced to the ground without the safe screen of height, the pilots, fleeing alongside their captors and the civilian population in 1995, become the very objects whose instantaneous yet agonizingly slow destruction Prokhanov records with remarkable eloquence. Ironically, the bombing is also the only time that the aviators actually escape from the cramped, derelict stone huts where they wait out long days shut off from the view of all but their captors. Air war thus becomes a force that compels the Russians to trade places with a people and a perspective that had remained literally invisible in both Soviet and post-Soviet discourses about Afghanistan. The terrifying moments when both captors and captives became an indistinguishable mass of disposable bodies on the ground also transform the 'body of men' flying the IL-76 into palimpsests mediating between the past and present.

The air attack achieves something that even the forced physical intimacy of captivity could not: it prompts a crucial interpersonal encounter between the captain and the navigator that begins to address the implicit questions raised both by the opening scene and by the Talib leader's acquaintance with Sharpatov. Once again, however, the film makes no pretense of reconstructing their memories. Instead, the camera stops at the surface of the body, which, like the aircraft, reveals itself to be both a screen and a layered repository of memory (Fig. 9.4).

Being the targets of air war brings about a literal merging of the skin with the screen in the sequence immediately following the aerial bombardment. Back in the courtyard of the walled compound, the navigator and the captain for the first time betray a sentiment that is inadmissible in any past or present narrative of patriotic heroism: fear. "I don't want to be the sacrificial lamb," begins Zdora, to

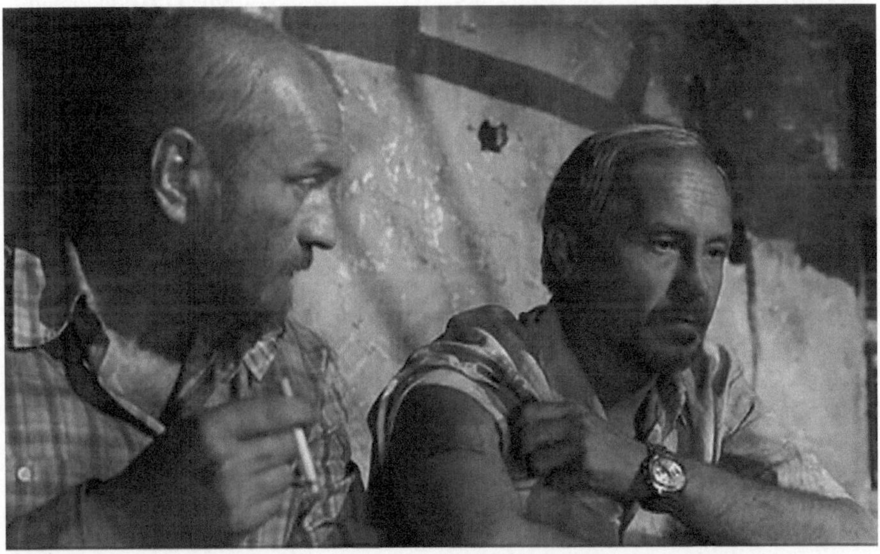

Figure 9.4 The airman's tattoo. Film still from Andrei Kavun's *Kandahar* (2010)

which Sharpatov responds with an understanding nod. The middle-aged navigator then rolls up his sleeve to reveal a tattoo representing the insignia of a Soviet air squadron. Although his interlocutor does not come forth with a reciprocal revelation, it is clear that he recognizes the symbol and acknowledges its connotations. Without penetrating the wall of silence—the scene ends at the moment when Sharpatov's eyes move from the tattoo to the navigator's lined face—the film imbues skin, the surface that conceals interiority and memory, with a narrative capability all its own.

Filling up the entire frame for a brief moment, Zdora's tattoo calls the viewer's attention to the rich social history of embodied inscription (Caplan 2000 and Gustafson 1997). The forbidden affect of fear that triggers its sighting in the film, however, also resonates with a more culturally specific dialectic of concealment and legibility. The tattoo's function in the film recalls the usage of body markings in Soviet penal communities, which Helena Goscilo has examined as a subversive alternative signifying system (Goscilo 2012). Even as the tattoo marks the airmen's place in the community of former Soviet combatants, its association with cowardice rather than heroism undermines the moral basis of their affiliation.

Screened bodies: reenacting and re/mediating Afghanistan

The implications of skin acting as a screen, defined by the properties of both projection and concealment, open up a much wider reading of the film that extends beyond this brief sequence and the characters of the two senior members of the crew. The trace of the past pushing into the present through the navigator's skin directs the viewer's gaze toward the particular ethno-racial identities that comprise the body of men who fly the iconic aircraft of Soviet operations in Afghanistan. It is not coincidental that only three of the five aviators have Russian names while both the co-pilot and the engineer are identified as Central Asians. The cast of characters seems not so much a matter of authenticity as a conscious choice on the part of the filmmakers. On the website, a list of the actors' names and pictures vis-à-vis the corresponding members of the actual crew shows that two additional Russian pilots were left out of the project. This particular mode of redaction transforms the community of kidnapped aviators in 1995 into a representative symbol of the Soviet bio-politics of empire.

Afghanistan provided the impetus for a particularly perverse deployment of the contradictory politics of identity in the Soviet Union. If one of the motivations of the war was to ameliorate the internal anxiety of secession by the Central Asian republics—as Adamishin, the military officer cited above, astutely notes—one of the principal ways of winning hearts and minds in Afghanistan was to recruit large numbers of Central Asian personnel who could ostensibly identify with the locals. The community of aviators in *Kandahar* thus constitutes not just an explicit throwback to the demographic makeup of Soviet forces in Afghanistan, but also the manifestation of a much more diffused politics of cultural production in the late-Soviet period. Katerina Clark notes that Afghanistan generated an imperative for literature to reflect the multiethnic, and specifically Central Asian, composition of Soviet citizenship more adequately (Clark 1984: 574). Although in Clark's account this mandate compelled the cultural elite to recognize a much more wider-ranging and globally oriented body of fiction, Kavun's film engages with its inherent contradictions and ultimate failure.

The co-pilot, Gazinur Hairullin, played by the macho screen icon Vladimir Mashkov, serves as a particularly eloquent medium for deconstructing the untenable logic of the Soviet empire. Whereas Afghanistan shocks Sharpatov and Zdora into silence and provokes only fear and anger from Abbayazov and Vshivtsev, Hairullin enthusiastically enacts exactly the kind of role that Soviet Central Asian soldiers were supposed to play. He is the only member of the crew who engages the landscape as well as the populace with an unswervingly direct gaze. Defying the verticality of aero-visual perception, Hairullin is never pictured from above or below. Whether framing Mashkov's character or seeing through his eyes, the camera itself remains at the character's eye-level. This perspective becomes particularly poignant in the scene when Hairullin encounters a mother and child in an alleyway of Kandahar (Fig. 9.5).

The deceptively tranquil triangulated encounter between the uniformed man, the stereotypically veiled woman, and the innocent child amid mud-baked houses could have been culled from Soviet television or, for that matter, from news footage of Iraq or Afghanistan in the past decade. In *Kandahar*, however, it represents a remarkable aberration from the dominant visual code through which the screen mediates between the past and the present.

A strictly binary configuration of bodies constitutes the only sense of order in the deterritorialized state of war that defines the city of Kandahar. Other than at the time of the air raid, the captives never occupy the same spatial plane as the locals. Even when they are subjected to an informal trial on the airfield after the IL-76 is forced to land, a large body of Talib fighters shields them from the audience baying for their blood—an audience largely comprising older men with mutilated bodies, who could very well have been among the *kishlaks* in Prokhanov's narrative and had miraculously survived the Soviet bombs, aerial mines, and chemical

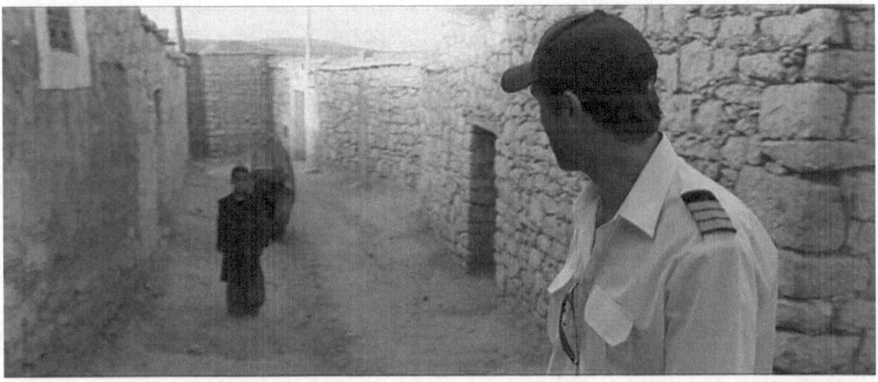

Figure 9.5 Encounter on the street. Film still from Andrei Kavun's *Kandahar* (2010)

attacks. The rare moments when the captives are transported from one informal prison to another on a decrepit Soviet-era military truck are the only times when they encounter the everyday life of Kandahar's citizens. Yet here, too, the captives perversely reenact the movement of troops from an earlier era, who remained not only elevated on the truck bed but also literally screened off from life at street level by a metal mesh. The compounds where the protagonists are held comprise another kind of screened tableau: in an inversion of the Soviet aviator's position vis-à-vis the Afghans, Talib guards sit, with carbines at the ready, high atop the surrounding wall of the courtyard, looking down on the unarmed pilots who are completely at their mercy.

The encounter with the woman and the boy is emblematic of the ways in which Hairullin resists, and up to a point manages to breach, the protocol of envisioning power through hierarchical screens erected between and by bodies. While the other airmen barely glance out of the truck during their periodic transitions, he calls out to the people thronging the streets and draws his companions' attention to both familiar and comforting scenes, such as an impromptu soccer game erupting on a dusty field, or to the horrific, when the same field becomes a theater where a woman is stoned to death for adultery. He is also the only captive who willingly participates in reading the Koran with the healer who comes to treat Abbayazov's wound. Hairullin vis-à-vis the mother and child is the only scene in the film that can be called authentically liberating, occurring soon after the co-pilot, unlike the others who passively remain in the compound where the insurgents first leave them without even bothering to lock the door, ventures out to "look around the place a bit." Yet, as the preceding analysis demonstrates, this picture of brief if wordless communion between fellow Muslims is also a palimpsest through which the past keeps manifesting itself anew across the unfolding present.

The body as a palimpsestic screen, ironically, becomes the agent of shattering Hairullin's naïve—and inimitably Soviet-style—faith in inter-ethnic dialogue and multiethnic citizenship. His brief foray into freedom, where the end of the narrow alley opens up into a sublime vista of receding mountain ranges, ends abruptly when the same little boy calls the Taliban to the scene. The innocent's figure, which has also served as a longstanding trope of empathy from a hundred zones of war all over the world, comes back to haunt the co-pilot in the prison compound as well. A slightly older reincarnation of the same boy joins the fighters guarding the walls and immediately bonds with the co-pilot through their shared love of soccer, the same game that had earlier provided the arena of execution glimpsed from the truck. What the child soldier lobs back at him, however, is a deadly hand grenade. Ultimately, it is the laconic Zdora who saves the Central Asian's life. He defuses the bomb with a professional detachment that attests to long experience in other un-nameable zones of conflict.

Conclusion: the subject of the screen

Even though the imprisoned crew and their grounded aircraft finally escape Kandahar in a grand coordinated effort, it remains questionable whether the film's

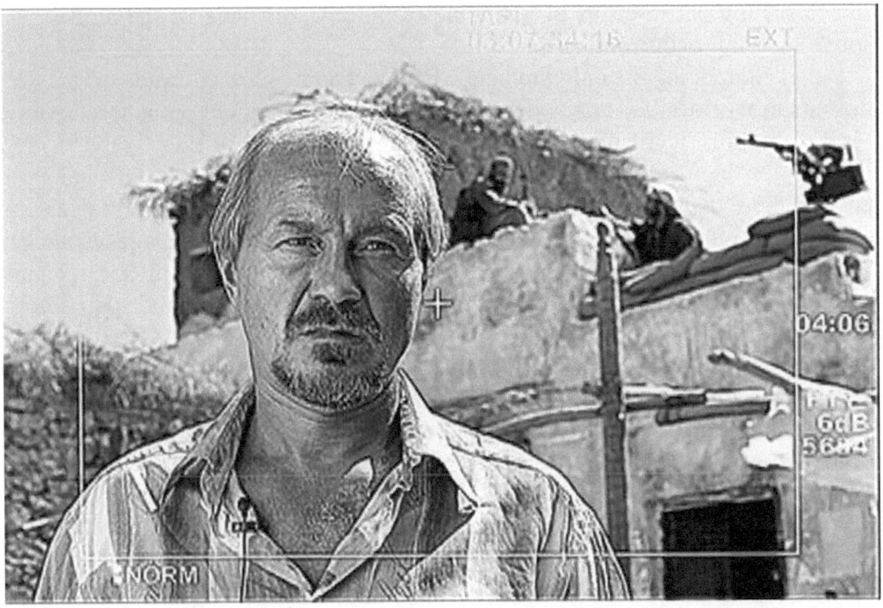

Figure 9.6 Screened silence. Film still from Andrei Kavun's *Kandahar* (2010)

deep reckoning with the past produces an authentic narrative of redemptive closure for the present and the future. *Kandahar* offers its own answer to the question that is far more disquieting than either a heroic or an elegiac reading of the film would suggest. And the answer, like other negotiations of the screen examined in this chapter, emerges explicitly framed in the instruments and languages of cinema (Fig. 9.6).

Midway through the film, navigator Zdora appears as if embedded within the lens of a video camera. It is soon revealed that the captors have engaged a crew to film the kidnapped aviators and transmit their personal testimonies to a putative 'international audience.' Even as the recording instrument etches an additional inner frame onto the larger cinematic one, it challenges the mediating element of the television screens against which the navigator appeared at the beginning. Instead of disembodied sites in Afghanistan projected across vast distances via televisual media, we see flesh-and-blood militants silhouetted high up in the sky behind and above him. The Taliban even take special pains to engage Russian-speaking women, presumably from post-Soviet Central Asia, for interviewing each of the protagonists. The first question posed to them, however, is designed to invoke precisely the kind of memory that Freud called a substitute or screen and Sebald characterized as an empty gesture. This question, which the interviewer puts to each of her interlocutors several times, is "Do you miss your motherland?" [Vy toskuete po rodine]—inviting one of two possible connotations of the word "motherland." The signifier itself is empty, however, as the airmen have been abandoned by both the USSR and post-Soviet Russia. Their clichéd response, correspondingly, fails to emerge each time. This first question is also one that forecloses the entire effort of gathering testimony, for none of the airmen can speak at all after confronting it.

Apart from being the only moment when the captives are allowed contact with the outside world, albeit only via the screen, this is also their only chance to become visible—not just in the eyes of the government and the public back home in 1995 but also for the imagined community of viewers in 2010. The hypothetical audience of the videos, like the real spectators of *Kandahar*, is, however, left facing a peculiarly eloquent screen of silence. This sequence, which engages explicitly with the materiality and apparatus of filming in conjunction with documentary testimony, also compels us to contemplate the limits of the relationship among visualization and memorialization in the age of the world target. By perverting the act of witnessing even as it deconstructs the aerial economy of visualizing distant places and forgotten wars, *Kandahar* seems to ultimately interrogate its own mission of presenting an authentic counter-narrative to the cinematic genre of the Afghanistan epic.

Notes

1 For a discussion of the continuities between *Battle of Algiers, Apocalypse Now*, and post-Soviet cinematic treatments of the Chechen wars, see Evangelista (2011).
2 I exclude from this list films that employ Afghanistan as a frame but do not represent it as a site, such as Aleksei Balabanov's *Cargo 200* [Gruz 200 2007], and Aleksandr Sokurov's lyrical documentary, *Spiritual Voices* [Dukhovnye golosa 1995], which references Afghanistan as a more proximate geographical horizon but does not attempt the historical reconstruction of soldiers' experiences.
3 It is both puzzling and significant that post-Soviet films have not found their way into the emerging work by American scholars on cinematic depictions of Afghanistan. Graham (2010), for example, covers a significant number of lesser-known international films alongside Hollywood productions, but not Bondarchuk's blockbuster cited by the *New York Times*.
4 Oushakine (2009) includes several visual examples of Black Tulip emblems in his study, albeit without any commentary about the role of aviation in Afghanistan.
5 Chow makes this point forcefully by establishing a continuum between the use of imaging in Hiroshima and Nagasaki and the progressive virtualization of aerial bombings for both the United States military and the American viewing public between the Gulf War of 1991 and the Afghanistan and Iraq wars after 9/11 (Chow 2006: 27–35).
6 For a discussion of the technologies of flight and computer gaming, see Chapter 10.

Bibliography

Banerjee, A. (2013) *We Modern People: Science Fiction and the Making of Russian Modernity*. Middletown: Wesleyan University Press.

Braithwaite, R. (2011) *Afgantsy: The Russians in Afghanistan*. Oxford: Oxford University Press.

Caplan, J. (Ed.) (2000) *Written on the Body: The Tattoo in European and American History*. Princeton: Princeton University Press.

Caruth, C. (1996) *Unclaimed Experience: Trauma, Narrative, and History*. Baltimore: Johns Hopkins University Press.

Chow, R. (2006) *The Age of the World Target: Self-Referentiality in War, Theory, and Comparative Work.* Durham, NC: Duke University Press.

Clark, K. (1984) "The Mutability of the Canon: Socialist Realism and Chingiz Aitmatov's *I dol'she veka dlitsia den'.*" *Slavic Review* 43. 4: 573–87.

De Certeau, M. (1984) *The Practice of Everyday Life.* Trans. S. Rendall. Berkeley: University of California Press.

Ehrenhaus, P. (1989) "Commemorating the Unwon War: On *Not* Remembering Vietnam." *Journal of Communication* 39. 1: 93–116.

Evangelista, M. (2011) *Gender, Nationalism, and War: Conflict on the Movie Screen.* Cambridge: Cambridge University Press.

Freud, S. (1996) *The Psychopathology of Everyday Life.* Trans. Anthea Bell. New York: Penguin.

Goscilo, H. (2012) "Texting the Body: Soviet Criminal Tattoos." In *A Cultural Cabaret*. Ed. D. Goldfrank and P. Lyssakov. Washington, DC: New Academia Publishing. 203–30.

Graham, M. (2010) *Afghanistan in the Cinema.* Chicago: University of Illinois Press.

Gregory, D. (2013) "American Military Imaginaries and Iraqi Cities." *The Visual Culture Reader*, 3rd ed. Ed. N. Mirzoeff. London: Routledge. 181–95.

Gustafson, W. M. (1997) "*Inscripta in Fronte:* Penal tattooing in Late Antiquity." *Classical Antiquity* 16.1: 79–105.

Kishkovsky, S. (2005) "From a Bitter War Defeat comes Russia's Latest Blockbuster Action Movie." *The New York Times.* October 29.

Loiko, S. and M. Stack. (2010) "Q and A with Film Director Andrei Kavun." *The Los Angeles Times.* 12 February.

Nelson, Lt. Col. D. R. (1985) "Soviet Air Power: Tactics and Weapons Used in Afghanistan." Available at <http://www.airpower.maxwell.af.mil/airchronicles/aureview/1985/jan-feb/nelson.htm> (accessed 30 March 2012).

Oushakine, S. A. (2009) *The Patriotism of Despair: Nation, War, and Loss in Russia.* Ithaca, NY: Cornell University Press.

Palmer, S. W. (2006) *Dictatorship of the Air: Aviation Culture and the Fate of Modern Russia.* Cambridge: Cambridge University Press.

Parks, L. (2013) "Zeroing In: Overhead Imagery, Infrastructure Ruins, and Datalands in Afghanistan and Iraq." *The Visual Culture Reader*, 3rd ed. Ed. N. Mirzoeff. London: Routledge. 196–206.

Prokhanov, A. (1991) *Tretii tost.* Moscow: Voennoe izdatel'stvo.

Putin, V. (2005) "Poslanie federal'nomu sobraniiu rossiiskoi federatsii." Available at <http://archive.kremlin.ru/text/appears/2005/04/87049/shtml> (accessed 30 March 2012).

Robins, K. (1996) *Into the Image: Culture and Politics in the Field of Vision.* London: Routledge.

Sebald, W. G. (2003) *On the Natural History of Destruction.* Trans. A. Bell. New York: Modern Library.

Virilio, P. (1997) *Pure War.* New York: Semiotext(e).

Westermann, E. B. (1999) "The Limits of Soviet Airpower: The Failure of Military Coercion in Afghanistan, 1979–89." *Journal of Conflict Studies* 19.2: 1–12.

10 Simulating *Sturm und Drang*

Theorizing digital historization, commemoration, and participation

Vlad Strukov

> You certainly have to admire the depth of the game and its sense of history. Successful completion of challenges unlocks archive footage, and the game itself moves from the Battle of Britain to the Battle of Berlin.
>
> Neil Davey's review of *Il-2 Sturmovik: Birds of Prey* in *The Guardian*[1]

> There above me was the black abyss of open space. The only thing between me and it was the thin Plexiglas of my motorcycle goggles. I was surrounded by total darkness.
>
> Viktor Pelevin *Omon Ra*[2]

Introduction

The term 'simulation' refers to a wide range of experiences and concepts. M.E. Gredler distinguishes between experiential and symbolic simulations, whereby the former place the learner/user in a professional role or social function (for example, a doctor) that requires deliberation to accomplish a task in a context rife with problems. The latter provide the learner/user with space for experimentation, in which the learner/user can try different strategies and observe and interact in environments that are typically not available or dangerous (for example, subatomic particles and radioactive materials) (Gredler 1996). Both experiential and symbolic simulations contain game features, such as a realization that actions taken by the learner/user will not have real-life implications. However, as Dennis Charsky argues, simulations require such characteristics as competition and fantasy to create an entertaining game experience (Charsky 2010). Simulations differ technologically: physical simulations use objects that substitute for the real thing or process, whereas computer simulations produce mathematical models that replicate real-life situations, or recreate hypothetical analogs.

In this chapter, the term 'simulation' is used as a metaphor for Russian and Soviet cultural practices that encompass experiential and symbolic contexts as well as physical and digital reconstructions. Furthermore, I use 'simulation' as a category that defines the condition of the human subject in relation to the contemporary economy of symbols as proposed in Jean Baudrillard's work (1983, 1994) and, in a wider context, as a manifestation of the industries of illusions, i.e. artifacts and processes that produce effects of reality as pleasure. Finally, the study exemplifies Henry Jenkins's concept of transmedial storytelling (2007) by employing transmedial analysis of the phenomenon. At the core of this study is an analysis of the ludic component of simulation, normally codified as narrative, representation, and performance, which accounts for the three main concerns of this chapter, with Russian aviation and the nation's history providing the cultural context for the discussion.

My chapter builds a discussion around flight simulation that approaches the subject from three perspectives: (A) simulation as a mechanical, or electronic, reproduction of a real-life experience (flying prosthetics and IL-2 Sturmovik);[3] (B) simulation as a form of an audiovisual apparatus that recreates reality in a specific temporal and spatial setting (panoramas and IL-2 Sturmovik); and (C) simulation as a cultural metaphor, i.e. as an element of cultural anthropology that conveys a culture's perception of its own existence (*Omon Ra* and game fan videos).[4] These three perspectives reveal the cultural potential of simulation as regards industries of symbolization and illusion that enable a hybridization of the perceptual apparatus and ideological imperatives. These concerns are brought together through the concept of cultural parallax, which defines not only strategies of vision but also strategies of cultural production in the era of global capitalism and digital modernity. The Russian case is not unique, and it enables me to conceptualize the development of aviation technology and simulation in the European context so as to theorize the relationship between gaming and simulation that is representative of a global trend.[5]

Writing a history of the use of flight simulators in the USSR is clearly beyond the scope of this chapter.[6] However, I aim to propose a framework for cultural exploration of flight training by analyzing two instances of (symbolic) use of flight simulation, one at the very start and the other at the end of the Soviet social experiment. Therefore, I approach the issue of flight simulation in Russian and Soviet empires from a temporal perspective, by analyzing two experiences that frame the use of flight simulation historically, as well as spatially, by examining post-Soviet dislocation of flight simulators that came about as a result of the centrifugal forces of the Soviet collapse. This implicit concern has to do with the notions of control (and/or its loss), which I analyze as a psychological phenomenon (as a desire and also gaming drive), physical (as a matter of gesture, and signification of body with its lost parts), and ideological (as a narrative of imperial expansion and contraction manifested in the works of historical panorama as well as their parodies in literature and fan-produced art).

From the theoretical perspective, I aim to reevaluate the existing reading of twentieth-century culture as the emergence and gradual proliferation of "computational" culture (see, for example, Aarseth 2004; Frasca 2001), in which "the flow of simulation into everyday life was led by the computerized implementation of the military-industrial complex" (Crogan 2007: 407). To counteract this computation-driven influence as well as a privileging of 'new' phenomena, in my analysis of simulational interactive media, I use a critical apparatus to explore simulation as a representational strategy as well as a cultural construct. Such an approach opens the door for a discussion of the post-Soviet Russian experience not only as the space of trauma (see, for example, Etkind 2009; Etkind 2013; Oushakine 2009; Oushakine and Trubina 2009), but also as a ludic field, whereby simulation mediates dominant semiotic domains and moves beyond restorative structures. Thus, the chapter posits my concept of cultural parallax as an experiential and symbolic context that defines contemporary Russian subjectivity in relation to the industries of vision, illusion, and simulation.

Narrative: flight simulators in the Soviet era

Before the launch of Sputnik in 1957, two copies of the satellite were made: one was built for the actual flight and the other for ground tests and simulations. The two hemispheres of the actual sputnik were covered with a highly polished 1-millimeter-thick heat shield made of an alloy of aluminium, magnesium, and titanium. Sergei Korolev (1907–66)[7] insisted that the test copy should be polished as well, so that in the future it could be displayed in museums.[8] The leading Soviet rocket engineer and space craft designer was apparently invested in the engineering and aesthetic qualities of the satellite as well as the historical legacy of the Soviet space enterprise. The two satellites—the sputnik that burned on entry into the earth's atmosphere and its earthbound sibling kept like jewellery on a piece of velvet—symbolize the Soviet space exploration as well as the conditions of research in Soviet aviation and the space program. While there is a wide range of studies of Soviet aviation and space (see, for example, Palmer 2006; Siddiqi 2010; Widdis 2003), virtually no research has been carried out on the use of flight simulators for training of both military and civil personnel in Russia and the USSR. There are practical and ideological reasons for such an oversight. First of all, Soviet aviation and space programs were shrouded in secrecy: aviators and cosmonauts always spoke at length about their emotions and commitment to the state, but rarely discussed any details of their training.[9] Being part of the Soviet military-industrial complex, engineers, trainers, and service people were absent from public view and official accounts of the space enterprise. Furthermore, because Soviet cosmonauts followed the precepts of Stalin-era glorification of aviators, as I demonstrate below, the emphasis was always on the ideological maturation of the pilots under the party leadership rather than on their actual training, thus focusing on the extraordinary achievement of flight and ignoring or forgetting the mundane circumstances of training. Since research on Soviet aviation and space, just like the space race, has been predominantly driven by ideology, a knowledge of the historical narrative focused on training helps overcome fossilized Cold War paradigms of knowledge.

The history of aviation simulators is a history of technological exchange and competition in post-war Europe, and it is intrinsically connected to our understanding of the human body and psyche. Flight simulators are built for training purposes as well as engineering and psychological research. Sanders Teacher introduced one of the early simulators in 1910. It was a modified airplane mounted on a framework linked to the ground, capable of responding to aerodynamic disturbances.[10] The need to train large numbers of pilots during WWI caused wider use of flight simulators; these started to include machines that would train pilots psychologically; for example, measuring reaction time in correcting disturbances. This period witnessed advances in understanding orientation in air, which largely depended on vision. This realization made the visual aspect of training most significant, and it explains the importance of flight simulators for our grasp on the national visual culture.

The history of flight simulators in Russia can be traced back to the imperial period. Iurii Vladimirovich Gil'sher (1894–1917) was an engineer and flying ace in the tsarist army, who in 1916, during a fighter patrol, lost control of his plane and crashed. Though Gil'sher survived, he lost his left foot. Determined to return to service, however, he started using a prosthetic.[11] In 1917, Gil'sher built an apparatus that enabled him to practice aiming and firing wing-mounted Lewis machine guns on the ground and to use the apparatus for training his fellow pilots.[12] In the 1920s and 1930s, the USSR relied heavily on technology imported from abroad. When, in 1925, a pilot training school was established in Lipetsk, central Russia, Soviet pilots and engineers gained access to Western aircraft as well as to flying experience and flight simulators (Sobolev and Khazanov 2000). During WWII, the USSR lacked the capacity to train pilots properly, normally drafting them from OSOAVIAKhIM[13] and providing only basic training before they took part in military action. Colossal war fatalities made it clear that pilot training was the key to success in military operations. Consequently, in 1946, on Stalin's orders, the *Elektroavtomatika* construction bureau was set up in Leningrad, with the purpose of developing flight simulators for the Soviet military and aviation industry. In other words, the trauma of human loss propelled the construction of simulators, which in its turn propagated the experience of trauma. A departure from this cycle became possible only after Stalin's death in 1953, with the increased use of electronics in building flight simulators (for example, PKBM[14] developed a simulator for MiG-19 in 1957 and for IL-14 in 1961). The electronic and, eventually, the digital realm provided a route for transmitting the national trauma with the help of technologies of visualization and participation.

Thus, in Russia, as anywhere else in the world, the use of flight simulators stemmed from military needs and was a means to achieve greater control of the plane on the ground. Early flight simulators provided an illusion of flying and focused on the aviator's need to operate the plane and simultaneously use weapons. Gil'sher's case also implies that—from the psychoanalytical point of view—a flight simulator is a form of prosthetic. In fact, Gil'sher's prosthetic limb enabled him to return to a career as a pilot, while the flight simulator helped him achieve military excellence.[15] Moreover, whereas in the first instance the pilot had no control over the production of the prosthetic, in the latter, he built the machine himself, exercising a greater degree of power and control. If in the early days of Russian aviation, the prosthetic/flight simulator provided the aviator with an opportunity to (re)gain authority—as is evident from historical accounts—at the end of the Soviet era, the prosthetic/flight simulator was used as a means of deconstructing the Soviet discourse of aviation, manhood, and power.[16]

Viktor Pelevin's novel *Omon Ra* (1992) is a postmodern parody of the Soviet cult of the aviator/cosmonaut and Soviet culture per se.[17] The protagonist, Omon Krivomazov, seemingly realizes his dreams of becoming a cosmonaut by joining other trainees in the Mares'ev flight school, where their bodies are mutilated to give the pilots a different symbolic status: "The next morning I was awakened by loud groans of pain and confusion right in my ear. The surrounding beds were alive with a strange squirming and muffled bellowing . . . . From the next bed I met the pain-filled eyes of Slava . . . where Slava's legs should have been, the blanket fell straight down in an abrupt step, and the freshly starched blanket cover was stained with red blotches like the marks left on cotton towels by watermelon juice" (Pelevin 1994: 33). Of primary importance here is not whether the maiming is real or imaginary, but that those enlisted take part in a sacrificial act that is associated with the loss of (male) body parts and their recuperation with the help of a machine, which enables the disturbance of cultural binaries. As Lilya Kaganovsky, following Roxanne Panchasi (1995) contends, "The disabled and patched-over male body, disrupting as it does a series of what otherwise appear to be stable binaries (animate-inanimate; organic/inorganic; human/mechanical, and in a different set: male/female; battle front/home front), points to a crisis in the dominant fiction, precariously balanced on the belief in the adequacy of the male subject and the commensurability of penis and phallus" (2008: 139). I argue that simulation provides a triangulation of trauma and memory, especially because of its performative nature and the possibility of ritualistic repetition; in other words, it enables a process that repeats and performs trauma outside the established boundaries of remembering and forgetting.

Indeed, the amputation of the feet is just one of many rituals in which Omon has to take part to achieve his goal. His narrative is presented as a loosely chronological account of past events: the narrator tends to mark gradual progression on his way to the dream. First of all, his experience is given spatial organization: the reader learns that as a child Omon lived "not far from the Cosmos cinema" (Pelevin 1994: 5). In other words, from early days he was involved in the two major spectacles of Soviet life—the flight and cinema—interlinked by visual strategies inherent to the medium. Second, Omon notes that the area was dominated by a sculptural piece—"a metal rocket standing on a tapering column of solid titanium smoke" (Pelevin 1994: 5)—that indicated the locus of power articulated with the help of the sculpture's form and politics of vision that seem to transform the very perception of the piece and its perspectival dominance.[18] Pelevin posits the prerogative of vision as an ideological and technological tool by making his character desire a spectacle of vision rather than an experience of flight itself: "If I'd just been able to glance at the screen and see the world from the cabin . . . flight is no more than a set of sensations" (Pelevin 1994: 7). However, it is a different structure that induced his enthusiasm: "But funny enough, it wasn't this rocket that marked the beginning of my personality, it was the wooden airplane in the small children's playground beside my block. It wasn't exactly an airplane, more a toy house with two windows, and during some repair work someone had nailed on a pair of wings and a tail made from the boards of a fence that had been pulled down, then covered the whole thing with green paint and decorated it with a few large reddish stars" (Pelevin 1994: 6). On one level, "the wooden plane" serves as a metaphor of makeshift, make-do Soviet economy; on another, it codifies the relationship between the child and aviation, between the subject and his nation. Structurally and intentionally, "the wooden plane" is a flight simulation apparatus set in the ludic environment of a child's game: "by unwritten agreement of the entire yard, this attic was regarded as the pilot's cabin, and when the plane was shot down, those who were sitting in the body of the fuselage jumped first, and only afterwards, when the earth was already roaring up toward the windows, could the pilot follow them" (Pelevin 1994: 6). Therefore, the plane provides children with a simulation experience, and the apparatus facilitates their (role) play, thus functioning as a prototype of an early simulator.

As these examples demonstrate, the formative relationship between the subject and flight via simulation is established in the opening sections of the novel. The impression of the flying machine and the simulative nature of life and experience in the Soviet Union are enhanced gradually as the novel progresses until the moment when Omon finds himself on a secret mission to the moon, which, as he discovers, is merely a gimmick, since instead of flying to the moon he is actually pedaling a bicycle in the murky tunnels of the Moscow metro. Omon's bicycle is an extension of Gil'sher's simulating machine in that it regressively transcends the entire course of Soviet high technological aspirations and achievements, and takes it back to the start of the twentieth century, i.e. pre-Soviet modernity. Pelevin continues to generalize when he compels his character to comprehend that, like all other citizens of the socialist state, he occupies a position in some sort of a simulation machine, i.e. an ideological virtual reality that provides everyone with an immersive experience of being. Such a realization forces Omon to abandon the project: he is literally catapulted into the grim reality of subterranean Moscow.

Peleven's novel registers a gradual debunking of the Soviet project embodied in its aviation and space programs. Omon Ra's final entry into the real signifies the collapse of the Soviet simulation machine. At the end of the novel, the hero boards a train on the Moscow metro,[19] where he finds himself staring at a fellow passenger's shopping bag: there he sees a packet of rice and a bag of noodles as well as a frozen chicken wrapped in Clingfilm. The dead bird symbolizes the death of the Soviet aspiration to "conquer space" [pokorenie kosmosa] and, evidently, an end of everything in itself. The train carries Omon Ra away, as bereft of his own identity, he requires coordinates for the construction of a new self: "I had to decide where to go. I looked up at the metro diagram on the wall beside the emergency-stop handle, and began to work out where exactly on the red line I was" (Pelevin 1994: 154). The hero is dislocated ideologically, metaphysically, socially, and historically, as well as spatially, and his disorientation epitomizes the state of ideology and technology in regard to the post-Soviet space.

Pelevin's novel documents the demise of the Soviet Union and its ideologemes as it depicts the claustrophobic contraction of space. In fact, Russian withdrawal from its former colonies in 1991 meant that important technological and military objects and sites were in danger of being left behind, i.e. would be an ideological and technological loss. Some of them were dismantled and put together in new places: for example, the 16th Soviet Air Force[20] was relocated from Eastern Germany to Russia in the manner of WWII industrial evacuation. Others—like the Baikonur space station—were leased back to Russia under new intergovernmental arrangements between Russia and newly independent states, in this case Kazakhstan.[21] In the third category are those phenomena that fell out of use and came to symbolize the demise of the Soviet military-industrial complex. (The loss of these territories and simulation machines based on them are similar to the loss of limbs in the symbolic imperial body.) A useful instance of technological deterioration and ideological bankruptcy is NITKA [Nazemnyi ispytatel'nyi trenirovochnyi kompleks aviatsionnyi], a ground-based simulator that enables the training of pilots operating planes aboard aircraft carriers. Built in the Crimea in the 1980s, it was the only training complex of its kind in the USSR. It became operational in 1983, when pilots of Cy-27 started training, but has been out of use since the dissolution of the Soviet Union. However, in 2010 Russia and Ukraine resumed talks about employing the simulator for training purposes ("Rossiiskie letchiki . . . "), and after the annexation of Crimea in 2014, the facility is part of Russia's military-industrial complex.

The loss of 'real-life' simulators in the early 1990s compelled Russia to invest strategically in digital simulators, thus following the trajectory of post-national trauma discussed above. Moreover, the deregulation and privatization processes of the 1990s resulted in the emergence of a number of privately owned companies that specialize in production of flight simulators; that is, transfer of the economy of symbols to private capital from state-controlled enterprises. In contemporary Russia, there are a few companies that make flight simulators—Dinamika (Zhukovskii 1989), Transas-Aviatsiia (Saint Petersburg 1995), Kronshtadt (Moscow 2000), and others. However, the crisis in pilot training is acute: for example, the participants of a round table discussion "Simulator Technologies as a Reserve to Increase Flight Safety," held in Moscow on 20 May 2011, concluded that in five to ten years the Russian Federation would face a shortage of pilots qualified for operating planes and helicopters in commercial aviation (Polov'ian 2011). Thus, the collapse of the USSR meant important technologies were left behind abroad, beyond the boundaries of the Russian Federation as a reminder of Russian scientific imperialism; it also propelled Russia's technological innovation into a new realm—the digital universe of computer gaming.

Game: post-Soviet ludic systems and cultural parallax

In an interview with Celia Pearce, the author of *Communities of Play: Emergent Cultures in Multiplayer Games and Virtual Worlds* (2009), Will Wright, a prominent American video game designer who developed *SimCity* (1989) and other games, made a connection between "modelling as a kid, building models," and interactive design: "When computers came along, I started learning programming and realizing the computer was this great tool for making things, making models, dynamic models, and behaviours, not just static models" (Pearce 2002). By extension, modeling and simulation are embodiments of Soviet and post-Soviet experiences, respectively, in that they shift between different types of architecture and design of flight as a symbol of Russian modernity.

As has been demonstrated in a number of studies (see, for example, Palmer 2006; Widdis 2003), in Russian cultural tradition, aviation and planes embody national aspirations and carry the logic of individual and collective ambition, whereby the perceived linear flight of history is undermined by the non-linear work of memory. Aviation, whether in its physical reality or in the form of simulation (i.e. plane models or plane sculptures, erected in all major Soviet cities—as depicted in Pelevin's novel), has facilitated the process of the nation's consumption of its own past, which since 1991 has undergone the process of commodification, with a great number of companies producing and selling plane models in Russian toy shops as well as online. The available merchandise typically includes models of contemporary planes as well as futuristic versions of flying machines. Mosigra,[22] Russia's largest company specializing in selling board games, with branches in the twenty-five largest Russian cities, offers models of IL-2, i.e. replicas of planes used in the Soviet army in 1942 (the average cost of a model is R270/$5). Produced by the Russian manufacturer Zvezda [star], this model is one of many items that commemorate the military history of the country via such items as the PE-8 bomber (1940) and the Katiusha rockets.[23] If such toys offer pleasure of a tangible nature that involves manipulation of a real object and the imagining of war episodes, flight simulators provide an immersive, imaginary experience of warfare and flying. Flight simulators are an established genre of video gaming, the number of games exceeding a hundred (nowadays over one hundred and fifty games are available, produced by various companies around the world).[24]

In his study of flight simulators as a genre of video games, Patrick Crogan demonstrates that "the flight simulator game genre . . . frequently takes military combat as its subject . . . and it is the direct descendant of computerized flight simulation developments that have been so crucial in the history of computer generated imaging and simulation" (Crogan 2003: 276). The pinnacle of Crogan's argument is the link between flight simulators and developments in the military-industrial complex. By using Paul Virilio's concepts of "pure war" and its reliance on logistics as the determining factor of contemporary warfare and subsequently of city planning and, generally, the organization of human life (Virilio 1997), Crogan demonstrates how flight simulators enable gamers to "repeat history" to develop their control over events (Crogan 2003: 296). He continues to assert that "flight simulation gaming offers a 'replay' of the passage of military-driven technological innovations into the heart of contemporary computer visualization and simulation" (Crogan 2003: 276). I argue that flight simulators provide a context in which historical events are experienced as a case of alterity, and that the gamer's desire is not so much to gain control, which in Russia's case would be to assert its position on the global scale and to replay historical contexts that constitute the core of Russian national mythology, i.e. WWII, but rather to experience an alternative configurational horizon.

To conceptualize my presupposition, I introduce the term 'cultural parallax.' It is derived from the term 'motion parallax' that connotes a difference in the direction of movement or speed viewed when the observer moves relative to his/her environment, i.e., when objects in the visual scene appear at different distances from the observer's line-of-sight. In other words, motion parallax is a phenomenon that a pilot or passenger of an aircraft regularly experiences during take-off and landing. In his critique of Western philosophy Slavoj Žižek writes about (2006) the parallax gap, or "the confrontation of two closely linked perspectives between which no neutral common ground is possible" (Žižek 2006: 4). By extending Kantian epistemological antinomies, Žižek examines his main concern, the notion of difference—ontological, scientific, and political (but not cultural)—and employs Derridean logic to deconstruct categories of defining and structuring. In Žižek's use, parallax accounts for gaps and tensions in the dualistic logic of capitalist culture; in my view, parallax enables simultaneous appreciation of multiple perspectives; that is, in visual terms, a shift from the centrality of the panorama to de-centered participation of multi-player online gaming. I use the term 'cultural parallax' to signify the phenomenon of cultural displacement that occurs when the events of the past are viewed in the ludic environments of computer gaming. In other words, cultural parallax defines an effect of divergence caused by the dislodgment of cultural coordinates. Simulation relates to the cultural parallax in that it fills in the gaps in the discourse—as identified by Žižek—and provides a ritualistic displacement of vision in non-predicated ludic environments.

A manifestation of such cultural parallax is IL-2 Sturmovik, a cluster of flight simulators that include the original WWII combat flight video game released in 2001 as well as a large number of add-ons, patches, variants, and sequels, such as *Forgotten Battles* (2003), *Pacific Fighters* (2004), *Manchuria* (2006), *1946* (2006), and *Cliffs of Dover* (2011) among many others.[25] In this chapter I chiefly analyze the original game; however, my discussion also tackles the game's sequels with the purpose of conceptualizing IL-2 Sturmovik not as a stand-alone occurrence, but as a complex cultural phenomenon that is representative of global trends in gaming. By focusing not so much on the game's structure and the interactive solicitations of the gamers, and more on the inclusion of the game into a cultural tradition of simulation, I trace how a certain perception of reality and national history is constructed. In the closing section of this chapter, I explore the meta-textual level of game play by analyzing videos produced by fans of IL-2 Sturmovik and distributed over networked video channels such as YouTube.

As Lev Manovich observes in his *Language of New Media*, in the ontology of the computer, "the world is reduced to two kinds of software objects that are complementary to each other—data structures and algorithms" (Manovich 2001: 223). Indeed, the makers of IL-2 Sturmovik have been preoccupied with the constructions of such databases and deploying algorithms as game-time tools.[26] In the cultural context, IL-Sturmovik since its inception has grown not only in terms of the number of playfields, missions, and interactive possibilities, but also in terms of the represented arsenal of warfare. If the original version of the game featured thirty-one flyable planes and forty additional non-flyable aircraft, the complete franchise now contains about three hundred and fifty planes, two hundred and fifty of them flyable: for example, Focke-Wulf Ta 152H, Kawasaki Ki-61, Lavochkin La-7, Yakovlev Yak-3, and so forth. Ultimately the game covers almost all major aircraft deployed during WWII, as well as those that never evolved beyond the prototype phase. Therefore, IL-2 Sturmovik has become an internationally acclaimed ludic encyclopaedia of world aviation during the war period, in which the growing database of aircrafts—in its morphological construction—has eclipsed the needs of the game in its game-time function. Such growth is a manifestation of the emergence of new pleasurable systems and modes that encompass an interpretive interest and a drive toward a manageable totality of vision and action. This is also evident in that IL-2 Sturmovik has advanced from a single-player game to include a multi-player feature that supports up to one hundred simultaneous players online. In addition to this, various independent multi-player hubs allow for IL-2 Sturmovik players to use their resources for hosting multi-player hubs. Thus, IL-2 Sturmovik increasingly involves cooperative forms of play and collaborative constructions of space.[27]

IL-2 Sturmovik's game-space is built using major historical events of WWII, focusing on air battles on the Eastern front (Moscow, the Crimea, Kursk, Smolensk, Berlin, etc.), as well as events elsewhere in the world (Normandy, Okinawa, Hawaii, Manchuria, Slovakia, etc.). The game reproduces the terrain of historical battlefields and employs such generic or imaginary landscapes as 'Pacific Islands' or 'Deserts.' IL-2 Sturmovik mixes reconstructed and imaginary settings, as well as different types of visual representation: the game's cinematics includes so-called 'cut-scenes': i.e. short, pre-recorded animated sequences that introduce missions as well as real wartime documentary footage that appears at the end of a successful campaign, thus combining creative adaptations of historical documents and fake imagery.[28] In its construction of space and its use of missions, IL-2 Sturmovik asserts the specifically Soviet narrative of WWII. The game begins with a video reel that shows German aircraft attacking Soviet troops on 22 June 1944. It also shows how German Bf-109Es are taking off from an airdrome, flying above roads filled with tanks and other military vehicles, and joining Ju-87s in their attack on a Soviet town. The city is ablaze with fighting, explosions, and fires; the game audio presents an auditory mix of sirens, aircraft engines, and falling bombs. The video now moves to the Soviet rear territories where the first IL-2s are being produced. The camera focuses on a plane that is being prepared for take-off and tracks the ascent of the fighter bomber into the sky. The ominous rumblings of the plane's engine serve as a premonition of the future fact of that the USSR will produce over thirty-six thousand such planes, which will determine the course of the war.

The video focuses on the immediate threat of the German attack, eliciting patriotism from a Russian gamer familiar with the sacrosanct Soviet account of the events: "On 22 June 1941, at exactly 4 o'clock in the morning, Germany invaded the Soviet Union." So, as one starts the game, the Great Patriotic War begins; this catastrophic moment has been immortalized in such diverse films as Mikhail Kalatozov's *The Cranes Are Flying* [Letiat zhuravli 1957] and Boris Galkin's *June 22, Exactly at 4 O'clock . . .* [22 iunia, rovno v 4 chasa . . . 1992], and is now exploited in the video game.

As the game evolved, so did its set of missions and campaigns. The game consists of air skirmishes and open combats, assault on targets on the ground, and missions of convoy and reconnaissance. It provides the gamers with options to fly not only as Soviets and Germans, but also as Americans, Finns, Hungarians, Japanese, etc. Such growth in gaming options signifies a clever marketing strategy with a global appeal, as well as a more complex view of history, with the Great Patriotic War represented not solely as a battle between the USSR and Germany, but as a combination of multiple forces.[29] The focus of the game, however, is not only on achieving various combat objectives, but also on mastering the flying skills that require learning about the planes, their engines, flap positions, and so forth, i.e. the complex mechanics of a real plane. Therefore, the game appeals to average gamers interested in combat and missions, on the one hand, and, on the other, aviation enthusiasts who cherish a great degree of detail in the representation of historical planes. Overall, IL-2 Sturmovik provides a unique sensorial experience that enables gamers to engage in historical reality by means of ludic interaction. One of the reviewers of the game notes the realism of representation in the following statement: "The sun reflects off rippling water, contrails snake across the sky when planes fly at high altitudes, and the clouds—particularly around nightfall—are almost photo-realistic at times" (Osborne 2003). The visual effects are of such high quality that they enable the gamer to distinguish between parts of the plane made of wood and metal, and the atmospheric optics appear realistic and provide a coordinated experience. So the player can enjoy pedantic reconstructions of landscapes and plane designs, including the cockpit with its numerous controls. The illusion of flight is enhanced by rich soundtrack and audio effects.

The realistic effects in the visual features of the game do not disable the simulation factor of the experience, which is enhanced by the participatory elements of the game. For example, the players can customize the appearance of their plane, and if they are successful as pilots, they will receive awards and military ranks. IL-2 Sturmovik simulates choice in that it offers a chance to fly on a Soviet plane as a pilot or gunner, and as an experienced aviator or a novice, thus defining various modes of gameplay, its scenarios, locales, and progression from one level to another. In terms of the game narrative, the game places special emphasis on the historical realism of combats rather than on a dynamic gaming campaign; in other words, IL-2 Sturmovik downplays the simplistic appeal of a first-shooter game and underscores the historical context: while special knowledge of historical events is not required, it is built into the very matrix of the flight simulator. Furthermore, the game does not celebrate the personal achievement of the gamer, i.e. the overall success of the gamers does not depend on their performance in specific sequences and missions. Instead, the game aims to make the gamers aware of the collective war effort, and in that sense it follows Soviet conventions while simultaneously advocating a new post-Soviet ideology of individualism. Finally, the players can exit pre-designed missions and create their own range of missions using provided tools, thus taking part in the collaborative reconstruction of the past.

IL-2 Sturmovik is simultaneously a flight simulator and an historical simulation, i.e. it simulates experiences that are distanced temporarily. Such distancing provides the player with a 'double vision' of historical events, whereby the historical perspective enhances the illusion of realism: the temporal gap verifies the employed narrative and its modifications while also exonerating any imperfections in simulation. In other words, IL-2 Sturmovik instances cultural parallax, for it oscillates among different visions of the past, multiple perspectives of gameplay, and cultural predispositions. In visual terms, IL-2 Sturmovik abandons a linear perspective in favor of an aerial one: it provides visual scenes that encompass great distances both horizontally and vertically, strong cues to distance and depth, and atmospheric particles that scatter and diffuse light emitted and reflected from objects. The effect of the aerial perspective is that it increases the perceived distance from the observer to the observed object[30] and makes objects appear in a desaturated, 'washed-out' manner. The 'washing out' effects of the game are symptoms of the cultural parallax and the instability of cultural memory, which the gamers strive to resolve via repetitive tasks of gameplay.

Simulation: repetitive action and a new field of play

> In striving to reproduce deceptively lifelike changes in their presentation of nature, the panoramas point ahead, beyond photography, to films and sound films . . . . The panoramas, which declare a revolution in the relation of art to technology, are at the same time an expression of a new feeling about life.
>
> Walter Benjamin[31]

The ultimate presentation of an aerial view is a panorama. Panoramas are a modern phenomenon: the first large-scale pictorial representations, often built in a circular auditorium, presented visual surveys of areas from an elevated vantage point, and were exhibited in Europe at the very end of the eighteenth century. Such a vantage point accounted for the new experience of observing landscape from hot air balloons; the connection to military use and representation of military action is particularly evident in that ballooning was often used for reconnaissance in the nineteenth century. "Over the next [nineteenth] century the panorama and its variants—cosmoramas . . . , dioramas, neoramas, cycloramas, and in particular, moving panoramas—came to epitomize an international hunger for physically, geographically, and historically extended vision" (Miller 1996: 34). As a child, Walter Benjamin admired the 'Kaiserpanorama' in Berlin: it employed over one hundred thousand photographs that passed before the viewers in circular fashion; he later grew to see the panorama as the precursor of photography in cinema because of its ability to capture and represent time (1978). In the age of digital reproduction, Google Earth has evoked the principles of the nineteenth-century exhibition and irrational desires for new vantage points, hence mastery over space. It does so by providing a vision of the planet as a set of interlaid and interconnected panoramas, thus raising the issue of illusionism and the blurring of boundaries between reality and art (in this case, digital reproduction of 'reality').

The panorama and IL-2 Sturmovik as an example of flight simulation video games display a range of common characteristics, including their mass appeal, i.e. the viewer/player is construed as a mass audience rather than as an individual; the viewer/player occupies the central position in a decentered, all-encompassing experience;[32] the illusion of reality is grounded in spatial design, whether architectural or algorithmic; the accuracy of representation 'automatically' (if erroneously) verifies truthfulness in the depiction of historical events and/or unusual terrains; and both the panorama and IL-2 Sturmovik display historicity of vision. The last concept refers both to an interest in reproduction of historical events (albeit in an edited, expedited, and telescoped manner) and to a passage of time that occurs thanks to the movement (of the gaze) in space: i.e., it adheres to the logic of cultural parallax. While in IL-2 Sturmovik the player can respond to visual stimuli by engaging in a combat or embarking on a specific mission, with panoramas, viewers are able to act in response only at the level of imagination, i.e. by following the display and (re-)assembling its interpretation in accordance with their previous viewing experience. However, it is possible to argue that the video game player is involved in a similar process of (re-)construction. As Will Wright, the architect of many simulation games, puts it, in an elaborate simulation of reality, a lot of what the player does is "reverse engineer the simulation" (Pearce 2002). Wright describes his objective as the game designer and the strategy of the player as follows: "The more accurately you can model that simulation in your head, the better your strategies are going to be going forward. So what we're trying to do as designers is build up these mental models in the player. The computer is just an incremental step, an intermediate model to the model in the player's head. The player has to be able to bootstrap themselves into understanding that model" (Pearce 2002). Therefore, the panorama and IL-2 Sturmovik provide the viewer/player with an opportunity to create mental maps of observed reality and its representation, with the latter gradually gaining the status of the former in terms of the cognitive response to the visual display and its internal design.

IL2-Sturmovik extends Russia's tradition of historic simulation originating in the so-called Potemkin villages, hollow facades of houses allegedly erected along the banks of the river Dnepr during the visit of Catherine the Great to the Crimea in 1787, and stretching to Sergei Eisenstein's *Battleship Potemkin* [Bronenosets Potemkin 1925], a filmic account of a naval mutiny in the Crimea and a resulting street demonstration that ended in a police massacre. While the former is largely a historical myth, the latter was one of the first cinematic attempts to reconstruct historical actuality with the help of sources and consultants, and of altering details to achieve the desirable effect. Therefore, the film is a fantastic compromise between historicity and the director's ingenuity, with Eisenstein's originality lying "in his skill in dramatizing an authentic element of the historical reality through the transformation of its detail" (Gerould 1988: 176). Eisenstein's later film *October* [Oktiabr' 1927] was derived from a different sort of spectacle/simulation—Nikolai Evreinov's re-enactments, on Dvortsovaia Square in St. Petersburg in 1918 and subsequent years, of the October revolution of 1917 that involved eight thousand participants and attracted thousands of spectators. In the twenty-first century, another Russian filmmaker likewise produced a historical simulation based on real events, personae, and artifacts: Aleksandr Sokurov's *Russian Ark* [Russkii kovcheg 2003] evokes scenes from Eisenstein's *October* and constructs an edifice of Russian history by letting his digital camera roll and produce footage for eighty-seven uninterrupted minutes inside the Hermitage. The result is a single continuous take that recreates the major events of Russian history of the past three hundred years, presented as elements of a large-scale panorama above which the camera glides effortlessly, thus constructing the viewers' gaze and their perception of history. Sokurov's film appears as a digital panorama whereby spatial montage functions akin to the architectural design of the nineteenth-century panoramas and dioramas, with the huge canvas of pre-Revolutionary national history hinging on the digital framework.[33]

In tsarist Russia, two actual panoramas were produced to celebrate the anniversaries of the major war battles, *The Defense of Sevastopol* [Oborona Sevastopolia][34] and *The Borodino Battle* [Borodinskaia bitva] panoramas. Produced by Frants Alekseevich Rubo (1856–1928), the first was completed in 1904 in time for the fiftieth anniversary of the Sevastopol siege and was displayed in the Crimea, and the other was completed in 1912, to commemorate the centenary of the battle of Borodino, and was shown in Moscow.[35] Under Stalin, *The Battle of Stalingrad* [Razgrom nemetsko-fashistskikh voisk pod Stalingradom] panorama was commissioned by the regime as early as 1944 and it involved a number of artists who produced a pictorial representation in spite of the availability of more modern tools of commemoration.[36] It was finished in 1950 and was displayed in Moscow until it was moved to Stalingrad in 1952.[37] In the second half of the twentieth century, cinema became the main medium of commemorating the nation's history as well as producing convincing simulations of past experiences and practices. In fact, it is possible to argue that the war film was one of the main genres of Soviet and post-Soviet cinema (see, for example, Smorodinskaya 2010; Youngblood 2007), and it served as the focal point of visual commemorative strategies until 2000s, when this function was delegated to Russian television[38] and eventually video game producers.[39] Therefore, the history of simulation apparatus in Russia is not only a case of critical cartography, with technical devices achieving various levels of illusion, but rather a means of mapping technology onto epistemology, and vice versa. Nineteenth-century panoramas and twenty-first-century flight simulation games provide a totalizing, transcendent gaze, and serve as an essential ideological component of control over subjectivity and spatial-temporal constructions of the nation.

It comes as no surprise that the Russian government of Vladimir Putin, ever conscious of the need to preserve and re-purpose national history, supported an IL-2 Sturmovik gaming tournament celebrating the sixty-fifth anniversary of Soviet victory in Great Patriotic War.[40] The tournament was a three-way joint venture of the State Duma Committee on Youth [Komitet Gosudarstvennoi Dumy po delam molodezhi], the 1C company, and the Russian National Association for Computer Sports, i.e. governmental, commercial, and grassroots agencies. Their emphasis on the educational aspect of the tournament and on what they call "aviation cyber sport" [aviakibersport] is strangely reminiscent of the early OSOAVIAKhIM, except that in 2010, unlike in the 1920s, they handed out not badges, but cash prizes of $1,000 and $2,000.

As with early Soviet military practices, the IL-Shturmovik games encourage participatory experiences in that they provide the possibility of recording flight missions. This has fostered activities of the game's fans, who post videos of their gaming achievements on YouTube.[41] Some posts are 'vanity videos,' boasting about the success of gamers; others are displayed with the pragmatic goal of providing video reviews or sharing gaming tips and strategies. However, some fans use their posts as a repository documenting the excitement of the gaming experience—a practice particularly common with multi-player sessions, videos of which showcaswe gamers' collaboration. Other fans use footage obtained from IL-2 Sturmovik to create their own original audiovisual works. For example, a video posted by vinsphoff on 26 January 2010 includes edited video footage from the game as well as the soundtrack from a popular Soviet animated film about Winnie-The-Pooh [1969–72; trilogy, dir. Fedor Khitruk]. The purpose of this video is to parody the Soviet obsession with weapons and military games: Piglet shoots a gun at a balloon that keeps Winnie-The-Pooh up in the air; however, because of how the video is edited, his shot brings down not his mischievous friend but a German plane. Another user, Gravesad, on 4 January 2007, posted a video that contains reworked and edited footage from the game accompanied by Vladimir Vysotskii's *Song of a Pilot* [Pesnia letchika 1968]. The video generated 152,211 views (as of 6 July 2010) and a heated discussion that thematically ranged from an analysis of the game to the role of the Soviet army in WWII and the work of cultural memory, thus signaling that IL-2 Sturmovik is not only an essential element in the assemblage of memories of the war—in this particular case, piecing together the game footage, Vysotskii's song, and the cinematic tradition of representation of war events and national trauma—but also a platform for grassroots debates over the Soviet past and particularly WWII.

The ironic engagement of gamers with the game content is post-Soviet modification of the Soviet past and the associated Grand Style of Socialist Realism. This is what Lipovetsky identifies as part of Russian postmodern poetics, which entails "a deconstruction of the binary opposition between the post-Soviet present and the Soviet past, which served as the basis for the post-Soviet mentality in the 1990s" (Lipovetsky 2004: 359). However, the user-generated content propagated with the help of social media signals for a more complex picture of cultural memory and practice, which moves beyond the binary structures proposed by Lipovetsky. The creative work of the fans signifies a disruption to existing media models as well as cultural discourses, with innovation emerging from the margins and facilitating a particular type of cultural citizenship, forging a new relationship between the individual and the community in the face of cultural difference and negotiation of cultural experiences, both present and past. These experiences enable parallel visions and interpretations of the past in the ludic environment of the game and socially networked domain of YouTube. In other words, the gamers adopt the logic of cultural parallax as a means of working on cultural displacement and effects of cultural divergence. In this process, visualization and participation maintain their primacy in terms of transmitting the national trauma and constructing alternative experiences. The expansion of the game and inclusion of various interconnected transmedial practices signifies an emergence of a new pleasurable system and a specific totality of vision. The latter relies on the technology of massive multi-player gaming and on the visual representations provided by IL-2 Shturmovik with its mixed cinematics of real and imaginary wartime footage. The game mechanics enable different gaming experiences as well as multiple cultural predispositions which enhance the interpretation of history. The multiple aerial perspectives allow new critical cartographies which shift from the totalizing gaze of the panorama to a variable, and often erratic, gaze of the war simulation game.

As a metaphor of cultural practice, simulation accounts for both experiential and symbolic contexts and defines the condition of the contemporary Russian subjectivity in relation to the industries of illusion as well as historicity. As this chapter has demonstrated, the Russian gamer narrates, represents, and performs the past by engaging simultaneously with multiple modes of cultural production. Simulation is a zone for ludic engagement with the national trauma; it is also a method of historicizing the trauma because simulation facilitates a dislocation of cultural coordinates and their free manipulation. In addition to that, simulation provides a framework for re-writing and re-considering the Russo-Soviet experience of flight, and also affords a new perspective on the grand narratives of Russian modernity and postmodernity, especially in respect to the means of psychological, physical, and ideological control, as well as spatiality, imperial expansion and contraction, and human enterprise. The (cultural) parallax is not only a visual mode of simulation but also an episteme of post-Soviet dislocation—spatial, ideological, aesthetic, and experiential—which deconstructs the restorative logic of trauma symbolized by the notion of loss—bodily, spatial, ideological, and stylistic—and propels ludic replications and reiterations. In relation to simulation, the cultural parallax supplies an opportunity to overcome the antinomies of experience, knowledge, and symbolic apparatus, as identified by Žižek, and injects the possibility of simultaneous location, dislocation, and relocation of subjectivity supported by a new type of visuality and participatory culture (Figs. 10.1 and 10.2).

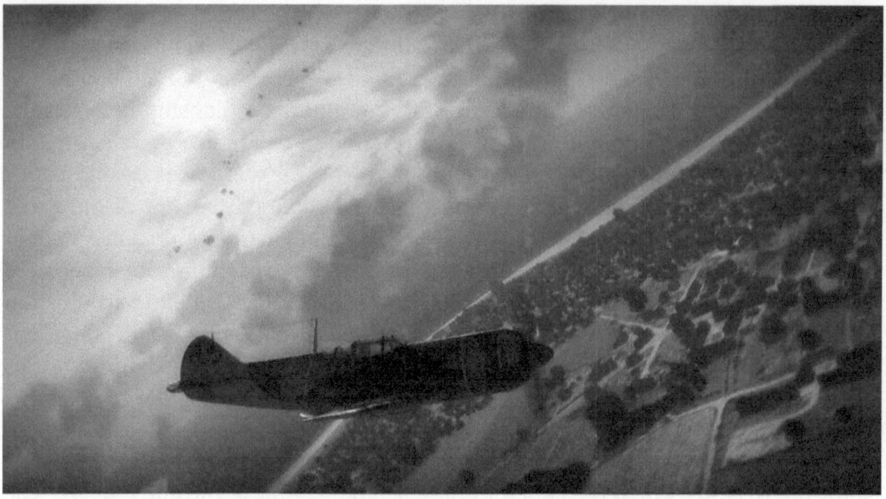

Figure 10.1 IL-86 Shturmovik gameplay (summer landscape). IL-86 Shturmovik

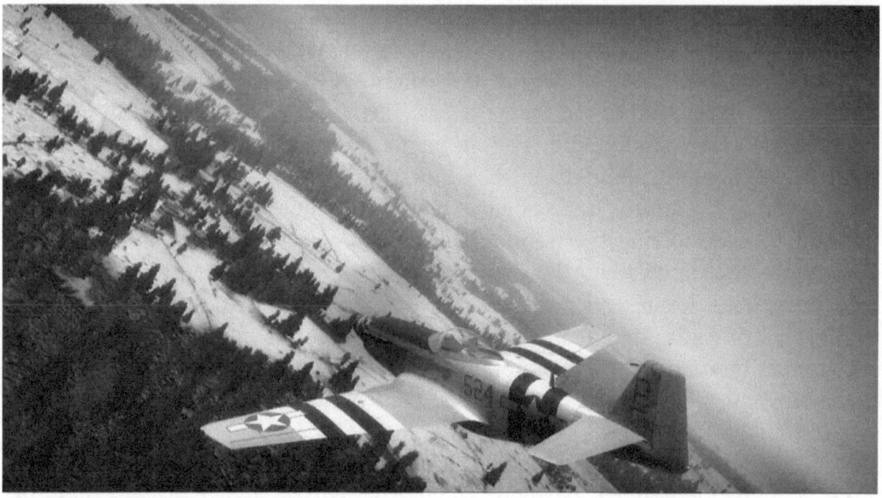

Figure 10.2 IL-86 Shturmovik gameplay (winter landscape). IL-86 Shturmovik

Notes

1 25 August 2009, http://m.guardian.co.uk/technology/gamesblog/2009/aug/25/sturmovik-ps3-xbox-psp?cat=technology&type=article (accessed 4 July 2011).

2 (1994: 138).

3 The game, praised for "attention to detail," "gorgeous graphics, and its interesting selection of planes" (Osborne 2003), was developed by 1C: Maddox games and released in Russia by 1C, and by Ubisoft in the rest of the world. Though the game was originality designed for PCs, versions for gaming consoles were developed later. The

original game was immensely popular with players around the world, though the actual number of players and revenue from sales are hard to estimate. It received extremely positive reviews in such publications as *The Guardian* (see the epigraph to the chapter) and had sundry expansions created by fans and third parties, such as *Plus Wave* and *X1 Software*, which are all eloquent indications of the game's success. (1C is the leading publisher, developer, localizer, and distributor of computer and video games in Russia, Eastern and Central Europe, http://www.1cpublishing.eu/.)

4 The politics of gaming in post-socialist countries is not my concern in this chapter. However, some discussion of the subject may be found in Strukov 2012a and Strukov 2012b.

5 Other successful international flight simulator games include *X-Plane* (Laminar Research), *Flight Unlimited* (Looking Glass Studios), *Pilot Academy* (PlayStation), and many others.

6 Possible directions for future research include studies of physical and psychological training of Soviet and Russian aviators and astronauts, simulators as technologies of warfare, and simulators as artefacts of Soviet design and ideology.

7 According to the Gregorian calendar that was in use at the time of his birth, Korolev was actually born in 1906.

8 For a full account of the event see Gerovitch 2011.

9 In Chapter 8 I also demonstrate how the animated film *Belka and Strelka,* which is a post-Soviet inversion of the Soviet myth of the aviator, emphasizes the dogs' training rather than the actual flight.

10 In Britain, Eardley Billing patented a purpose-built airplane for elementary training and amusement, which relied on the instructor to provide the disturbances while the student tried to maintain equilibrium (Rolfe and Staples 1986).

11 Gil'sher's experience would be serialized and celebrated in the cultural canon of Socialist Realism, culminating in Boris Polvevoi's *Tale about a Real Man* [Povest' o nastoiashchem cheloveke 1946].

12 Allegedly (Mitiurin 2006), Gil'sher's machine was the first ground-based cockpit machine gun trainer in aviation history.

13 Union of Societies of Assistance to Defence and Aviation-Chemical Construction of the USSR.

14 [Penzenskoe konstruktorskoe biuro modelirovaniia].

15 This example provides an illustration of Marshall McLuhan's theories of media as extensions of man (1964), whereby the prosthetic is a site of trauma and a symbol of the symbolic connectedness of the psyche to the imaginary world.

16 A range of studies have explored the cult of the aviator in Stalin's Russia and numerous publications have examined the legacy of flight and Soviet masculinity, including an analysis of the famous story of Aleksei Mares'ev and Boris Polevoi's novel *A Tale about a Real Man* (1946) based on real-life events. See, for example, Kaganovsky 2004. The story of Mare'sev has a parallel in the British war hero, Douglas Bader (1910–82), whose biography was written by Paul Brickhill, an Australian journalist and WWII fighter pilot, and was later adapted into a film titled *Reach for the Sky*, 1954. A London-based artist, Pavla Alchin, remastered the two films for her 2008 piece of video art entitled *Legless (East-West)*, in which she delights in the similarity between the Soviet and British films, their visual styles and ethos. The narrative of the (assumed) male physical inadequacy, space heroism, and identity building is not specific to Soviet and Russian contexts; examples may be found in other cultures, including, for instance, Andrew Niccol's 1997 film *Gattaca* about a genetically inferior man who assumes the identity of a superior one to pursue his lifelong dream of space travel.

17 The following are some of the sources that provide a general analysis of the novel: Anisimova 2011; Dalton-Brown 1997; Kaganovsky 2004; Lipovetsky 2004; McCausland 2000.

18 Sally Dalton-Brown notes, "Pelevin's textual type or mode of writing can be called that of the 'reverse image' text. His work is highly visual, and his texts usually focus on a few well-chosen metaphors, often only one extended metaphor per text . . . which

lose their status as created images and become 'real', reversing the status of image/phenomenon (or signifier/signified). Reality, however, is never ultimately signified in this simulated world" (1997: 227). Also see Chapter 1 on the discussion of flight and sculpture.

19 The train accidentally signifies the return to pre-Soviet symbols of modernity—see Lev Tolstoy's *Anna Karenina.*

20 [Shestnadsataia vozdushnaia Krasnoznamennaia armiia].

21 Kazakhstan is now part of the Eurasian Economic Union, conceived and promoted by the Kremlin, which means that in the long run Russia may gain free access to the facilities.

22 http://www.mosigrushka.ru/ (accessed 22 July 2011).

23 This new range of toys is in stark contrast to the experience of a Soviet child, who, according to Pelevin, had access to a few wooden toys.

24 For a comprehensive list of available games, see http://gameguru.ru/pc/genre/flight/page9/ (accessed 22 July 2011).

25 The game also inspired gaming projects produced by other companies: for example, CodeExotics, based in the provincial town of Orel in central Russia, debuted in 2009 with a 2D-arcade called *Stalin's Falcons* (Stalinskie sokoly, http://www.codeexotics.com/?id=project [accessed 20 June 2013]).

26 "The algorithm consumes the topographic and turns it into topological" (Wark 2007: 69).

27 For a discussion of collaborative aesthetics of space, see Strukov 2012.

28 On the relationship between flight and memory see Chapters 8 and 9.

29 An alternative reading of this phenomenon is that the game advances Russia's imperialist ambitions in the digital realm by enabling the player to take part in the annexation of new territories in Europe. These opposing interpretations are grounded in a specific 'Post-Sots' sensibility. In his analysis of contemporary Russian literature and film, Mark Lipovetsky notes that "the paradox of Post-Sots is thus based on the juxtaposition of two opposing modalities: the affirmation of the absent collective identity (through various signifiers of military discourse) combined with a desperate nostalgia for it" (Lipovetsky 2004: 365). Computer gaming provide users with opportunities to maintain the identified desires and affiliation as well as to construct new networked forms of belonging in ephemeral digital environments.

30 In linguistic terms, it destabilizes the signifier and signified.

31 1978: 149–50.

32 Such a central position facilitates the work of the perspective system, which in its turn creates an illusion.

33 The film generated a large corpus of research; see, for example, Harte 2005; Hashamova 2006; Kujundzic 2004; Ravetto-Biagolli 2005. For a detailed discussion of the gaze in *Russian Ark,* see Strukov 2009.

34 The website of the panorama (http://sevpanorama.ru/) lists a photograph of Iurii Gagarin viewing the exhibition, a historical anecdote which allows for a comparison of different strategies of vision and representation, the Tsarist and Soviet, with the latter attaining greater authority due to the technological advantage. At the same time, the scene documents the humility of the Soviet subject in light of the past national trauma.

35 *The Defense of Sevastopol* panorama was rescued during the second siege in 1942. Though large sections of the panorama were lost and damaged, it was restored after WWII and put on display in 1952. It became an extremely important commemorative tool in post-war USSR, as it provided a link between events of the nineteenth and twentieth centuries, depicting the heroism of the 'Russian' soldier. During the Stagnation period, the panorama gained extra prominence due to large-scale memorialization projects carried out in the Crimea because of Leonid Brezhnev's service on the Crimean front. The panorama eventually became part of his cult of personality until it was stripped of such connotations after the dissolution of the USSR. *The Battle of Borodino* panorama was removed from public display in 1918 and stored in various locations in Moscow until 1948, when complex restoration work commenced; in 1962 the panorama reopened for visitors in a new museum built on Kutuzov Avenue in Moscow.

Strictly speaking, both original panoramas were lost in the 1930s and 1940s, i.e. the existing panoramas are products of Soviet restoration efforts resulting in 'a recreation of a recreation' of the historical events: they result as much from the nineteenth-century obsession with pictorial illusion as from the Soviet propaganda machine.

36 If these three panoramas signify the role of the state in controlling the nation's memory, i.e. time, with the help of visual spectacles put on public display, then another large-scale panorama produced by Pavel Iakovlevich Piasetskii in 1897–1903 and called *Trans-Siberian Railway* [Velikii Sibirskii put': Transsibirskaia zheleznodorozhnaia magistral'] documents imperial expansion and tsarist appropriation of the nation's space.

37 Interestingly, in Stalingrad it was initially displayed in a cinema until 1958, when a special building to house the panorama was constructed.

38 For example, Aleksei Feoktistov's *And the War Commenced* [I byla voina 2009]; Vitalii Vorob'ev's *Ballad about a Bomber* [Ballada o bombere 2011]; Ivan Shurkhovetskii's *Fog* [Tuman 2012]; Tigran Kiosian's *Yalta 1942* [Ialta 1942; 2012]; and many others.

39 For example, *Art of War: Kursk 1943* [Iskusstvo voiny: Kurskaia Duga] produced in 2009 by 1C, which is also responsible for localizing such games as *Front Line: Battle for Kharkov* [Liniia fronta: Bitva za Khar'kov], released by Graviteam in 2009.

40 The official website of the tournament is http://il2pobeda.ru/ (accessed 19 July 2011).

41 Eighty-four fan videos were chosen and analyzed for the purpose of this chapter; they were selected using the Google search engine and keyword search function as well as search paths created by user-generated interlinks.

Bibliography

Aarseth, E. (2004) "Genre Trouble: Narrativism and the Art of Simulation," *First Person: New Media as Story, Performance, and Game.* Ed. N. Wardrip-Fruin and P. Harrigan. Cambridge, MA: MIT Press. 45–55.

Anisimova, I. (2011) "From Consumption to Objectification in Viktor Pelevin's 'Akiko.'" *Studies in Slavic Cultures* 10: 60–74. Available at: http://www.pitt.edu/~slavic/sisc/SISC10/index.html (last accessed 7 July 2013).

Baudrillard, J. (1983) *Simulations.* Paris: Semiotext(e) Inc.

—. (1994) *Simulacra and Simulation.* Ann Arbor, MI: Michigan University Press [first edition 1981].

Benjamin, W. (1978) *Reflections: Essays, Aphorisms, Autobiographical Writing.* Trans. E. Jephcott. New York: Harcourt, Brace, Jovanovich.

Bolter, J.D. and R. Grusin. (2000) *Remediation: Understanding New Media.* Cambridge, MA: MIT Press.

Charsky, D. (2010) "From Edutainment to Serious Games: A Change in the Use of Game Characteristics." *Games and Culture* 5.2: 177–98.

Crogan, P. (2003) "Gametime: History, Narrative, and Temporality in *Combat Flight Simulator 2.*" *The Video Game Theory Reader.* Ed. M.J.P. Wolf and B. Perron. New York, London: Routledge. 275–303.

—. (2007) "Remembering (Forgetting) Baudrillard." *Games and Culture* 2.4: 405–13.

Dalton-Brown, S. (1997) "Ludic Nonchalance or Ludicrous Despair? Viktor Pelevin and Russian Postmodernist Prose." *The Slavonic and East European Review* 75.2: 216–33.

Etkind, A. (2009) "Post-Soviet Hauntology: Cultural Memory of the Soviet Terror." *Constellations* 16.1: 182–200.

—. (2013) *Warped Mourning: Stories of the Undead in the Land of the Unburied.* Stanford, CA: Stanford UP.

Frasca, G. (2011) "Simulation 101: Simulation versus Representation" [Blog entry]. Available at: http://www.ludology.org/articles/sim1/simulation101.html (last accessed 20 July 2011).

Gerould, D. (1988) "Historical Simulation and Popular Entertainment: The 'Potemkin' Mutiny from Reconstructed Newsreel to Black Sea Stunt Men." *TDR* 33.2: 161–84.

Gerovitch, S. (2011) "'Why are We Telling Lies?' The Creation of Soviet Space History Myths." *Russian Review* 70: 460–84.

Gravesad. "IL-2 Shturmovik (o vozdushnom boe) [fan video]. (2007) *YouTube*. 4 January. Available at: http://www.youtube.com/watch?v=ekZdGxb-jTA (last accessed 4 July 2011).

Gredler, M.E. (1996) "Educational Games and Simulations: A Technology in Search of a Research Paradigm." *Handbook of Research for Educational Communications and Technology*. Ed. D.H. Jonassen. New York: MacMillan. 521–39.

Harte, T (2005) "A Visit to the Museum: Aleksandr Sokurov's *Russian Ark* and the Framing of the Eternal." *Slavic Review* 64.1: 43–58.

Hashamova, Y. (2006) "Two Visions of a Usable Past in (Op)position to the West: Mikhalkov's *The Barber of Siberia* and Sokurov's *Russian Ark*." *Russian Review* 65.2: 250–7.

Jenkins, H. (2007) "Transmedial Storytelling" [Blog posting]. Available at: http://henryjenkins.org/2007/03/transmedia_storytelling_101.html (last accessed 20 July 2013).

Kaganovsky, L. (2004) "How the Soviet Man Was (Un)Made." *Slavic Review* 63.3: 577–96.

—. (2008) *How the Soviet Man was Unmade: Cultural Fantasy and Male Subjectivity under Stalin*. Pittsburgh, PA: Pittsburgh University Press.

Kujundzic, D. (2004) "After 'After': The *Ark*ive Fever of Alexander Sokurov." *Quarterly Review of Film and Video* 21.3: 219–39.

Lindley, C.A. (2005) "The Semiotics of Time Structure in Ludic Space As a Foundation for Analysis and Design." *Game Studies: An International Journal of Computer Game Research* 5.1. Available at: http://www.gamestudies.org/0501/lindley/ (last accessed 15 July 2011).

Lipovetsky, M. (2004) "Post-Sots: Transformations of Socialist Realism in the Popular Culture of the Recent Period." *SEEJ* 48.3: 365–77.

McCausland, G. (2000) "Viktor Pelevin and the End of Sots-Art." *Endquote: Sots-Art Literature and Soviet Grand Style*. Eds. M. Balina, N. Condee and E. Dobrenko. Evanston, IL: Northwestern University Press. 225–38.

Manovich, L. (2001) *The Language of New Media*. Cambridge, MA: MIT Press.

McLuhan, M. (1964) *Understanding Media: The Extensions of Man*. New York: McGraw-Hill.

Miller, A.L. (1996) "The Panorama, the Cinema, and the Emergence of the Spectacular." *Wide Angle* 18.2: 34–69.

Osborne, S. (2003) "*IL-2 Sturmovik: Forgotten Battles* Review." *GamespotUK*. 12 March. Available at: http://uk.gamespot.com/pc/sim/il2sturmovikthefb/review.html (last accessed 2 July 2011).

Oushakine, S.A. (2009) *The Patriotism of Despair: Nation, War, and Loss in Russia*. Ithaca: Cornell University Press.

Oushakine, S. and E. Trubina. (2009) *Travma: Punkty [Trauma: Points]*. Moscow: Novoe literaturnoe obozrenie.

Palmer, S.W. (2006) *Dictatorship of the Air: Aviation Culture and the Fate of Modern Russia*. Cambridge: Cambridge University Press.

Panchasi, R. (1995) "Reconstructions: Prosthetics and the Rehabilitation of the Male Body in World War I France." *Differences* 7.3: 109–40.

Pearce, C. (2002) "Sims, BattleBots, Cellular Automata God and *Go*: A Conversation with Will Wright." *Game Studies* 2.1. Available at: http://www.gamestudies.org/0102/pearce/ (last accessed 19 July 2011).

Pelevin, V. (1994) *Omon Ra*. Trans. Andrew Bromfield. London: Faber & Faber.

Polevoi, B. (1946) *Povest' o nastoiashchem cheloveke.* Moscow: Khudozhestvennaia literatura.

Polov'ian, S. (2011) "Kruglyi stol 'Trenazhernye tekhnologii—eserv povysheniia bezopasnosti poletov.'" Available at: http://www.dinamika-avia.ru/mcenter/exhibition/hr2011.php (last accessed 8 July 2011).

Ravetto-Biagolli, K. (2005) "Floating on the Borders of Europe: Sokurov's *Russian Ark.*" *Film Quarterly* 59.1: 18–26.

Rolfe, J.M., and K.J. Staples. (1986) *Flight Simulation.* Cambridge: Cambridge University Press.

"Rossiiskie letchiki vnov' poluchili vozmozhnost' trenirovat'sia na unikal'nom trenazhernom komplekse NITKA." (2010). 31 March. Available at: http://sevpatriot.ru/novosti/rossiyskie-letchiki-na-trenazhernom-komplekse-nitka (last accessed 8 July 2011).

Siddiqi, A.A. (2010) *The Red Rockets' Glare: Spaceflight and the Soviet Imagination 1857–1957.* Cambridge: Cambridge University Press.

Sobolev, D.A., and D.B. Khazanov. (2000) *Nemetskii sled v istorii otechestvennoi aviatsii.* Moscow: RUSAVIA.

Smorodinskaya, T. (2010) "The Fathers' War through the Sons' Lens." *Cinepaternity: Fathers and Sons in Soviet and Post-Soviet Film.* Eds. H. Goscilo and Y. Hashamova. Bloomington: Indiana University Press. 89–113.

Strukov, V. (2009) "A Journey through Time: Alexander Sokurov's *Russian Ark* and Theories of Mimesis." *Realism and Audio-Visual Media.* Eds. L. Nagib and C. Mello. London: Palgrave. 119–32.

—. (2012a) "Spatial Imagining and Ideology of Digital Commemoration (Russian Online Gaming)." *Europe-Asia Studies* 64.8: 1584–604.

—. (2012b) "Ludic Digitality: A. Sokurov's *Russian Ark* and A. Popogrebskii's *How I Ended This Summer* as Cinegames." *Studies in Russian, Eurasian and Central European New Media* 8: 19–45.

Virilio, P. (1997) *Pure War: Interviews with Sylvere Lotringer.* Paris: Semiotext(e).

Wark, McK. (2007) *Game Theory.* Cambridge, MA: Harvard University Press.

Widdis, E. (2003) *Visions of a New Land: Soviet Film from the Revolution to the Second World War.* New Haven, CT: Yale University Press.

Youngblood, D.J. (2007) *Russian War Films: On the Cinema Front, 1914–2005.* Lawrence, KS: Kansas University Press.

Žižek, S. (2006) *The Parallax View.* Cambridge, MA: MIT Press.

Index